About Island Press

Island Press is the only nonprofit organization in the United States whose principal purpose is the publication of books on environmental issues and natural resource management. We provide solutions-oriented information to professionals, public officials, business and community leaders and concerned citizens who are shaping responses to environmental problems.

In 1994, Island Press celebrated its tenth anniversary as the leading provider of timely and practical books that take a multidisciplinary approach to critical environmental concerns. Our growing list of titles reflects our commitment to bringing the best of an expanding body of literature to the environmental community throughout North America and the world.

Support for Island Press is provided by The Geraldine R. Dodge Foundation, The Energy Foundation, The Ford Foundation, The George Gund Foundation, William and Flora Hewlett Foundation, The James Irvine Foundation, The John D. and Catherine T. MacArthur Foundation, The Andrew W. Mellon Foundation, The Joyce Mertz-Gilmore Foundation, The New-Land Foundation, The Pew Charitable Trusts, The Rockefeller Brothers Fund, The Tides Foundation, Turner Foundation, Inc., The Rockefeller Philanthropic Collaborative, Inc., and individual donors.

Wildlife and Recreationists

Wildlife and Recreationists

COEXISTENCE THROUGH
MANAGEMENT AND RESEARCH

EDITED BY

Richard L. Knight and Kevin J. Gutzwiller

ISLAND PRESS

Washington, D.C.   Covelo, California

Library of Congress Cataloging-in-Publication Data
Wildlife and recreationists : coexistence through management and research / edited by Richard L. Knight & Kevin J. Gutzwiller.
    p.   cm.
    Includes bibliographical references and index.
    ISBN 1-55963-257-7 (cloth). — ISBN 1-55963-258-5 (paper)
    1. Outdoor recreation—Environmental aspects. 2. Animal ecology.
3. Human-animal relationships. 4. Wildlife conservation—Social aspects.
I. Knight, Richard L. II. Gutzwiller, Kevin J.
QH545.O87W54   1995
33.78—dc20   94-30142
                                               CIP

Printed on recycled, acid-free paper ∞ ♲

Manufactured in the United States of America

10 9 8 7 6 5 4 3 2 1

*To those who place land health
before its utilitarian values*

Contents

Part III. Case Studies   221

Part IV. Ethics and Answers   325

Figures

Tables

Preface

As we near the end of the twentieth century, a paradigm shift in land management is nearing fruition. Historically, human activities on public and private lands were extractive (e.g., logging, grazing, hunting). Today, we live in a society that speaks nearly as one in its objections to primarily utilitarian uses of our once-vast natural resources. With an ever-expanding world population, a shift from rural to urban societies, and a concomitant increase in leisure time in developed countries, there has been a decades-long increase in outdoor recreational activity, and there is no decline in sight. In addition, ecotourism, a form of recreation focused on bringing tourists to biologically rich ecosystems, is becoming increasingly popular. Our society seems to have requested that the primary use of public lands be that of outdoor recreation, and to the degree that utilitarian uses conflict with recreation, they should be of secondary importance.

One upshot of this "recreation boom" is the recognition that wildlife and recreational activities are not compatible without some form of overall management. Researchers continue to report damaging effects of recreationists on wildlife. Whether it be a shift in age and sex ratios of a hunted population of big game or displacement of a species sensitive to off-road-vehicle activity, recreational disturbance can have previously unappreciated impacts on wildlife. Legislation protecting wildlife (e.g., the Endangered Species Act, National Forest Management Act) may conflict with expanding recreational disturbance in shrinking wildlands, so the need for understanding and managing recreational impacts on wildlife will continue.

We believe that the principal natural-resource management issue for the remainder of this century, and into the next, will revolve around conflicts between outdoor recreationists and wildlife (and between different types of outdoor recreationists!). Today, natural-resource agencies are preoccupied with changing their focus from largely extractive uses of the land to increasing outdoor recreation opportunities; we envision the next step in agency evolution will be toward land health and stewardship. We feel certain, however, that these agencies will not be able to accomplish this transition without first passing through a painful period of introspection—one caused by unchecked outdoor recreation. Accordingly, the primary audience for our book is the natural-resource manager. We hope that the information contained in these pages will help define and resolve the increasingly important issue of outdoor recreation and its impacts on wildlife and ecosystems.

In this book, we focus on the direct effects of recreationists on wildlife (e.g., behavior change, energetic imbalances, death); however, we also consider indirect effects such as habitat modification (see Chapter 11). We recognize that there is a profound need to understand the human dimensions of outdoor-recreation management, and we have included two chapters (Chapters 2 and 3) as a primer on this important topic.

The first part of the book (General Issues) explores topics common to most wildlife-recreation interactions, ranging from how wildlife responds to disturbance, to the origin of these responses. In Part II (Specific Issues), we examine detailed points relevant to wildlife-recreation interactions, which span the gamut from physiological responses of wildlife to disturbance, to the effects of ecotourism. The third part (Case Studies) presents seven case studies that provide insights into how specific recreational activities affect diverse types of wildlife, from manatees to rattlesnakes. The final part (Ethics and Answers) looks to the future, addresses how wildlife and recreationists might coexist, and explores ethical issues relevant to this field.

Our goal in compiling this book was for it to have practical value in helping minimize or ameliorate the negative impacts of recreation on wildlife. Therefore, each chapter in the second and third parts has a segment describing realistic options for avoiding or managing detrimental effects of recreation. In addition, for all but the final two chapters, we have asked the authors to conclude with a section that identifies major gaps in our knowledge on their topic. We hope this information can play a part in prioritizing the research necessary for solving the wildlife-recreation conflicts that are sure to occur. Each chapter should provide managers and researchers alike with information that is useful in directing and integrating their activities toward a particular type of impact.

RLK wishes to acknowledge the support of students and colleagues at Colorado State University who assisted in many ways with the preparation of this book. He is in debt to Heather A. L. Knight for her support and encouragement. He also acknowledges the financial support of the Intermountain Research Station, U.S. Department of Agriculture, Forest Service, which enabled him to prepare a review of recreational impacts on wildlife. KJG appreciates all aid received for this effort from Baylor University. He is especially grateful to Jerome and Rita Gutzwiller for the life-long support that made his role in this work possible, and to Pam and Robert Gutzwiller for their assistance and enduring patience during this project. Amy D. Lauderdale served as our editorial assistant on this project. Were it not for her, the book would be but a shadow of its present form. Finally, we owe a very special thanks to the authors. Their enthusiasm and commitment to this project made it not only enjoyable for us, but will help ensure better stewardship of our wildlife in the years to come.

PART I

General Issues

CHAPTER 1

Outdoor Recreation: Historical and Anticipated Trends

Curtis H. Flather and H. Ken Cordell

One attribute common to the diverse array of outdoor recreational pursuits is the need for a spacious land- or water-base (Clawson and Harrington 1991). The fact that outdoor recreation is dispersed over large areas has undoubtedly contributed to the perception that it has little environmental impact compared to extractive uses of natural resources such as timber harvesting or livestock grazing. Some have concluded that outdoor recreation is benign, or at worst neutral, in its environmental consequences (see Wilkes 1977; Duffus and Dearden 1990).

Recreational Influences

Given the growing number of outdoor recreationists, and an emerging disposition among some public land management agencies to shift their emphasis from commodity to amenity uses (Brown and Harris 1992), the notion that recreation has no environmental impacts is no longer tenable. Recreationists often degrade the land, water, and wildlife resources that support their activities by simplifying plant communities, increasing animal mortality, displacing and disturbing wildlife, and distributing refuse (Boyle and Samson 1985). These impacts can be particularly extensive for the very reason that many outdoor recreational impacts were initially thought to be diluted—namely, recreationists are dispersed over large areas (Cole and Knight 1991).

Management strategies for regulating recreational impacts on wildlife often involve restricting access to public lands, and, proactive management would benefit from an analysis of historical and anticipated trends in outdoor recreation. In this chapter we review these trends for outdoor recreation in the

United States, speculate on the potential causes of these trends, and suggest research to extend the effectiveness of recreation forecasting in resource management planning.

A Typology of Wildland Outdoor Recreation Activities

Common criteria used to distinguish levels of potential interaction with wildlife among recreational activities include consumptive versus nonconsumptive motivations, species harvested, and whether wildlife is a purposeful or incidental component of the experience. We have categorized activities on the basis of these criteria to distinguish among the potential impacts on wildlife habitats and populations.

We have made a primary distinction between activities that directly depend on wildlife and those that do not. Participation in wildlife-dependent activities is contingent on the expected occurrence of wildlife in the area. In contrast, the enjoyment of nondependent activity is often enhanced by, but participation is not conditioned on, the presence of wildlife. Among wildlife-dependent activities we distinguish between consumptive and nonconsumptive recreation. A final level of distinction broadly groups consumptive activities according to species harvested. Species groupings correspond to state licensing categories and include big game, small game, and migratory bird hunting, and freshwater and saltwater fishing. In this chapter, references to "wildlife" include fish.

Historical Context

Public demand for outdoor recreation opportunities grew rapidly with the revival of the U.S. economy following World War II. Annual growth rates in the use of public parks and recreation facilities often exceeded 10% from the early postwar period through the mid-1960s (Walsh 1986). This period of rapid growth was coincident with a general rise in affluence as indicated by increased disposable income, increased leisure time, institutionalization of paid vacations, and transportation improvements that facilitated mobility (Clawson and Harrington 1991).

Increased affluence has also been associated with the formation of a conservation ethic (see Myers 1985; Brady 1988), so it is not surprising that the outdoor recreation boom paralleled the growth of the conservation movement; its beginnings were marked by the establishment of The Conservation Foundation in 1948, the Sport Fishing Institute in 1949, and The Nature Conservancy in 1951 (Clawson and Harrington 1991). Although the concurrent evolution of conservation ethics and outdoor recreation was initially regarded as mutually beneficial, many of the goals of natural resource conservation, wilderness preservation, and provision of outdoor recreation are now viewed as conflicting (Nash 1982:316). Quantifying the magnitude and the nature of the conflict requires, in part, a review of recent historical trends in the number of outdoor recreationists.

Recent Historical Trends in Outdoor Recreation

To characterize the temporal data series relevant to outdoor recreation, growth rate for a given activity is usually compared to population growth (see Snepenger and Ditton 1985). Although growth rates enable one to infer trends in the popularity of an activity, statistics on participation (e.g., the number of people, number of visits, and aggregate time on the recreation landscape) are necessary to judge the potential impacts on wildlife resources.

Number of participants, as an indicator of potential impact, is particularly important in light of the decreasing availability of places for outdoor recreation. Although the United States has a substantial public land base to support outdoor recreation (300 million ha), much of the potential land and water recreation base is under rural private ownership (60% of United States land area) with restricted access to outdoor recreationists. In 1987, only 23% of rural private lands were open to the public without restriction, a decline of nearly 30 million ha since 1977 (Cordell et al. 1990:14). More recent evidence indicates that the trend toward greater closure and exclusive leasing of private land is continuing (Cordell et al. 1993).

TRENDS IN WILDLIFE-DEPENDENT ACTIVITIES

Trends in wildlife-dependent activities have been mixed over the last three decades (Fig. 1.1). More people participated in fishing (freshwater fishing in particular) than in any other wildlife-dependent activity. In 1985, nearly 25% of this country's inhabitants fished.

In contrast to the monotonic increase in the number of anglers, hunter numbers have remained essentially unchanged since 1975. The stability in the number of total hunters, however, is misleading. The number of small game and migratory bird hunters has declined substantially since 1975, while the number of hunters pursuing big game species has increased during every survey period since 1955.

The divergence in participation trends by species category may be explained, in part, by trends in game populations. Small game species associated with agricultural habitats, including ring-necked pheasant, northern bobwhite, and cottontail have shown declines in abundance (Flather and Hoekstra 1989:33). Similarly, breeding duck populations declined by 30% from the early 1970s to the mid 1980s (U.S. Department of the Interior, Fish and Wildlife Service and Canadian Wildlife Service 1992). Conversely, big game populations have increased in most states (Flather and Hoekstra 1989:28).

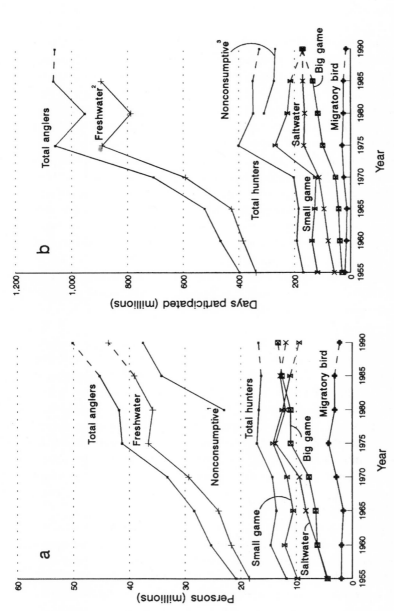

Figure 1.1 Trends in (a) the number of persons (≥12 years old) and (b) the and numbers of days (for persons ≥16 years old) spent participating in recreational activities dependent on wildlife ([1]number of persons ≥ 6 years old; [2] 1985–1990 freshwater angler days could not be extrapolated because of disaggregation into "Great Lake" and "Other freshwater" categories; [3]Based on number of trips but not adjusted for difference in survey design across years). (U.S. Department of the Interior, Fish and Wildlife Service, and U.S. Department of Commerce, Bureau of the Census 1993: Appendix Table B-3; dashed lines represent participants or days adjusted to be consistent with 1955–1985 trend based on estimates presented in Appendix Table B-4.)

Other factors that may be contributing to declines in migratory bird and small game hunters include restricted access, crowding, and less leisure time available for hunting (Smith et al. 1992; Enck et al. 1993). Participation in small game and migratory bird hunting appear to be more tied to land access than big game hunting. In 1985, 63% of small game hunters and 62% of migratory bird hunters hunted on private lands only, compared with 51% of big game hunters (U.S. Department of the Interior, Fish and Wildlife Service 1988).

Nonconsumptive recreational activities are growing in popularity relative to traditional wildlife and fish recreational pursuits (Duffus and Dearden 1990). The number of persons that actually traveled more than 1.6 km from their residence to observe, photograph, or feed wildlife increased from 22.9 to 37.5 million from 1980 to 1990 (Fig. 1.1a)—an average annual rate of increase that exceeds all other wildlife-oriented recreation. Based on 1991 survey results, most of the participants in nonconsumptive activities simply observed wildlife (96%); substantially fewer people photographed (47%) or fed wild animals (44%) (U.S. Department of the Interior, Fish and Wildlife Service and U.S. Department of Commerce, Bureau of the Census 1993).

Trends in the total number of days devoted to wildlife-dependent recreation (Fig. 1.1b) tend to mirror trends in participants, with two notable exceptions. First, despite increased numbers of participants, the number of nonconsumptive trips declined between 1980 and 1990. Second, since 1975, the number of days spent angling and hunting have deviated from participation trends. Despite consistent increases in the total number of anglers, the number of days spent fishing declined conspicuously in the 1980 survey and has only recently recovered to 1975 levels. Similarly, the total days spent hunting has continued to decline since 1975 despite a nearly constant number of participants. Although definitive studies are lacking, data do indicate that the amount and focus of leisure time can restrict the level of participation in all types of wildlife-oriented outdoor recreation (see Goodale 1991; Schor 1991).

TRENDS IN ACTIVITIES NOT DEPENDENT ON WILDLIFE

Trends in outdoor recreational activities not dependent on wildlife were established from several sources (Table 1.1). Although the estimates were developed using different methods, the results are sufficiently comparable to indicate general trends over a ten-year period.

Among the land-based activities, those occurring within developed recreation sites or near roads had the highest number of persons 12 years old or older participating. The consistent improvement in bicycle technology has, in large part, been responsible for biking's strong growth. The increasing popu-

larity in motorized off-road vehicles also seems to be in response to techno-
logical advancements that make it easier for less experienced persons to participate. Day hiking, photography, and nature study have shown moderate
growth as Americans seek educationally oriented outdoor experiences. Horseback riding and backpacking are growing at slower rates than many of the
other land-based activities; they are expanding approximately at the rate of
population growth.

Among water-based activities, swimming in natural water bodies has continued to rise in popularity and has tended to concentrate at a limited number
of relatively small access points. Motorboating and waterskiing, however,
cover large stretches of water, and with development of jet-boat technology,
few water bodies are inaccessible.

Participation in downhill skiing and the concurrent ski-resort development are growing at moderate rates. Although the actual area modified and
developed for ski slopes is relatively small, they tend to be in high-elevation
ecosystems and thus are concentrated within a relatively narrow band of
habitat types. More significant to wildlife are the concomitant developments
and modifications to the natural landscape resulting from the services and facilities that complement downhill skiing.

Avidity, or frequency of participation, is measured in number of different
days on which participation occurred. Biking, swimming, motorboating,
off-road driving, day hiking, and developed-site camping have the highest
avidity among the activities listed in Table 1.1. Greater frequencies of participation combined with significant numbers of people participating translate
into greater pressures on the resources and ecosystems where these activities
occur.

Several activities have only recently emerged as popular avocations, including trail (mountain) biking, mountain climbing, rock climbing, caving,
orienteering, rafting and tubing, and jet skiing (Cordell et al. 1990). Although
these activities are still relatively novel, 1% to 11% of the population 12 years
and older participate. Participation often occurs in fragile environments, including alpine tundra, caves and on cliff faces.

Several socioeconomic and resource management factors seem to have
shaped recent trends in recreation not dependent on wildlife. Possible causes
include an aging population, population growth and redistribution to warmer
regions, immigration, increasing numbers of dual-income households,
smaller percentages of two-parent households, greater educational attain-
ment, and economic instability as indicated by more frequent recessions
(Cordell et al. 1990). Other factors empirically shown to affect participation
include reductions in social and physical barriers to participation, advances in

Table 1.1

Recent Participation Trends and 1992 Avidity for
Outdoor Recreational Activities Not Dependent on Wildlife

Activities	Millions of persons ≥ 12 years old participating			Millions of persons participating ≥ 10 days in 1992
	1982–83	1985–87	1992	
Land-based				
Bicycling	60	72	86	60.3
Camping in developed campgrounds	32	40	48	13.2
Day hiking	26	32	50	13.4
Nature study/ photography	22	24/26	—[a]	—[a]
Driving motorized vehicles off-road	21	24	38	16.2
Camping in primitive campgrounds	19	22	25	6.7
Horseback riding	17	20	19	6.7
Backpacking	9	10	8	3.6
Water-based				
Swimming in lakes, streams, ocean	60	70	90	41.4
Motorboating	36	42	65	24.6
Waterskiing	17	20	21	7.6
Canoeing/kayaking	15	18	19	3.2
Snow- and/or ice-based				
Downhill skiing	11	14	21	3.6
Snowmobiling	6	8	8	1.7
Cross-country skiing	6	8	8	2.9

Sources: 1982–1983 National Recreation Survey, USDI National Park Service; 1985–1987 Public Area Recreation Visitors Survey, USDA Forest Service; and the 1992 Pilot of the National Survey on Recreation and the Environment (unweighted data), USDA Forest Service.

[a]Participation in nature study/photography not estimated in 1992.

recreation equipment technology, expanded availability of information and transportation, accessibility to private and public lands, types and location of recreational facilities, and accessibility of remote areas (Cordell et al. 1990; Cordell and Bergstrom 1991).

Anticipated Trends in Outdoor Recreation

Projecting participation in outdoor recreation activities has often involved a simple extrapolation of historical trends. Although this approach has been used widely, it assumes that past factors affecting recreation participation will continue immutably into the future. This assumption may be suitable for short-term forecasts but is likely to result in biased projections in the long-term (>5 years) (Walsh 1986:353).

A more realistic approach for longer-term projections is to model participation as a function of factors known or hypothesized to affect personal decisions on whether or not to engage in outdoor recreational activities. Under this approach, participation is often modeled a a function of (1) price (e.g., average per capita costs of transportation, food, lodging, fees, distance traveled); (2) socioeconomic factors acting as surrogates for differences in tastes and preferences (e.g., per capita income, age, education, ethnicity, marital status); and (3) resource availability (e.g., big game populations, harvest success rates, proximity and capacity of camping facilities), including substitute opportunities. Our projections of participation in outdoor recreation are based on this modeling approach, the details of which are reviewed by Hof and Kaiser (1983), Walsh et al. (1989), and Cordell and Bergstrom (1991).

PROJECTED PARTICIPATION IN WILDLIFE-DEPENDENT ACTIVITIES

The anticipated trends in the number of people participating in wildlife-oriented recreation presented here are based on an analysis of the 1985 National Survey of Fishing, Hunting, and Wildlife-Associated Recreation (U.S. Department of the Interior, Fish and Wildlife Service 1988) completed by Walsh et al. (1989). In general, future participation patterns are consistent with those observed since 1980 (Fig. 1.2). Fishing and nonconsumptive participation are projected to increase 63% to 142% over the next 50 years. Conversely, participation in big game hunting is projected to remain relatively stable, while participation in small game hunting is expected to decline.

The empirical relations accounting for the difference in future participation among these activities primarily involve income, education, and residence (i.e., urban vs. rural). The expected increase in household income is associated with increased probability of participation in nonconsumptive activities, fishing, and migratory bird hunting. Lower participation rates in big game hunting are associated with increasing income. With increasing education level, the likelihood of participating in coldwater fishing and migratory bird hunting increases, and the likelihood of participating in small game hunting decreases. Finally, as the proportion of the population that lives in an urban setting increases, participation rates rise for coldwater fishing and mi-

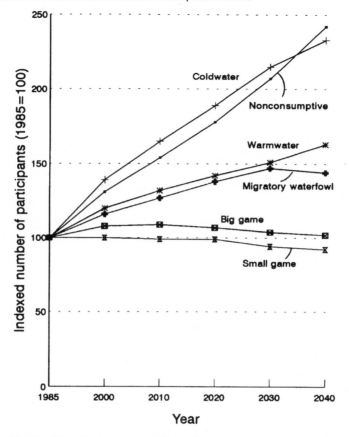

Figure 1.2   Projected trends in the number of persons participating in recreational activities dependent on wildlife. To facilitate comparisons among activities, participation is indexed to levels observed in 1985 (Walsh et al. 1989).

gratory bird hunting, and rates decline for big game hunting and warmwater fishing.

The most significant projected shift from recent historical trends concerns migratory waterfowl hunting, which is projected to increase over time. This deviation from past trends is due primarily to an assumption that per capita waterfowl numbers will remain stable over the projection period, allowing the growth-stimulating effects of greater income and education to drive up predicted participation (Walsh et al. 1989:346). The future demand for migratory waterfowl hunting will probably not be met unless recent declines in duck populations and wetland habitats are reversed and access to hunting areas is increased. Although discrepancies between future and recent historical trends in participation may indicate uncertainties with model specification and estimation, they also may indicate outdoor recreational activities in which significant gains could be made to meet demand through resource management and policy adjustments.

PROJECTED PARTICIPATION IN ACTIVITIES NOT DEPENDENT ON WILDLIFE

Activities that do not directly depend on wildlife constitute the vast majority of outdoor recreation participation now and in the future. Projected participation in these activities (Table 1.2) is based on an analysis of the 1985 to 1987 Public Area Recreation Visitors Survey (Cordell and Bergstrom 1991) and on an updated analysis that includes additional data from surveys conducted across numerous national forests in the United States (English et al. 1993).

Table 1.2

Projected Indices of Growth[a] in Recreation Trips
for Activities Not Directly Dependent on Wildlife, 2000 to 2040

Activities	Projected participation index by year (1987 = 100)				
	2000	2010	2020	2030	2040
Land-based					
Day hiking	123	144	168	198	229
Bicycling	124	146	170	197	218
Developed camping	120	138	158	178	195
Horseback riding	114	125	135	144	149
Primitive camping	108	115	122	130	134
Off-road driving	104	108	112	118	121
Nature study	99	101	103	107	108
Water-based					
Rafting/other floating	123	151	182	229	267
Canoeing/kayaking	113	126	138	153	163
Swimming in streams/lakes	108	118	128	140	152
Motorboating	107	114	122	131	138
Snow- and ice-based					
Snowmobiling	120	131	137	141	137
Cross-country skiing	125	136	142	141	126

Source: 1992 Pilot of the National Survey on Recreation and the Environment (unweighted data), USDA Forest Service.

[a] These projections assume that recent trends in facility development, access, and services for outdoor recreation will continue into the future.

In the last 15 to 20 years, per capita availability of land-, water-, snow-, and ice-based resource opportunities have been decreasing, but the trends in demographic variables and increases in population generally have had more influence on participation changes than the effect of increasingly scarce wildland recreation opportunities.

Activities projected to grow most rapidly include day hiking, bicycling, developed camping, and rafting and tubing. By the year 2000, we expect these activities to grow by 23%, 24%, 20%, and 23%, respectively, relative to their 1987 levels. We anticipate that off-road driving, motorboating, and snowmobiling will grow at respective rates of 4%, 7%, and 20% by the year 2000. Growth in the above activities will mean greater use of and need for trail and stream access. Thus, whether by horseback, on foot, by bicycle, by ski, or by motorized vehicle, we predict that people will increasingly enter wildland areas to enjoy their natural beauty and to experience less crowded environs. We are likely also to see more participation in high-technology activities such as rock climbing, white-water recreation, cave exploration, diving, cross-country snow travel, and ice climbing. As the frequency and spatial scale of these activities grow, so too will the pressures they place on heretofore undisturbed or lightly disturbed areas and resources.

Knowledge Gaps

Although historical trends in participation are useful for simple retrospective description, greater understanding of what motivates people to participate or forgo participation is needed to advance our ability to predict trends. Although much empirical investigation has been directed at this problem, the predictive capability of models has remained low and few theory-building papers have emerged (Kelly 1991:399). Areas of research that could broaden our understanding include: improved specification of behavioral models, validation of empirical results, and improved grounding in theory.

Model Specification

The goal of most empirical studies of outdoor recreation participation has been the specification and estimation of an econometric prediction model. These studies have stressed statistical analysis of cross-sectional survey data. By including nontraditional factors in model specification, we may better understand recreational choice behaviors. Along with measures of price and socioeconomic factors that are thought to reflect preference for certain activities, refined measures of access should also be incorporated (e.g., geographic proximity and participation barriers), opportunity quality (e.g., visual attractiveness, availability of services, and social atmosphere), and participant attitudes.

Quantifying some of these factors will likely require nonstandard survey procedures including direct participant observation or projective techniques (e.g., imaginative role playing) (Kelly 1991).

Validation of Empirical Results

Barnett (1991:343) asks: "Should correlational findings based on nonexperimental designs be taken seriously?" Certainly, patterns of association between participants and their socioeconomic profiles have offered important, if only exploratory, insights into factors that potentially affect outdoor recreation behavior. However, little has been done to establish a concrete base of evidence to validate the estimated relations between recreation participation and changes in human and resource conditions.

Experiments offer a powerful protocol for inferring cause-and-effect relationships that is not possible through observational studies. Although the applicability of findings generated in a highly controlled environment is a legitimate concern, this same environment offers a much needed opportunity for analyzing the complex causes of recreational choice behavior.

Additional validation is needed to test the temporal stability of model specification and estimated parameters. To date, little work has been done on these frontiers. Consistency in measures across time, periodic re-estimations of model parameters, and comparisons between forecast participation and actual participation at common points in time would all be useful.

Improved Grounding in Theory

Research can address specific solutions to applied problems, or it can focus on development of general theory about phenomena. Because participation research has focused on empirical relations, some investigators feel that this area has failed to develop adequate theory. As noted by Kelly (1991), facts, as represented by empirical findings, do not themselves lead to generalization, but the hypotheses they either support or refute do. Strong inference (sensu Platt 1964), however, will be constrained unless theory is sufficiently defined to permit its interaction with empiricism. That some are calling for higher-order hypotheses to guide and summarize leisure research (Barnett 1991:348) suggests that the interaction between theory and empirical study is currently weak. It also suggests that progress toward understanding recreationist behavior could be made by synthesizing the numerous empirical efforts, to distill a theoretical framework for guiding future modeling research.

Literature Cited

Barnett, L.A. 1991. Leisure research is a very good thing. In *Recreation and Leisure: Issues in an Era of Change*, eds., T.L. Goodale and P.A. Witt, 343–358. State College, Pennsylvania: Venture Publishing.

Boyle, S.A. and F.B. Samson. 1985. Effects of nonconsumptive recreation on wildlife: a review. *Wildlife Society Bulletin* 13:110–116.

Brady, N.C. 1988. International development and the protection of biological diversity. In *Biodiversity,* ed., E.O. Wilson, 409–418. Washington, D.C.: National Academy Press.

Brown, G. and C.C. Harris. 1992. The United States Forest Service: changing of the guard. *Natural Resources Journal* 32:449–466.

Clawson, M. and W. Harrington. 1991. The growing role of outdoor recreation. In *America's Renewable Resources: Historical and Current Challenges,* eds., K.D. Frederick and R.A. Sedjo, 249–282. Washington, D.C.: Resources for the Future.

Cole, D.N. and R.L. Knight. 1991. Wildlife preservation and recreational use: conflicting goals of wildland management. *Transactions of the North American Wildlife and Natural Resources Conference* 56:233–237.

Cordell, H.K. and J.C. Bergstrom. 1991. A methodology for assessing national outdoor recreation demand and supply trends. *Leisure Sciences* 13:1–20.

Cordell, H.K., J.C. Bergstrom, L.A. Hartmann, and D.B.K. English. 1990. An analysis of the outdoor recreation and wilderness situation in the United States: 1989–2040. USDA Forest Service General Technical Report RM-189. Fort Collins, Colorado: Rocky Mountain Forest and Range Experiment Station.

Cordell, H.K., D.B.K. English, and S.A. Randall. 1993. Effects of subdivision and access restrictions on private land recreation opportunities. USDA Forest Service General Technical Report RM-231. Fort Collins, Colorado: Rocky Mountain Forest and Range Experiment Station.

Duffus, D.A. and P. Dearden. 1990. Non-consumptive wildlife-oriented recreation: a conceptual framework. *Biological Conservation* 53:213–231.

Enck, J.W., B.L. Swift, and D.J. Decker. 1993. Reasons for decline in duck hunting: insights from New York. *Wildlife Society Bulletin* 21:10–21.

English, D.B.K., C.J. Betz, J.M. Young, J.C. Bergstrom, and H.K. Cordell. 1993. Regional demand and supply projections for outdoor recreation. USDA Forest Service General Technical Report RM-230. Fort Collins, Colorado: Rocky Mountain Forest and Range Experiment Station.

Flather, C.H. and T.W. Hoekstra. 1989. An analysis of the wildlife and fish situation in the United States: 1989–2040. USDA Forest Service General Technical Report RM-178. Fort Collins, Colorado: Rocky Mountain Forest and Range Experiment Station.

Goodale, T.L. 1991. Is there enough time? In *Recreation and Leisure: Issues in an Era of Change,* eds., T.L. Goodale and P.A. Witt, 33–46. State College, Pennsylvania: Venture Publishing.

Hof, J.G. and H.F. Kaiser. 1983. Long-term outdoor recreation participation projections for public land management agencies. *Journal of Leisure Research* 15:1–14.

Kelly, J.R. 1991. Leisure and quality: beyond the quantitative barrier in research. In *Recreation and Leisure: Issues in an Era of Change*, eds., T.L. Goodale and P.A. Witt, 397–411. State College, Pennsylvania: Venture Publishing.

Myers N. 1985. Endangered species and the north-south dialogue. In *Economics of Ecosystem Management*, eds., D.O. Hall, N. Myers, and N. S. Margaris, 139–148. Boston, Massachusetts: Dr. W. Junk Publishers.

Nash, R. 1982. *Wilderness and the American Mind*. New Haven, Connecticut: Yale University Press.

Platt, J.R. 1964. Strong inference. *Science* 146:347–353.

Schor, J. 1991. *The Overworked American: The Unexpected Decline of Leisure.* New York: Basic Books.

Smith, J.L.D., A.H. Berner, F.J. Cuthbert, and J.A. Kitts. 1992. Interest in fee hunting by Minnesota small-game hunters. *Wildlife Society Bulletin* 20:20–26.

Snepenger, D.J. and R.B. Ditton. 1985. A longitudinal analysis of nationwide hunting and fishing indicators: 1955–1980. *Leisure Sciences* 7:297–319.

U.S. Department of the Interior, Fish and Wildlife Service. 1988. *1985 National Survey of Fishing, Hunting, and Wildlife Associated Recreation.* Washington, D.C.: U.S. Department of the Interior, Fish and Wildlife Service.

U.S. Department of the Interior, Fish and Wildlife Service, and Canadian Wildlife Service. 1992. *Status of Waterfowl and Fall Flight Forecast.* Washington, D.C.: U.S. Department of the Interior, Fish and Wildlife Service.

U.S. Department of the Interior, Fish and Wildlife Service, and U.S. Department of Commerce, Bureau of the Census. 1993. *1991 National Survey of Fishing, Hunting, and Wildlife-Associated Recreation.* Washington, D.C.: U.S Government Printing Office.

Walsh, R.G. 1986. *Recreation Economic Decisions: Comparing Benefits and Costs.* State College, Pennsylvania: Venture Publishing.

Walsh, R.G., D.A. Harpman, K.H. John, J.R. McKean, and L. LeCroy. 1987. Wildlife and fish use assessment: long-run forecasts of participation in fishing, hunting, and nonconsumptive recreation. Technical Report 50. Fort Collins, Colorado: Water Resource Research Institute, Colorado State University.

Walsh, R.G., K.H. John, J.R. McKean, and J.G. Hof. 1989. Comparing long-run forecasts of demand for fish and wildlife recreation. *Leisure Sciences* 11:337–351.

Wilkes, B. 1977. The myth of the non-consumptive user. *Canadian Field-Naturalist* 91:343–349.

CHAPTER 2

Human Dimensions of Wildlife Management: Basic Concepts

Michael J. Manfredo, Jerry J. Vaske, and Daniel J. Decker

In the past decade there has been increased interest in what has been described as a "human dimensions" approach to wildlife management. Human dimensions offers promise in efforts to make decisions that are more responsive to the public and that, in the long term, increase the effectiveness of decision making (Decker et al. 1989, 1992).

In its simplest form, a human dimensions approach can be described in two parts. The first emphasizes acquisition of sound information that explains human thought and action regarding wildlife using the concepts and methods of social science. The second part is determining how to use that information in wildlife decision making. Social information is just one consideration among many (e.g., biological, legal, political) in the decision-making process. Deciding how to interpret and weigh that information offers challenges to researchers and managers alike.

While each of these parts is equally important, the utility of a human-dimensions approach ultimately rests on the quality of the information provided. The quality of information is further determined by the integrity of an investigation's conceptual basis. Theory and concepts provide a storehouse for knowledge obtained from studies on the same topic. Theory allows studies to build upon one another, successively teaching more about human-wildlife interactions as more studies are conducted.

Recreational Influences

Given the importance of a human-dimensions approach to wildlife management, and specifically its importance to examining recreation-wildlife interactions, our chapter provides an overview of prominent conceptual approaches used in human-dimensions investigations, which can be used by managers to minimize conflicts between wildlife and recreationists. Ours is only a very broad treatment of the human-dimensions topics that have emerged from social psychology and that are those most frequently used in natural resource investigations.

Wildlife managers are typically faced with three broad mandates. The first is to ensure the conservation and protection of wildlife and their habitats. The second is to provide opportunities for people to enjoy and learn about wildlife, and the third is to protect people from potential hazards caused by wildlife. Wildlife managers dealing with the challenge of coexistence must mitigate and balance the impacts of these somewhat conflicting mandates. To assist managers in this endeavor, human-dimensions inquiry can be used to examine two broad types of questions. One asks what human behaviors lead to undesirable impacts on wildlife or humans, why do they occur, and how might human behavior be controlled to minimize impacts? Other questions ask what types of interactions with wildlife provide for positive and desirable human responses (e.g., what kinds and types of recreation lead to happy customers)? Each of these questions touches upon an even more basic question: Why do humans behave the way that they do? Some of the concepts that have evolved to explain human phenomena are described in the next section.

The prominent conceptual bases in human-dimensions inquiry fit into one of six broad categories: attitudes, attitude change, norms, values, motivation, and satisfaction. In this section we provide generic definitions of each and offer examples of research from our experience that illustrate the utility of the concept.

Attitudes

Attitude is a central concept in social psychology. Attitudinal studies are also quite common in human dimensions of natural resources. Human-dimensions research often purports to measure "preferences," "opinions," "perceptions," or "images," yet these studies employ methods that would more appropriately classify them as attitudinal investigations.

Attitudes have been defined in a variety of ways, but common to the definitions is the understanding that attitudes are an evaluation or a feeling state about a person, object, or action. For example, verbs with opposite meanings that capture the essence of evaluation include like-dislike, positive-negative, and good-bad. Statements such as "I like fishing" and "My taking a trip to go hunting would be a positive experience" are attitudinal in nature.

Most social psychologists agree that attitudes have a *cognitive* basis. Cognitions or *beliefs* refer to the factual information we hold about a person, object, or action. This is also referred to as knowledge we hold, but in this context, knowledge is not equated with a level of formal or scientific learning. People possess knowledge that may be right or wrong when measured by other people's standards, but that knowledge serves as the basis for their attitude. The essence of beliefs can be captured in opposite verbs such as "agree-disagree," "true-false," and "likely-unlikely." Statements like "Fishing takes a great deal of time" or "Hunting requires much physical exertion" are belief statements. A special class of particularly important beliefs are known as *behavioral intentions*. These are beliefs about how one will behave in a given situation at a particular time.

While it is known that attitudes have a cognitive basis, it can be difficult to make inferences about people's attitudes based on partial information about what they believe, or based on related attitudinal positions. For example, a person may have a positive attitude toward wolves and be expected to support wolf reintroduction. The person may actually oppose reintroduction, however, because his/her feelings toward wolves are outweighed by the belief that reintroduction would have a negative effect on ranchers in the area of reintroduction. When studying a controversial natural resource issue, it is important not only to know a person's attitudinal position, but why they hold that position. In this situation, measures are taken of attitudes and related beliefs that are salient to the population of interest.

It is important to measure attitudes because they are believed to cause human behavior. If people evaluate "going fishing on the Poudre River this weekend" very positively, it would be expected that they would go fishing on the Poudre River this weekend. The relationship between attitudes and behavior has been debated. Studies in the late 1960s and 1970s indicated that people do not act in ways that are consistent with their attitudes. Studies in the last decade, however, have shown that, with careful conceptualization and implementation, attitudes *are* consistent with behavior. (Fishbein and Ajzen 1975; Ajzen and Fishbein 1980; Fishbein and Manfredo 1992).

ATTITUDE EXAMPLE

The forest fires that occurred in much of the West during the summer of 1988 caused a major debate over controlled-burn policies (also referred to as "prescribed fire" or "let-burn" policies). The fire policies of the land management agencies were re-evaluated following the fires of 1988. Part of the review involved a social analysis that included a survey regarding public attitudes toward controlled-burn fire policies (Manfredo et al. 1990).

A national survey was conducted to learn about attitudes toward the policy and the beliefs that were the basis of those attitudes. Subjects were also given a short true-false test to determine the extent of their technical knowledge about fires. Findings were broken down for the region where fires occurred

(Wyoming-Montana), and for the rest of the country. Overall, 54% of the national group and 57% of the regional group had positive attitudes toward the policy, while 46% and 43% had negative or neutral attitudes.

Those with negative attitudes believed the policy results in the loss of animals' homes, the destruction of natural settings and scenery, out-of-control fires, and damage to private property and human life. Those with a positive attitude were more likely to believe that the policy would improve conditions for wildlife, allow natural events to occur, and remove dead vegetation.

The study revealed a high degree of polarization over the fire issue. A clear majority of people in the region of the fire supported a controlled-burn policy. Interestingly, people who had negative attitudes scored lower on the true-false test of technical knowledge. The study indicated a need for public education about controlled-burn fires, and wildfire in general.

Attitude Change

Important tools for managers attempting to mitigate recreation-wildlife interactions are those that interpret, inform, and educate the public. From a theoretical standpoint, these management functions can be addressed within an attitude-change framework. Changing people's behavior or attitudes can be accomplished by changing the beliefs that form the foundations for the attitudes. In the example above, therefore, if we could change people's belief that the controlled-burn fire policy leads to animals losing their homes, then we might change their attitude toward policy itself.

Unfortunately, attitude change is notoriously difficult. Research has suggested that if you can change the behavior of 5% of your targeted audience, your campaign can be considered successful. Why is it so difficult? Four classes of variables are known to affect attitude change. Source variables relate to who is delivering a message and whether they are perceived to be trustworthy or credible. Receiver variables deal with people targeted by the message; what are their current attitudes and how involved are they in the topic? Message variables describe the characteristics of the appeal, including its content and style of its presentation. Finally, the medium deals with the channel through which delivery occurs—via personal appeals, brochures, radio, speeches, or television. The interaction among source, receiver, message, and medium is complex, so simple decision rules on how to change attitudes are non-existent.

Important advancements in attitude-change theory have occurred in the last decade. One suggests a need for a distinction between situations of persuasion with high elaboration and situations of persuasion with low elaboration. High elaboration persuasion occurs when the recipient of information attends to the information, comprehends it, and integrates it into his/her existing thoughts. Persuasion can also occur with low elaboration when people do not attend to the persuasive message, but are influenced by something tangential to the message content. This might include variables such as the attractiveness or credibility of the person delivering the message.

This has important implications for natural resource managers, who often develop public information that requires a high degree of elaboration. Hunt and Brown (1971) revealed that among a sample of brochures developed by public land managers, most were evaluated at the college level of reading difficulty and were dull in terms of human interest. Resource managers clearly need to begin considering the use of low-elaboration persuasion attempts (simpler, easier to understand messages, more attractive presentation).

ATTITUDE CHANGE EXAMPLE

When developing persuasive appeals, managers should first attempt to understand the current beliefs that serve as the basis for attitudes and behaviors, and target them in developing the content of messages.

This approach is illustrated in a follow-up to the study described in the first example (Fishbein and Manfredo 1992; Bright et al. 1993). A sample of people visiting Yellowstone National Park were interviewed on-site to determine their attitude toward a controlled-burn fire policy. The sample was divided into two groups: those with positive attitudes and those with negative attitudes. Differences in beliefs between these two groups served as the basis for two messages that were developed to change the attitudes of people surveyed. One message was intended to change those with negative attitudes (to make them more positive); the other message was to change positive attitudes (to make them more negative). The strategy of message development was to target beliefs held by each group (e.g., the negative group believed that the controlled-burn policy caused animals to lose their homes, so the message presented evidence to the contrary). A follow-up survey showed that targeted beliefs changed as expected. However, while the degree of change was sufficient to alter the group with positive attitudes (making them more negative), it was not sufficient to change the group with negative attitudes. Results revealed the complexity of attitude-change attempts, and underscored the importance of developing persuasion strategies that consider multiple aspects of the persuasion process.

Values

Another commonly used term in human-dimensions research is *values* (Decker and Goff 1987). The term *value* has been used in two very different ways. The distinction between the two uses can be thought of as the distinction between *assigned* and *held* values. Assigned values are an indication of how important something is to us. This usage is commonplace in economics. For example, economists often attempt to identify the monetary value of nonmarket goods (how much we would pay for scenes or preservation of wildlife species).

Held values are "basic evaluative beliefs" and serve as the building blocks of attitudinal positions and behaviors. For example, one's attitude toward specific issues such as recycling, littering, contributing to environmental groups, or rain forests, may all have a common thread based on values regarding environmentalism.

Values form gradually within an individual. They are shaped largely during one's youth through the environmental surroundings and the people in our lives (parents, peers, and teachers). A potential solution to management problems is to initiate programs to change people's environmental values. It is important to realize, however, that this would require a most extraordinary undertaking. Changing people's attitudes is possible, but difficult. Changing their values, at least using short-term, narrowly focused approaches, is quite improbable.

VALUE EXAMPLE

In the wake of 1992 ballot initiatives the Colorado Division of Wildlife embarked on a revision of its long-range plan. One key initiative placed a ban on spring bear hunting in the state, while another provided money from lottery proceeds for wildlife, parks, open spaces, and trails. These initiatives signaled a change in the public's values regarding wildlife. Consequently, as part of the planning process, a survey was conducted to describe Coloradans' wildlife values.

A research team composed of agency managers and university researchers developed and used measurement scales to assess the following values:

Value topic	Short description
Bequest and existence	Knowing healthy populations exist and ensuring them for future generations.
Recreational	Importance of wildlife in leisure and recreation.
Residential	The importance of wildlife in the neighborhood and around the home.
Educational	Importance of people learning about wildlife.
Animal use	Feeling about whether wildlife should be used in a way that provides benefits to society.
Animal rights	Feeling about treatment and rights of individual animals.
Hunting and fishing uses	Focus on whether or not fishing and hunting are positive and humane leisure pursuits.

Results indicated that, overall, the public was quite positive toward bequest and educational values, but was diverse on other values important to them. Cluster analysis allowed the sample to be divided into subgroups. One group, estimated to be 27% of all Coloradans over 18 years of age, was distinguished by strong values toward animal rights. A second group (31% of Coloradans) was marked by strong values toward animal use, hunting, and fishing. A third (32% of Coloradans) held a moderate position between these two extremes. The fourth group (9%) showed low interest in recreational and residential uses of wildlife, but strong animal-use values. The large proportion of people who place importance on animal rights was surprising and suggests a unique clientele for Colorado's Division of Wildlife.

Norms

The normative approach has been developed as a useful way to conceptualize, collect, and organize empirical data representing value judgments about issues in human dimensions (Vaske et al. 1986; Shelby and Vaske 1991). Norms are standards that individuals use for evaluating activities, environments, or management proposals as good or bad, better or worse. They define what people think behavior should be.

Cialdini et al. (1990) identify two types of norms. Injunctive norms are rules about what behaviors you should exhibit, while descriptive norms reflect what behavior actually occurs. The distinction is illustrated by a recent investigation we conducted to examine illegal feeding of wildlife at Rocky Mountain National Park. People realized the appropriate behavior was to refrain from feeding wildlife at overlooks (injunctive norm), yet more than 50% engaged in feeding behavior (descriptive norm).

Formal sanctions (e.g., laws, regulations) and informal sanctions (dirty looks, comments from others, etc.) are often associated with norms. Different groups of people have different norms about appropriate behavior. Inexperienced wildlife viewers, for example, may feel it is appropriate to get close to others who are attempting to photograph wildlife, while experienced viewers may consider that a violation of unwritten rules of conduct among wildlife enthusiasts. Norm violations often form the basis for recreation conflict.

Much of the current human-dimensions, normative research is based on the work of Jackson (1965), who proposed a model to describe norms (evaluative standards) by means of a graphic device that we have referred to as an impact-acceptability curve (see Vaske et al. 1986, for a review). The curve can be analyzed for various normative characteristics including optimum conditions, the range of tolerable conditions, the intensity or strength of the norm, and the crystallization or level of group agreement about the norm. Evaluative standards for encounters in a wilderness setting, for example, might have an optimum of zero encounters, a low range of tolerable contacts, high intensity, and high crystallization, while norms for a wildlife-viewing area might show a greater tolerable range, lower intensity, and less group agreement.

NORM EXAMPLE

The Brule River study was the first attempt to use Jackson's model to describe encounter norm curves (Vaske 1977). Visitors were asked how they felt about seeing 0, 1, 2, 3, 5, 7, 9, 15, 20, and 25 canoers, tubers and anglers. Zero was the optimum encounter level, with higher levels rated as relatively less pleasant. For canoers, the mean range of tolerable contacts was smallest for encounters with tubers (0 to 2.3); increasing to 0 to 5.7 for contacts with other canoers, and 0 to 7.2 for encounters with anglers. These contact norms were held with different intensities. Encounters with tubers provoked the most extreme reactions, resulting in the highest intensity. Encounters with anglers provoked reactions nearer to the neutral line, so intensity here was lowest, and intensity for encounters with canoes was in between. Crystallization, or the level of group agreement, for all three encounter norms was fairly comparable, although there was most consensus about encounters with canoes and least about those with tubers.

Using the upper limits of the encounter tolerance ranges (three for tubers, seven for canoers, and nine for anglers), data on encounter reports were used to examine how often these standards were exceeded. Canoers reported having more than three encounters with tubers 37% of the time, more than seven canoes 76% of the time, and more than nine anglers only 5% of the time. Thus, the canoers' standard for meeting canoers was exceeded most of the time and some of the time for tubers. The standard for anglers was rarely exceeded.

During the year following the Brule River study (1976), a management plan was introduced that restricted the number of canoes and eliminated tubers from the river. Although the restrictions on canoeing were consistent with the normative tolerance limits, understanding the ban on tubing requires an examination of the intensity and crystallization indicators. Because tubing represented a nontraditional activity on the Brule, encounters with even a few individuals engaged in this new form of recreation evoked the most intense negative reactions. Crystallization of an encounter norm may also be a function of an activity's heritage on a river. Trout fishing, for example, began as a recreational pursuit on the Brule before the turn of the century. This history of usage increases the probability that anglers were relatively more aware of a norm regarding appropriate encounters.

Conceptually, the Jackson analytical approach applies equally to normative evaluations of wildlife management practices (Vaske et al. in press). New

Hampshire, for example, had not had a moose-hunting season since the early 1900s, but in recent years a moose population has become re-established. The state's Fish and Game Department proposed a hunting season for the fall of 1988. Data showing the norms held by different groups of New Hampshire residents (conservation groups, hunters, and the general public) for the proposed hunt illustrate the application of the model to wildlife management.

Respondents to a mailed survey rated their overall reactions to re-introducing the controlled moose hunt; their tolerances for acceptable statewide harvest limits; and their norms regarding an acceptable season length, a fair price for a resident moose license, and the percentage of licenses that should be allowed for nonresidents.

Although the different publics varied in their level of support, a majority of each group were favorable to the hunt (general public, 60%, conservation groups, 51%, hunters, 81%). Hunters favored a higher harvest limit than the other two publics, although their norm was consistent with the limit proposed by the state's Fish and Game Department. Norms for season length varied statistically between the hunters and the other two groups, with hunters specifying a slightly longer season. The norms, however, were again within the range suggested by Fish and Game. A similar pattern of findings emerged for the three publics' norms for the fair price of a moose hunting license. The norms for percentage of nonresident licenses did not vary significantly among the three groups and were consistent with the Fish and Game proposal.

Motivation

The fourth social psychological concept applied in the human-dimensions area is motivation and the related topics: needs, satisfactions, and desired psychological outcomes. This area of theory addresses why people behave the way they do. It is similar to attitude theory, but represents a unique and distinct theoretical tradition in the literature.

Motivation has been addressed in the work of need classification theorists who suggest that humans have five levels of need, including physiological needs, safety needs, belongingness and love needs, esteem needs, and the need for self-actualization. While researchers have debated the utility of the term "need," they have continued to focus effort on a classification of factors that motivate people in their day-to-day behavior. This has received considerable attention in the area of human dimensions.

Decker et al. (1989) proposed three general motivational orientations for wildlife recreation including affiliative, achievement, and appreciative orientations. Driver et al. (1991) developed and refined a classification of motivational factors labeled "desired psychological outcomes" through a series of more than 35 studies involving more than 30,000 survey participants.

The conceptual foundation of this approach was based primarily on expectancy-valence motivation theory. This theory suggests that behavior is a function of a person's motivation and ability. Motivation to engage in recreational activity can stem from two different expectancies: (1) the expectation that expended effort (travel costs, time) will lead to certain occurrences (catching fish, being alone, seeing a high degree of naturalness in the scenery), and (2) the expectation that these occurrences will lead to valued psychological outcomes (stress reduction, achievement, privacy).

The psychological outcomes examined in outdoor recreation include: enjoyment of nature, introspection, physical fitness, social security, reduced tension, achievement/stimulation, escape from physical stressors, physical rest, outdoor learning, risk-taking, sharing similar values, risk reduction, independence, meeting new people, family relations, creativity, and nostalgia. Questionnaire items have been developed that can be used to measure these outcomes (Driver et al. 1991; Manfredo and Driver, in press).

The concept of motivation has been useful in explaining the basis of conflict among recreationists, in determining user segments, and in the planning process for making inferences to the benefits of leisure.

MOTIVATION EXAMPLE

Manfredo and Larson (1993) illustrate how the concept of motivation helps develop wildlife-viewing user segments. The study provided information that could be used in planning for a variety of desirable wildlife experiences. A survey, which included desired psychological outcome items (Driver et al. 1991), was administered to a sample of residents in the Denver metro area of Colorado. Using responses to these items, cluster analysis was conducted to divide the sample into subgroups, based on motivations for taking trips to view wildlife. Four types of preferred wildlife-viewing experiences were described and recommended for consideration in the planning process. High involvement experiences provide participants with many types of outcomes. Opportunities to teach and lead others were particularly important outcomes. Creative experiences allowed participants the opportunity to fulfill creativity motivations through activities such as painting, photography, and sketching. Generalist experiences focused on outcomes related to change of pace, affiliation with friends and family, and change of scenery. Occasionalist experiences were for those who participate in wildlife viewing rather infrequently and who do so for change of pace. Study results are being used by Colorado Division of Wildlife to assist in developing a planning system for providing wildlife-viewing recreation.

Satisfaction

There have been two major uses of the satisfaction concept. The first identifies the types of satisfaction a person receives from recreation. In concept, motivation and satisfaction are similar to cause and effect. While motivation focuses on what initiates behavior, satisfaction focuses on the result of action (i.e., the satiation of motivations). Satisfaction deals with the extent to which the motivational forces people acted upon were actually fulfilled. While conceptually distinct, applications of these two concepts in the human-dimensions literature have been interchangeable.

A second use of satisfaction is as an evaluative concept to measure how "happy" people are, overall, with a recreation trip. This use of the satisfaction concept is quite similar to the notion of attitude discussed earlier. Global measures of satisfaction should be regarded with some suspicion since they are likely only to identify major changes in the quality of service delivery. Research has indicated that there is little variability in measures of overall satisfaction (Shelby and Heberlein 1986; Kuss et al. 1990). These literature reviews suggest that overall measures may not accurately reveal levels of satisfaction with numbers and types of other recreationists at a setting.

A holistic approach to assessing satisfaction is offered in the multifaceted-discrepancy model. The multifaceted component of the model suggests that overall satisfaction is a function of satisfaction with various components of a recreation experience. Satisfaction with a recreation trip, for example, may be a function of satisfaction with the natural environment, the social surroundings, the extent of effort expended, and the presence of management. Satisfaction within a particular facet will be a function of the discrepancy between what one expected of, and what one received from, the desired aspects of the trip. If you expected, and wanted, to see two other people when fishing and you saw twenty, your level of satisfaction would be low.

Satisfaction Evaluation Example

One of the most widely recognized uses of the satisfaction concept was proposed by Hendee (1974) who suggested that hunting recreation should be managed for multiple satisfactions, as opposed to more traditional methods of management. These include the "game bagged approach" (a problem because it suggests only hunters who bagged game receive benefits) and the "days afield approach" (which suggests everyone derives the same benefits from time spent). Hendee proposed hunting be managed for the satisfactions derived from it using the following typology:

Backcountry hunt	Focuses on solitude, companionship, escapism, nature appreciation, outdoor skill, trophy, exercise.
General-season party hunt	Focuses sociability, camp life, escapism, equipment, harvesting game, exercise.
Meat hunt	Focuses on harvesting game, small party, skill, road hunting.
Special skill hunts	Focuses on outdoor skill and equipment.

The concepts introduced by Hendee (1974) are quite similar to those proposed by Driver et al. (1991) in our previous section on motivation.

Colorado Division of Wildlife used an evaluative measure of satisfaction when examining new buck-deer hunting regulations effective for the 1992 hunting season. These regulations drastically reduced the number of days available for buck-deer hunting. The regulations were intended to provide better buck escapement and, in the long run, provide larger buck deer in the state. As part of an effort to assess the impacts of these regulations, a longitudinal study was conducted to examine shifts in satisfaction with deer hunting in the state.

Two surveys were conducted. The first assessed satisfaction with the 1991 deer hunting season, which was held prior to the new regulations. The survey was administered to a sample, stratified by residence, of 1991 rifle buck-deer hunters. One of the questions asked respondents to rate satisfaction with their 1991 deer-hunting experience. These same people were contacted a second time after the 1992 season. Subjects were asked whether or not they participated in buck-deer hunting in Colorado, and for those who hunted, their satisfaction with the 1992 deer hunting experience.

Results indicated that buck-deer hunting satisfaction dropped dramatically between 1991 and 1992 for those in the study. These findings will allow the Colorado wildlife commission to re-examine and adjust the regulations in a way that preserves buck escapement, while reducing negative impacts to recreationists.

Knowledge Gaps

While interest in human dimensions of wildlife has grown considerably in recent years, it should be remembered that this is a rather recent area of scientific inquiry, with involvement from a broad range of disciplines. Consequently, the gaps in knowledge are considerable, reflecting an incipient stage of scientific development.

One problem is the relatively small number of studies in human dimensions of wildlife; even fewer examine human/wildlife interactions. A related problem is that new ideas or new issues are introduced to the literature and subsequently abandoned, so few substantive generalizations can be drawn from the literature. Moreover, because human-dimensions research is conducted where funding is available, we know a great deal more about certain recreation activities such as hunting and fishing than about other types of outdoor recreation, such as rock climbing and canoeing.

A great deal of research has been largely descriptive. Too often, this research offers no conceptual foundation and makes no suggestion about theory development (see Chapter 1). While we recognize that human dimensions is an applied field of study, theoretical development is critical if we are to predict and manage wildlife/human interactions.

A comprehensive review of knowledge gaps is beyond the scope of the present chapter. It is worth noting, however, a few broad generalizations about future research needs. Future studies of recreation-wildlife interactions would benefit from an examination of, from the human perspective, whether the interaction is purposive (taking trips to view wildlife), incidental (taking trips in which encounters with wildlife would be a possible outcome), or unintended.

A comprehensive approach to studying recreation-wildlife interactions should examine four areas. The first deals with understanding the factors that lead to human-wildlife interactions. For example, what are people's motivations for taking wildlife viewing trips, for approaching wildlife in parks, or for feeding wildlife (all purposive behaviors)? Also important would be attempts to explore the relationship between knowledge about wildlife and unintended impacts to wildlife.

A second area of investigation deals with the factors that dictate the flow and nature of interactions. For example, what norms of behavior or beliefs about wildlife are associated with people's responses during specific types of wildlife encounters? A third area of questioning concerns the types of short- and long-term effects resulting from interactions. To what extent do interactions with wildlife affect knowledge about wildlife? What effect do interactions with wildlife, particularly at early ages in humans, have on wildlife values and on attitudes toward wildlife uses?

A fourth area should address the extent to which we can influence and control recreation-wildlife interactions. For example, when impacts to wildlife are unacceptable, what types of messages and modalities are effective in changing behavior? Can educational programs in our grade schools and high schools be effective in making recreationists more sensitive to their impacts on wildlife? What are ways we can increase people's enjoyment of wildlife while minimizing the adverse consequences?

Human-dimensions represents a new and broad field that holds tremendous promise to wildlife managers (Decker et al. 1989). As illustrated in this chapter there are a host of conceptual tools that can be useful in dealing with the multiple objectives of recreation-wildlife interactions. The integrity of a human-dimensions approach in natural resources, however, rests on our ability to use and extend concepts for describing, predicting, explaining, and controlling recreation-wildlife interactions.

Literature Cited

Ajzen, I. and M. Fishbein. 1980. *Understanding Attitudes and Predicting Social Behavior.* Englewood Cliffs, New Jersey: Prentice-Hall.

Bright, A.D., M.J. Manfredo, M. Fishbein, and A. Bath. 1993. The theory of reasoned action as a model of persuasion: a case study of public perceptions of the National Park Service's controlled burn policy. *Journal of Leisure Research* 25(3):263–280.

Cialdini, R.B., R.R. Reno, and C.A. Kallgren. 1990. A focus theory of normative conflict: recycling the concept of norms to reduce littering in public places. *Journal of Personality and Social Psychology* 58:1015–1026.

Decker, D.J. and G.R. Goff. eds. 1987. *Valuing Wildlife: Economic and Social Perspectives.* Boulder, Colorado: Westview Press.

Decker, D.J., T.L. Brown, and G.F. Mattfeld. 1989. The future of human dimensions of wildlife management: can we fulfill the promise? *Transactions of the North American Wildlife and Natural Resources Conference.* 54:415–425.

Decker, D.J., T.L. Brown, N.A. Connelly, J.W. Enck, G.A. Pomerantz, K.G. Purdy, and W.F. Siemer. 1992. Toward a comprehensive paradigm of wildlife management: integrating the human and biological dimensions. In *American Fish and Wildlife Policy: The Human Dimension,* ed., W.R. Mangun, 33–54. Carbondale, Illinois: Southern Illinois Press.

Driver, B.L., H.E. Tinsley, and M.J. Manfredo. 1991. Leisure and recreation experience preference scales: results from two inventories designed to assess the breadth of the perceived benefits of leisure. In *The Benefits of Leisure,* eds., B.L. Driver, P.J. Brown, and G. Peterson, 263–287. State College, Pennsylvania: Venture Publishing.

Fishbein, M. and I. Ajzen. 1975. *Belief, Attitude, Intention and Behavior: An Introduction to Theory and Research.* Reading, Massachusetts: Addison-Wesley.

Fishbein, M. and M.J. Manfredo. 1992. A theory of behavior change. In *Influencing Human Behavior,* ed., M.J. Manfredo. Champaign, Illinois: Sagamore Press.

Hendee, J.C. 1974. A multiple-satisfaction approach to game management. *Wildlife Society Bulletin* 2:104–113.

Hunt, J.D. and P.J. Brown. 1971. Who can read our writing? *Journal of Environmental Education* 2:4(Summer):27–29.

Jackson, J.M. 1965. Structural characteristics of norms. In *Current Studies in Social Psychology,* eds., I.D. Steiner and M.F. Fishbein, 301–309. New York: Holt, Rinehart, and Winston.

Kuss, F.R., A.R., Graefe, and J.J. Vaske. 1990. *Visitor Impact Management: A Review of Research.* Washington, D.C.: National Parks and Conservation Association. 362 pp.

Manfredo, M.J. ed. 1992. *Influencing Human Behavior.* Champaign, Illinois: Sagamore Press.

Manfredo, M.J. and B.L. Driver. A meta analysis of the recreation experience preference scales. *Journal of Leisure Research,* in press.

Manfredo, M.J. and R. Larson. 1993. Managing for wildlife viewing recreation experiences: a case study in Colorado. *Wildlife Society Bulletin* 21:226–236.

Manfredo, M.J., M. Fishbein, G.E. Haas, and A.E. Watson. 1990. Attitudes toward prescribed fire policies. *Journal of Forestry* 88:19–23.

Shelby, B. and T.A. Heberlein. 1986. *Carrying Capacity in Recreation Settings.* Corvallis, Oregon: Oregon State University Press.

Shelby, B. and J.J. Vaske. 1991. Using normative data to develop evaluative standards for resource management: a comment on three recent papers. *Journal of Leisure Research* 23:173–187.

Vaske, J.J. 1977. The relationship between personal norms, social norms and reported contacts in Brule River visitors perception of crowding. Masters thesis. Madison, Wisconsin: University of Wisconsin.

Vaske, J.J., B. Shelby, A. Graefe, and T.A. Heberlein. 1986. Backcountry encounter norms: theory method and empirical evidence. *Journal of Leisure Research* 18:137–153.

Vaske, J.J., M.P. Donnelly, and B. Shelby. Establishing management standards: selected examples of the normative approach. *Environmental Management,* in press.

CHAPTER 3

Human Dimensions of Wildlife Management: An Integrated Framework for Coexistence

Jerry J. Vaske, Daniel J. Decker, and Michael J. Manfredo

Outdoor recreation may influence wildlife behavior and populations, the amount and diversity of vegetation and soil, and the quality of the visitor's experience. Each of these impacts has its own body of literature and the existing knowledge pertinent to each particular discipline is well documented (see Chapter 4; Hammitt and Cole 1987; Shelby and Heberlein 1986). Relatively little attention, however, has focused on integrating the findings across ecological and social research (Decker et al. 1992; Kuss et al. 1990). This failure to integrate the available empirical evidence has limited the application of research data to visitor impact management, because natural resource planners must contend with both ecological and social issues; not one or the other. Moreover, ecological and social impacts are often interrelated. Perceptions of ecological disturbance, for example, may influence the quality of a visitor's experience in much the same way as do conflicts arising from interactions among user groups.

Recreational Influences

Although much has been written about the advantages of an interdisciplinary approach to visitor impact management and the need to improve the researcher-practitioner relationship, resolution of these issues remains a major stumbling block. Our chapter seeks to address these problems by combining a Visitor Impact Management (VIM) process (Graefe et al. 1990) with findings from previous research on visitor impacts (Kuss et al. 1990). Similarities and differences inherent in the ecological and social impact literature are discussed first.

In general, the differences between the two areas of emphasis center around different research procedures and associated difficulties and limitations. Ecological impacts occur within ecosystems characterized by complex interactions between plant and animal species. Wall and Wright (1977) suggest four factors that limit ecological studies and introduce difficulties in identifying human impact: (1) there are often no baseline data for comparison to natural conditions; (2) it is difficult to disentangle the roles of humans and nature; (3) there are spatial and temporal discontinuities between cause and effect; and (4) in light of complex ecosystem interactions, it is difficult to isolate individual components. Some impacts take the form of naturally occurring processes that have been accelerated by human interference. Even without human activity, however, severe impacts can occur because of natural fluctuations and disturbances that render effects associated with recreational use insignificant (Schreyer 1976).

Impacts on wildlife are perhaps most difficult to identify. Research findings are often mixed and animal responses to human intruders are divergent, even within a single species (see Chapter 6). For example, although many studies describe avoidance behavior by animals as a result of human interaction, less research has focused on the long term effects of this behavior. Likewise, little attention has been given to the relationships between numbers of visitors and wildlife behavior and population changes.

Research on social impacts avoids the problem of multiple species and concentrates on only the human response to other visitors. As a result, many studies can be identified that deal specifically with relationships between use levels and visitor experience parameters. The understanding of social impacts, however, remains incomplete because of the complexity of human values and behavior (see Chapter 2). In addition, some types of social impacts are difficult to evaluate due to logistical constraints. Displaced visitors who no longer use a given area, for example, cannot be located easily. To measure the psychological adjustments that visitors make when confronted with too many people requires elaborate procedures that are usually beyond the time and budget constraints of most field studies.

Despite these differences and difficulties associated with ecological and social research, there appear to be several general areas where they overlap. A recent review of the scientific literature concerning visitor impact management (Kuss et al. 1990) concluded that there are five major sets of considerations that are critical to understanding the nature of ecological and social impacts.

Impact Interrelationships

There is no single, predictable environmental or behavioral response to recreational use. Instead, an interrelated set of impact indicators can be identified. Some forms of impact are more direct or obvious than others (e.g., displacement of wildlife species or altered visitor experiences). The direct impacts of recreation on wildlife are best described by the term "harassment," and include the various responses of wildlife to interactions with people. Indirect impacts result from changes in vegetation, habitat, or other environmental variables. To understand how wildlife and the visitor's experience in a given area are affected by recreational use, it is necessary to consider a range of possible impact variables.

Use-Impact Relationships

The relationships between use levels and impact variables are neither simple nor uniform. Most impacts do *not* exhibit a direct linear relationship with user density. It appears that the number of people using a given area plays a smaller role in human-wildlife relationships than selected characteristics of recreational use, such as frequency of use, type of use, and the behavior of visitors (see Chapter 5).

Varying Tolerance to Impacts

Different types of wildlife and user groups have differing tolerances for interactions with people. Some wildlife species have declined in response to increasing use levels, while other species have benefitted. Animal responses to human intruders are often divergent even within a single wildlife species.

Activity-Specific Influences

Some recreational activities create impacts faster or to a greater degree than other types of activity. Impacts can vary even within a given activity according to type of transportation or equipment used and visitor characteristics such as party size and group behavior (see Chapters 4 and 5).

Site-Specific Influences

The impacts of recreation are influenced by a variety of site-specific and seasonal variables. Given a basic tolerance level for a particular type of recreation, the outcome of recreational use may still depend on the time and place of the human activity. Recreational users may create critical situations for given species during some seasons and have no effect on the same animals under other conditions.

These five issues represent important management considerations regardless of the type of impact—ecological or social.

The Visitor Impact Management (VIM) process builds upon the widely accepted notion that effective management involves both scientific and judgment considerations (Hendee et al. 1978; Stankey 1980; Shelby and Heberlein 1986). The scientific component focuses on documenting the relationships within the system and thereby provides the data needed to predict the impacts of different planning alternatives (e.g., doubling recreational use of a wildlife viewing area will increase the average number of human-wildlife encounters by some number). The evaluative (judgment) component is concerned with the desirability of different management alternatives and the associated levels and types of impact on the experience.

Effective management is also more than carrying capacities and use limits (Washburne 1982). While use quotas represent one possible strategy for reducing the impacts of visitors, research has repeatedly demonstrated that only a weak relationship exists between impacts to the experience and overall use levels (Graefe et al. 1984; Kuss et al. 1990; Manning 1985). In such instances, establishing capacities and limits may do little to reduce the impact problems they were intended to solve, whereas other potential management strategies may be quite effective for reducing the impact conditions.

The VIM framework includes an eight-step sequential process for assessing and managing recreation impacts (Fig. 3.1). The steps in this process are designed to help managers in dealing with three basic issues inherent to impact management: (1) the identification of problem or unacceptable visitor impacts; (2) the determination of potential causal factors affecting the occurrence and severity of unacceptable impacts; and (3) the selection of potential management strategies for ameliorating the unacceptable conditions. Special attention is paid to illustrating the integration of management and research, including making use of existing data sources as well as identifying the most critical research questions.

The first five steps are devoted to the important, yet often slighted, task of documenting problem conditions. This basic issue is separated into several steps to highlight the decisions that must be made to assess the current situation. These steps use the concepts of objectives, indicators, and standards, which are also the central elements of other current frameworks for outdoor recreation resource management (Shelby and Heberlein 1986; Stankey et al. 1985).

STEP 1

The objective of Step 1 is to identify and summarize what is already known about the situation, so that existing information can be put to its best use as

Basic Approach - Systematic process for identification of impact problems, their causes, and effective
management strategies for reduction of visitor impacts.
Conditions for Use - Integrated with other planning frameworks or as management tool for localized impact
problems.

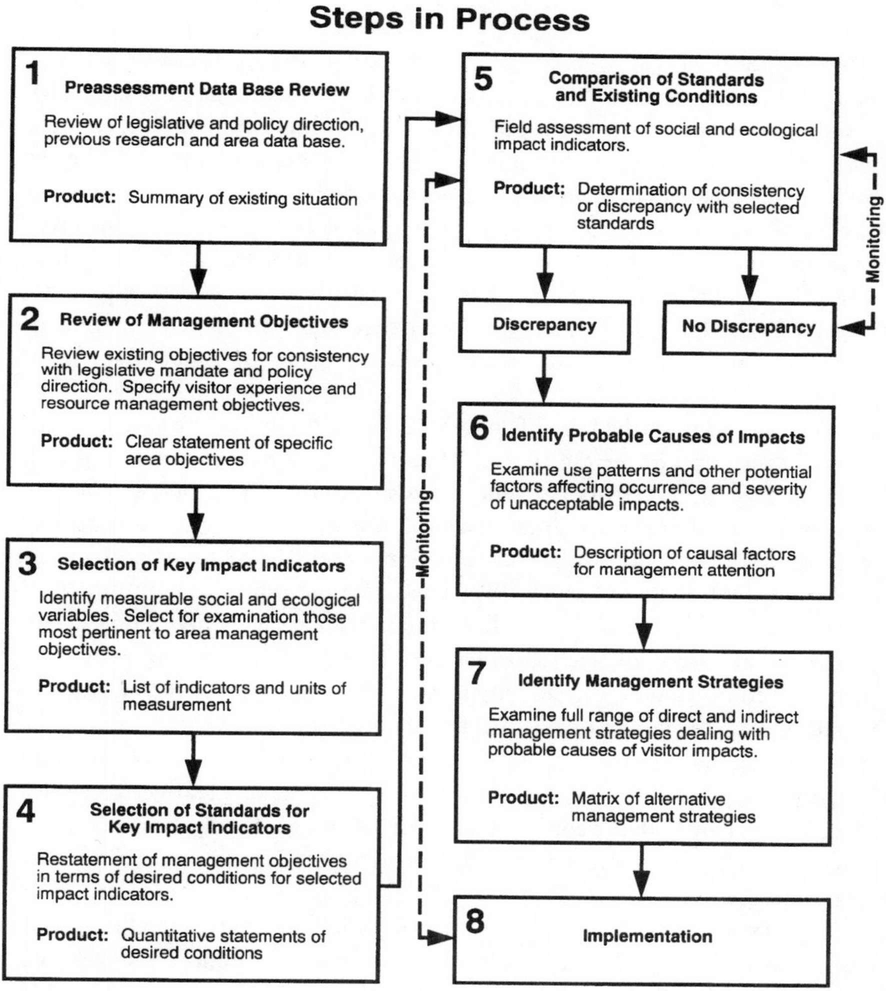

Steps in Process

1 Preassessment Data Base Review

Review of legislative and policy direction,
previous research and area data base.

Product: Summary of existing situation

**5 Comparison of Standards
and Existing Conditions**

Field assessment of social and ecological
impact indicators.

Product: Determination of consistency
or discrepancy with selected
standards

2 Review of Management Objectives

Review existing objectives for consistency
with legislative mandate and policy
direction. Specify visitor experience and
resource management objectives.

Product: Clear statement of specific
area objectives

Discrepancy   **No Discrepancy**

6 Identify Probable Causes of Impacts

Examine use patterns and other potential
factors affecting occurrence and severity
of unacceptable impacts.

Product: Description of causal factors
for management attention

3 Selection of Key Impact Indicators

Identify measurable social and ecological
variables. Select for examination those
most pertinent to area management
objectives.

Product: List of indicators and units of
measurement

7 Identify Management Strategies

Examine full range of direct and indirect
management strategies dealing with
probable causes of visitor impacts.

Product: Matrix of alternative
management strategies

**4 Selection of Standards for
Key Impact Indicators**

Restatement of management objectives
in terms of desired conditions for selected
impact indicators.

Product: Quantitative statements of
desired conditions

8 Implementation

— Monitoring —

Figure 3.1   Management/planning process.

the process continues. Management decisions can then be based on the best
available understanding of the overall recreation system. The amount of relevant material will vary from situation to situation, but there will always be
some background information that can be used to establish an initial perspective on the problem. Planning documents, for example, may include useful
data as well as management guidance or constraints. Area records and pre-

vious surveys may provide baseline data on recreationist characteristics, motivations, and participation patterns within the area.

STEP 2

The second step in the VIM process is to review planning objectives pertinent to the situation at hand. The importance of clear and specific planning objectives is a common theme in the literature (Shelby and Heberlein 1986; Stankey et al. 1985). "A major shortcoming in most . . . management plans is the lack of specific objectives that allow managers to explicitly state the conditions they seek and to measure performance with regard to achieving these objectives" (Hendee et al. 1978:80). The objectives should be prioritized, since any single objective may lead to potentially conflicting goals. To be effective, planning objectives need to define the type of experience to be provided in terms of appropriate environmental and social conditions (Stankey 1980; Graefe et al. 1990).

The definition of the type of experience to be provided in a given area requires a selection of certain type(s) of recreationists over competing groups seeking different types of experiences. Such decisions go beyond a choice between allowing hunting/angling versus wildlife viewing in a given setting. Distinct subgroups of both hunters (Hautaloma and Brown 1978) and wildlife viewers (Manfredo and Larson 1993) have been identified. The hunting literature, for example, has shown that hunters can be classified along a continuum that reflects differences in levels of participation, types of equipment used, the importance of the activity to the individual, and the motivations for hunting (e.g., bringing home game, being outdoors). The wildlife viewing literature suggests a somewhat similar continuum. Manfredo and Larson (1993) identified four distinct experiences (high involvement, creative, generalist, and occasionalist) that people prefer when viewing wildlife. In general, this "experienced-based management" approach is based on the recognition that traditional activity classifications (e.g., hunting, angling, wildlife viewing) are inadequate in guiding management because they do not reflect what people seek (what motivates behavior) and receive (the satisfaction and benefits) from a recreation engagement (Driver and Brown 1975; Manfredo et al. 1983).

While planners are often reluctant to make explicit decisions for distinct experience types, it is important to recognize that this judgment is inherent in people management and will occur regardless if it is not made deliberately. Avoidance of a specific experience-definition essentially allows those activities that can preempt other opportunities to determine the recreational character of the area (Schreyer 1976). Determining the type of experience to be provided can be difficult. The decision, however, can be guided and defended by a variety of criteria compiled during Step 1, including institutional or policy mandates, alternative opportunities available in the area, and recreationist preferences.

STEP 3

The third step in the VIM process involves identifying measurable indicators for the planning objectives. Once objectives have described the particular type of experience to be provided, this step prescribes how the specified experience will be measured. The planners must select the most important impacts or experiential attributes to serve as indicators of the targeted experience.

There is no single indicator or set of indicators that is appropriate for all situations. The choice of indicators depends on the particular type of impact under consideration and the specific characteristics of the situation. Useful indicators include those that are directly observable, relatively easy to measure, directly related to the objectives for the area, sensitive to changing use conditions, and amenable to management.

STEP 4

Step 4 adds another layer of specificity by providing standards for the previously selected impact or experience indicators. This step calls for a restatement of planning objectives in quantitative terms. Standards differ from planning objectives because they specify the appropriate levels or acceptable limits for the impact indicators designated in Step 3. The selected standards become the basis for evaluating the existing situation. This step serves the function of describing the type of experience to be provided in units of measurement compatible with available measures of the current situation.

Indicators and standards specify the type of experience one is trying to provide. The standards should be expressed using impact indicators like number of encounters with groups of a particular size or type, or number of people encountered at attraction sites (Shelby and Heberlein 1986). For the Flathead Wild and Scenic River Management Plan, Stokes et al. (1984) suggested such standards as 80% probability of encountering no more than two floater parties (or six shore parties) per day, and no more than an average of four occurrences of litter viewable from the watercraft per management unit. Standards of this nature demonstrate the importance of selecting units of analysis and indicators that are tangible, observable, and measurable qualities of a recreation experience. Such statements also reflect the site- and activity-specific influences that have been shown to influence wildlife populations and the quality of the recreation experience.

Normative information makes a vital contribution to Steps 2 through 4 (see Chapter 2). At a general level, it is necessary to decide what type of experience is to be offered. Findings reported by Vaske et al. (1993) illustrate how the norms specified by different user groups assist in this definition process. At a more specific level, norms are evaluative standards that define the important aspects of a particular experience. Data reported by Vaske et al. (1986) identify numerous ways by which norms define experiences in terms of encounters at various locations with various kinds of groups.

STEP 5

After the first four steps have clarified the desired conditions for a given area, the existing situation can be compared to this desired state (Step 5). This step requires an assessment of current conditions for the impact indicators selected in Step 3. Step 5 does not necessarily involve elaborate and costly studies. What is necessary, however, is a level of observation and measurement that provides for a reasonable comparison of existing conditions and their corresponding standards. If current measures of pertinent indicators are consistent with standards, one needs only to monitor the situation for future changes. In this instance, the site is currently providing the type of experience defined as appropriate for the area. The monitoring that is done, however, should include both the impact indicators that are most susceptible to future changes and the use patterns that may lead to changes in the status of these indicators. Such a monitoring program would provide a basis for evaluating probable causes of impacts, were they to reach an unacceptable level at some future time.

If measures for certain indicators do not meet the standards for the area, a problem situation is documented. It is then appropriate to move on to the identification of probable causes of the unacceptable impacts.

STEP 6

Because of the many potential factors that may contribute to impact conditions, the challenge of Step 6 is to isolate the most important cause(s) of the problem situation. This involves examining the relationships between recreation use patterns and the impact indicators that have exceeded their respective standards. When evaluating potential causal factors, it is important to consider the full range of aspects of use that may influence the situation, including type of use, length of stay, size of groups, time of use, concentration of use, frequency of peak use periods, overall amount of use, and behavior of recreationists. Such use/impact relationships are often complex, mediated by site characteristics, and consequently will vary with time and place.

STEP 7

With some understanding of how the number, type, and distribution of people using a given area affect wildlife or the quality of the experience, it is possible to identify a range of alternative management strategies (Step 7). Just as many aspects of use may contribute to the problem, numerous alternatives are available for dealing with the problem. At this phase, the focus is on the probable causes of the impacts rather than on the impact conditions themselves. It is important to recognize that one may never have a complete understanding of the causes underlying certain recreation impacts, nor can one predict exactly how a given planning action will affect a particular problem situation.

A classification of management strategies that might be used as tools for controlling visitor impacts is shown in Table 3.1. The strategies include direct approaches that regulate or restrict activities and indirect approaches that attempt to achieve the desired outcome by influencing recreationist behavior. Selecting the best approach for any given situation is difficult because of the wide range of potential alternatives and the difficulty of predicting all of the outcomes that can result from the options.

Table 3.1
Classification of Visitor Management Strategies

Indirect strategies	Direct strategies
Physical alterations	*Enforcement*
Improve or neglect access	Increase surveillance
Improve or neglect campsites	Impose fines
Information dispersal	*Zoning*
Advertise area attributes	Separate users by experience level
Identify surrounding opportunities	Separate incompatible uses
Provide minimum impact education	
Economic constraints	*Rationing use intensity*
Charge constant fees	Limit use via access point
Charge differential prices	Limit use via campsite
	Rotate use
	Require reservations
	Restricting activities
	Restrict type of use
	Limit size of group
	Limit length of stay
	Restrict camping practices
	Prohibit use at certain times

Source: Adapted from Hendee et al. 1978.

Planning techniques aimed at reducing a particular impact problem may adversely affect the recreational experience or may introduce other problems. For this reason, a matrix approach for evaluating alternative management strategies is recommended (Table 3.2). This approach provides a vehicle for evaluating the range of planning alternatives against a set of selection criteria. The suggested criteria cover issues related to any implementation program. A given option may seem desirable for some criteria, but less appropriate from other perspectives. A strategy with high probability of producing the desired outcome may be impractical due to the difficulty or cost of implementation, or because it causes problems related to recreationist freedom and/or other experience indicators. In general, there are no right or wrong answers for dealing with recreation impacts. The goal is to strive for a balance among criteria when selecting a particular technique.

STEP 8

The selected management strategies should be implemented as soon as possible for those situations where impacts to wildlife and/or the quality of the experience are severe. Because the nature and causes of impacts are highly variable, management programs designed to deal with these impacts should be flexible and able to respond quickly to changing conditions.

The task of managing visitor impacts is not over when a management program has been implemented. Monitoring of key impact indicators is critical to determine whether the management actions are producing the desired outcomes without altering other experiential characteristics.

APPLYING THE VIM FRAMEWORK

The Trustees of Reservations (TTOR) is the world's oldest private, non-profit land conservation organization. TTOR's fundamental mission is to preserve for public enjoyment places of exceptional scenic, historic, or ecological value throughout Massachusetts. More specifically, the goal is to protect sensitive shorebird populations and fragile dune systems from the impact of human use, while providing opportunities for recreation activities (Deblinger et al. 1989).

In 1986, TTOR initiated a program of ecological and social research to assist in the development of management plans for three barrier beaches managed by the organization: Cape Poge Wildlife Refuge and Wasque Reservation, Edgartown; Coskata-Coatue Wildlife Refuge, Nantucket; and Crane Beach, Ipswich (see Vaske et al. 1992 for a description of the study areas).

Ecological research focused on the effects of trampling sand dune vegetation (Carlson and Godfrey 1989a, 1989b), and human/predator impacts on shorebirds (Rimmer and Deblinger 1990; Deblinger et al. 1991). A shorebird

Table 3.2

Matrix for Evaluation of Alternative Management Strategies

Management strategy	Consistency with management objectives	Difficulty to implement	Probability of achieving desired outcome	Effects on visitor freedom	Effects on other impact indicators
Indirect strategies					
Physical alterations					
Information dispersal					
Economic constraints					
Direct strategies					
Enforcement					
Zoning					
Use rationing					
Activity restrictions					

Source: Adapted from Kuss et al. 1990.

research and protection program was instituted between 1986 and 1989. Impacts to nesting piping plovers were measured through direct observation and predator population surveys. Finally, a study of human disturbance on migratory shorebirds at Crane Beach was conducted during 1990 (Deblinger et al. 1991).

The social research consisted of a series of on-site visitor surveys at each beach. Surveys were conducted at Cape Poge Wildlife Refuge and Wasque Reservation during 1987 and 1988, at Coskata-Coatue during the summer of 1990, and at Crane Beach during 1990 (Vaske et al. 1992).

Results from the vegetation studies at all three beaches revealed that human foot traffic as well as off-road vehicle (ORV) use was having a detrimental effect on the dunes. Where people accessed dunes, vegetation cover and dune height were significantly lower than areas unimpacted by visitors.

At Crane Beach, the mean percent plant cover (an ecological impact indicator) was, on average, 45% lower at disturbed sites than undisturbed sites (Carlson and Godfrey 1989a, 1989b). Management techniques such as fencing and vehicle ramps significantly increased plant cover after three years along these same transects. Dune damage was greatest when caused by ORVs, next by human foot traffic (20% more plant cover), and least by deer (40% more plant cover). Similar results were observed for ORVs and pedestrians at Cape Poge/Wasque.

Research results regarding nesting birds indicated that predator impacts were more serious than those caused by visitors. Single-strand wire fences at Crane Beach effectively segregated visitors from nesting areas (Rimmer and Deblinger 1990). At Cape Poge/Wasque, pedestrian visitors were easily kept out of nesting areas, but off-road vehicles were much more difficult to segregate as they did not always remain outside of fenced areas, especially at night. Because the vehicles tended to remain within a single two-track roadway, it was more difficult to spatially restrict vehicles when they occurred on the beach. Pedestrians tended to create many different pathways through the beachgrass at Crane Beach and Cape Poge/Wasque. Pedestrians were successfully channeled along paths using ramps, boardwalks, or fencing.

Results from the Crane Beach migratory shorebird research indicated that while four different bird species were not disturbed by boater's off-shore activities, they were disturbed by pedestrian activities. Shorebird behavior and distribution patterns were related to tidal action and beach topography (Deblinger et al. 1991). At high tide, the shallow, muddy feeding flats became inundated with water and were inaccessible to shorebirds. At that time they rested above the high tide line in the dunes. Prior to and after high tide, the birds preferred the mud flats. Because of good anchorage, the boaters preferred another section of the beach, thus creating a natural zoning between the birds and visitors.

Efforts to reduce human-environmental impacts involved a number of management strategies. The ecological data highlighted the sensitivity of beachgrass to human trampling. To minimize the impact of foot traffic at Crane Beach, three boardwalks were constructed to move people from the parking area to the beach. At Wasque, pedestrian traffic was channeled through the dunes using fences, signs, and boardwalks. The toe of the primary dune at the Wasque swimming beach was fenced with three-strand, smooth wire to prevent pedestrian trespass on the primary dune. Snow fencing, wire fencing, and beachgrass planting was used to restore the blow-outs at Cape Poge. To mitigate the impact of ORVs on the dunes at all three locations, some existing access locations were closed, and special vehicle ramps were constructed where access was allowed to the beach.

The management technique selected to protect plovers from predators and eliminate disturbance by visitors depended on nest location and visitor activity. Findings from the shorebird research at Cape Poge/Wasque and Crane Beach indicated that predators were causing greater impact than either pedestrians or ORVs. Two types of protection were applied to mitigate these impacts. Small wire-mesh fences were installed around nests to protect piping plovers from striped skunks, raccoons, red fox, gulls, and crows. Outside of these exclosures, symbolic fencing composed of a single strand of twine was erected to eliminate disturbance by visitors. These areas were posted with signs to educate visitors about nesting shorebirds. Because Coskata-Coatue is devoid of mammalian predators such as skunks and red foxes, protection only from human disturbance was implemented.

These methods, when combined with visitor education programs, proved effective for pedestrian visitors. When ORVs are allowed in an area, single-strand fences are neither effective nor popular. Management techniques for ORV use areas were designed to physically deter vehicles and their operators. The wooden snow-fencing used at Wasque has effectively eliminated ORV traffic from selected wildlife management areas. By restricting use of nesting areas during critical seasons, as opposed to prohibiting use altogether, both plovers and humans can exist sympatrically. Support for these spatial and temporal restrictions is further enhanced when beach closures are kept relatively small and recreationists understand the rationale for the closure. There is, however, a brief period of time after chicks hatch and before they fledge, when they leave protected areas and can be in conflict with ORVs.

The Crane Beach shorebird disturbance study indicated a natural zoning existed between visitors and the birds. Most boaters were attracted to a portion of the beach that, due to habitat considerations, the birds did not use for either feeding or resting. By designating restricted boat-landing areas outside of the feeding flats and restricting visitors from bird-nesting areas in dunes, migratory shorebirds could stop over at Crane Beach undisturbed.

The social research indicated that the beliefs held by some of the visitor groups conflicted with the management goals. Comparisons of questionnaire responses between pedestrians and boaters at Crane Beach, for example, revealed a clear distinction. Boaters were less educated about property regulations, ecological issues, and human impact. Each of the visitor-impact management plans had to deal with public education. At Crane Beach, pedestrian visitors enter via a gatehouse where they receive educational information. Conversely, boaters land at many sites along the beach where educational information is unavailable. The management plan designated boat-landing areas where boaters would receive educational information, be segregated from swimmers to promote safe recreation, and be segregated from wildlife and dunes.

The social research results also indicated that some visitors at TTOR beaches were beginning to feel crowded (a social impact indicator). To eliminate potential conflicts, the Wasque swimming-beach area was closed to ORV traffic year-round. At Crane Beach, designated boat landing areas were specified to segregate the boaters from the swimmers. While current use levels do not necessarily warrant restricting visitor numbers at this time, future research and planning will monitor shifts in these baseline data.

Knowledge Gaps

The studies conducted at the three barrier beaches managed by TTOR illustrate the value of integrating ecological and social data within the VIM framework. TTOR's objectives of balancing preservation with recreation (Step 2) influenced the selection of indicators (Step 3) for piping plover impacts (disturbance from predators and humans), dune impacts (loss of vegetation), and the visitors' recreational experience (crowding). The criteria for selecting impact indicators, however, may not always be obvious. For example, if the goal is to provide wildlife viewing opportunities, research on what constitutes a quality viewing experience is far from complete. Investigations that examine a range of indicators for different experiences will be useful in this regard.

Step 4 in the VIM process calls for the creation of specific standards for the selected impact or experience indicators. When the impact involves threatened or endangered species, a standard of no impact may be the only alternative. Failure to adhere to such a standard may imply total elimination of the species. When the impact involves a social indicator like crowding, a sufficient amount of research exists to select among desired experience opportunities (e.g., Shelby et al. 1989). Standards for other indicators such as vegetation loss may require more of a judgment call. The limited amount of normative research in this area (Shelby et al. 1988) may provide some guidance for future studies.

The comparison of existing conditions against the desired state (Step 5) as well as the continual monitoring (Step 8) requires a commitment of resources. Given the time and monetary constraints faced by natural resource agencies, research is needed to identify cost-effective procedures for acquiring the necessary data.

Finally, the studies at the TTOR barrier beaches were successful in identifying the probable causes of the problem (Step 6) and workable solutions (Step 7) in part because the assembled research team included both ecologists and social scientists. The effective application of the VIM process requires collaboration among the involved disciplines. Achievement of this goal will be enhanced when universities recognize the value of an interdisciplinary curriculum, and encourage students to broaden their scope of study.

Literature Cited

Carlson, L.H. and P.J. Godfrey. 1989a. Human impact management in a coastal recreation and natural area. *Biological Conservation* 49:141–156.

Carlson, L.H. and P.J. Godfrey. 1989b. An evaluation of human impacts on the Wasque Reservation and Cape Poge Wildlife Refuge, Edgartown, Massachusetts. Phase III. The Trustees of Reservations, Beverly, Massachusetts, 69 pp.

Deblinger, R.D., J.J. Vaske, and M.P. Donnelly. 1989. Integrating ecological and social impacts into barrier beach management. *Proceedings of the Northeast Recreation Research Symposium,* Northeast Forest Experiment Station Technical Report NE-132:49–56, Saratoga Springs, New York.

Deblinger, R.D., J.J. Vaske, M.P. Donnelly, and R. Hopping. 1991. Boater impact management planning. *Proceedings of the Northeast Recreation Research Symposium,* Northeast Forest Experiment Station, Saratoga Springs, New York.

Decker, D.J., T.L. Brown, N.A. Connelly, J.W. Enck, G.A. Pomerantz, K.G. Purdy, and W.F. Siemer. 1992. Toward a comprehensive paradigm of wildlife management: integrating the human and biological dimensions. In *American Fish and Wildlife Policy: The Human Dimension,* ed., W.R. Mangun, 33–54. Carbondale, Illinois: Southern Illinois Press.

Driver, B.L. and P.J. Brown. 1975. A socio-psychological definition of recreation demand with implications for recreation resource planning. In *Assessing Demand for Outdoor Recreation,* 64–68. Washington, D.C.: National Academy of Science.

Graefe, A.R., J.J. Vaske, and F.R. Kuss. 1984. Social carrying capacity: an integration and synthesis of twenty years of research. *Leisure Sciences* 6:395–431.

Graefe, A.R., F.R. Kuss, and J.J. Vaske. 1990. *Visitor Impact Management: The Planning Framework.* Washington, D.C.: National Parks and Conservation Association. 105 pp.

Hammitt, W.E. and D. Cole. 1987. *Wildland Recreation.* New York: John Wiley.

Hautaloma, J.E. and P.J. Brown. 1978. Attributes of the deer hunting experience: a cluster analytic study. *Journal of Leisure Research* 10:271–287.

Hendee, J.C., G.H. Stankey, and R.C. Lucas. 1978. Wilderness management. Publication No. 1365. Washington, D.C.: USDA Forest Service.

Kuss, F.R., A.R. Graefe, and J.J. Vaske. 1990. *Visitor Impact Management: A Review of Research.* Washington, D.C.: National Parks and Conservation Association. 362 pp.

Manfredo, M.J. and R. Larson. 1993. Managing for wildlife viewing recreation experiences: a case study in Colorado. *Wildlife Society Bulletin* 21:226–236.

Manfredo, M.J., B.L. Driver, and P.J. Brown. 1983. A test of concepts inherent in experienced-based setting management for outdoor recreation areas. *Journal of Leisure Research* 15:263–283.

Manning, R.E. 1985. Crowding norms in backcountry settings: a review and synthesis. *Journal of Leisure Research* 17:75–89.

Rimmer, D.W. and R.D. Deblinger. 1990. Use of predator exclosures to protect piping plover nests. *Journal of Field Ornithology* 61:217–223.

Schreyer, R. 1976. Sociological and political factors in carrying capacity decision making. In *Proceedings of the Third Resources Management Conference,* 228–258. Ft. Worth, Texas: USDI National Park Service.

Shelby, B. and T.A. Heberlein. 1986. *Carrying Capacity in Recreation Settings.* Corvallis, Oregon: Oregon State University Press.

Shelby, B., J.J. Vaske, and R. Harris. 1988. User standards for ecological impacts at wilderness campsites. *Journal of Leisure Research* 20:245–256.

Shelby, B., J.J. Vaske, and T.A. Heberlein. 1989. Comparative analysis of crowding in multiple locations: results from fifteen years of research. *Leisure Sciences* 11:269–291.

Stankey, G.H. 1980. A comparison of carrying capacity perceptions among visitors to two wildernesses. Report INT-242. Ogden, Utah: USDA Forest Service.

Stankey, G.H., D.N. Cole, R.C. Lucas, M.E. Petersen, and S.S. Frissell. 1985. The limits of acceptable change (LAC) system for wilderness planning. Report INT-176. Ogden, Utah: USDA Forest Service.

Stokes, G.L., R. Des Jardins, G. Walker, R. Hurd, F. Flint, M. Childress, M. Weesner, and L. Reesman. 1984. Planning and maintaining a river recreation spectrum: the Flathead Wild and Scenic River. In *Proceedings, 1984 National River Recreation Symposium, Baton Rouge, Louisiana.*

Vaske, J.J., B. Shelby, A.R. Graefe, and T.A. Heberlein. 1986. Backcountry encounter norms: theory, method and empirical evidence. *Journal of Leisure Research* 18:137–153.

Vaske, J.J., R.D. Deblinger, and M.P. Donnelly. 1992. Barrier beach impact management planning: findings from three locations in Massachusetts. *Canadian Water Resources Association Journal* 17:278–290.

Vaske, J.J., M.P. Donnelly, and B. Shelby. 1993. Establishing management standards: selected examples of the normative approach. *Environmental Management* 17:629–643.

Wall, G. and C. Wright. 1977. The environmental impact of outdoor recreation. Publication Series No. 11. Ontario, Canada: University of Waterloo, Department of Geography.

Washburne, R.F. 1982. Wilderness recreation carrying capacity: are numbers necessary? *Journal of Forestry* 80:726–728.

CHAPTER 4

Wildlife Responses to Recreationists

Richard L. Knight and David N. Cole

Recreational activities are widespread, yet our understanding of their effects on wildlife is rudimentary. Although numerous studies of recreational impacts have been conducted, the knowledge gained is disparate and seldom definitive. Preliminary evidence, however, suggests that recreational activities can harm wildlife. For example, Boyle and Samson (1985) reviewed 166 articles that contained original data on the effects of nonconsumptive outdoor recreation on wildlife. In 81% of them, the effects were considered negative. In this chapter, we have organized the information relative to wildlife and recreation. First, we present a conceptual model that distinguishes the most common causes of recreational impacts, as well as a hierarchy of wildlife responses to recreational disturbance. Alternative typologies of impacts are offered by Pomerantz et al. (1988) and Kuss et al. (1990). Second, we describe what is known about the effects of a wide variety of recreational pursuits on wildlife.

Recreational Influences

Human activities can impact animals through four primary routes—exploitation, disturbance, habitat modification, and pollution (Fig. 4.1). The first two are direct impacts. Exploitation involves immediate death from hunting, trapping, or collection; disturbance can be either intentional (i.e., harassment) or unintentional. Activities such as photographing wildlife, bird watching, or simply hiking through an animal's territory can cause unintentional disturbance (e.g., Gutzwiller et al. 1994). Alternatively, habitat modification and pollution are indirect forms of impact (see Chapter 11). Recreationists can modify vegetation, soil, water, and even microclimates, which in turn can impact species dependent on these habitats. Wildlife are indirectly affected when their habitats are contaminated with discarded human food or foreign objects, such as tangled fishing line or plastic six-pack tops.

Wildlife Responses

Most studies have documented immediate, rather than long-term, responses to disturbance (Fig. 4.1). Those immediate responses are either death or behavioral changes including nest abandonment, change in food habits, and physiological changes, such as elevated heart rates due to flight. These imme-

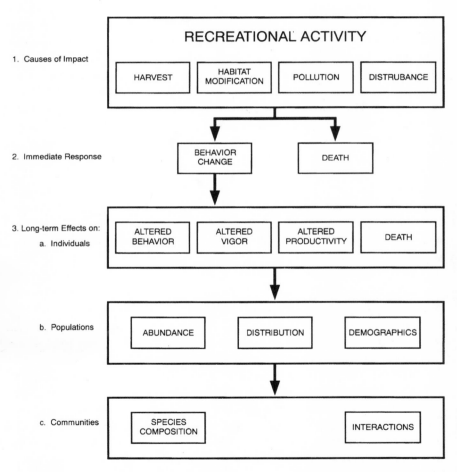

Figure 4.1 A conceptual model of responses of wildlife to recreational activities.

diate responses generally apply to individuals rather than populations or communities.

Many behavioral responses are of short duration (Cassirer et al. 1992). For example, deer that were disturbed by snowmobiles returned within several hours to areas they had fled (Dorrance et al. 1975). Long-term behavioral changes, however, also occur. Examples include abandonment of preferred foraging areas (Geist 1978) and changes in food sources (Klein 1971). Vigor is reduced due to elevated heart rates (MacArthur et al. 1982), an increase in energy expended during flight, or a decrease in energy acquired because foraging is disrupted (see Chapters 7, 13, and 14). Alterations in animal energy budgets may cause death or decreased productivity (Yarmoloy et al. 1988).

Consumptive recreational activities are known to have an impact on the abundance, distribution, and demographics of some populations (e.g., Wood 1993). Nonconsumptive activities are likely to have similar effects because they can increase mortality, reduce productivity, and result in population displacement. Research showing impacts at the guild or community level is equally sketchy. A few studies have identified changes in species composition and diversity that can be attributed to recreational impact. In some cases diversity increased (e.g., Guth 1978), while in other situations it decreased (e.g., Robertson and Flood 1980; Skagen et al. 1991).

Wildlife Responses to Recreational Activity Types

The type and magnitude of impact on wildlife varies with type of recreational activity. For example, consumptive activities such as hunting have a different impact than nonconsumptive activities such as nature viewing. Motorized and nonmotorized recreation, likewise, have distinctive effects. Extensive activities such as mountain biking differ in impact from intensive ones such as rock climbing. Below, we summarize wildlife responses to a variety of recreational types, beginning with consumptive activities, followed by nonconsumptive events, then motorized activities, and ending with recreational development.

Hunting

Hunting can alter behavior, population structure, and distribution patterns of wildlife (Cole and Knight 1990), and unhunted populations function differently from hunted ones (e.g., Wood 1993). The primary rationale for hunting is that wildlife populations demonstrate a compensatory response to predation, diseases and parasites, and competition (Bartmann et al. 1992). Excess individuals in a population will succumb to any one of these natural mortality factors, so hunters can remove a segment of these "expendable" animals without harming the population. Batcheler (1968) failed to find compensatory responses in hunted populations of introduced red deer, chamois, and opossums in New Zealand. He explained that the normal responses to reduction in numbers (i.e., improved physical conditions, higher reproductive rates) were suppressed by the inability of the remaining individuals to use certain habitat components.

Likewise, Small et al. (1992) found that hunting was at least partially, if not completely, additive to natural mortality of ruffed grouse populations on public lands in Wisconsin. Hunting mortality of both adult and juvenile grouse was significantly higher on public hunting areas than on private lands, and grouse numbers on public lands were sustained only by immigration from private lands.

Hunting may alter reproductive behavior in wildlife. It has been suggested that heavy hunting pressure on elk in Colorado interfered with normal breeding by influencing the date of conception (Squibb et al. 1986). Reproductive tracts recovered from female elk indicated a bimodal distribution of conception dates. Although this bimodal distribution could not be explained by the age or lactating condition of the cows, it did coincide with hunting season dates.

The behavior of wildlife can be influenced by hunting. Populations of red deer and chamois in New Zealand began feeding at night in response to hunting, and they returned to diurnal feeding only after hunting ceased (Douglas 1971). Others have found that hunted populations of ungulates became more shy and wary, stayed closer to cover, and fed in the open mainly at night (King and Workman 1986; Jeppesen 1987a, 1987b, 1987c).

Hunting may affect spatial and temporal patterns of wildlife populations (Bell and Austin 1985). Hunted geese in Denmark avoided traditional feeding sites and fed more during the night (Madsen 1988). When shooting had not occurred during the previous afternoon, there were six times as many geese present than when there had been shooting the day before. Newton and Campbell (1973) noted a similar response by hunted geese in Scotland; geese usually avoided feeding in fields where shooting had occurred. Thornburg (1973) investigated the relationship between hunter activity and diving duck movements along the Mississippi River, and concluded that movements and distribution were determined by two factors: hunting activity and food availability. Altered spatial patterns were associated with the opening of duck hunting season.

FISHING

Fishing is considered to be less disturbing to terrestrial wildlife than either hunting or motorized boating (Tuite et al. 1983). This is attributed to the fact that fishing, when not done from a motorboat, is both silent and relatively stationary.

Anglers had no apparent effect on presence or absence of three avian scavengers along a river in Washington; however, the anglers did influence both the numbers and behavior of these species (Knight et al. 1991). For each species, there were fewer individuals on the river at the same time as the anglers. For bald eagles and ravens, the anglers' presence also caused decreased feeding, but crows, considered human commensals, were more likely to be found feeding when anglers were present.

NATURE VIEWING

Nature viewing, by its very definition, has great potential to negatively affect wildlife. Avid wildlife viewers intentionally seek out rare or spectacular species. Some types of wildlife viewers have a reputation for striving for the most viewing opportunities in the least amount of time (e.g., bird listing). Because these activities may occur during sensitive times of the year (e.g., nesting), and because they often involve close approaches to wildlife for purposes of identification or photography, the potential for negative effects is large. Of five different recreationist-user groups at a wildlife refuge in Florida, photographers were the most disruptive, since they were most likely to stop, leave their vehicles and approach wildlife (Klein 1993).

Songbirds may alter their behavior after repeated interactions with humans. Red-winged blackbirds, goldfinches, and American robins became much more aggressive toward humans who repeatedly visited their nests (Knight and Temple 1986a, 1986b, 1986c). In addition, nesting red-winged blackbirds learned to distinguish between people who visited their nests often and people not seen previously; the blackbirds responded much more aggressively to the familiar people.

Wildlife may alter nest placement based on prior experience with humans (Marzluff 1988). Black-billed magpies, in response to people climbing to their nests, altered nest placement in subsequent years in an attempt to make their nests less accessible to human beings (Knight and Fitzner 1985). There were no changes in nest placement at unclimbed nests.

Predators learn to follow human scent trails to nest sites (e.g., Earl 1950; Hammond and Forward 1956; but see MacIvor et al. 1990). Avian predators apparently learn to forage in the vicinity of people who are visiting bird nests (MacInnes and Misra 1972; Strang 1980). Likewise, people who are visiting nests may decrease nest or nestling survivorship, provoke nest abandonment, or discourage renesting (reviewed by Götmark 1992).

Mortality rates of wildlife can be elevated through association with humans. Nestlings of Cooper's hawks that were handled by humans or studied from blinds suffered a significantly higher mortality rate after fledging than nestlings from nests that were not exposed to frequent human contact (Snyder and Snyder 1974).

BACKPACKING/HIKING/ CROSS-COUNTRY SKIING/HORSEBACK RIDING

These activities have two factors in common: they involve nonmotorized human travel, and all have increased in popularity, which, in turn, has caused the creation of more hiking, skiing, and riding trails in wildlands (see Chapter 1). These activities are extensive in nature and have the ability to disrupt wildlife in many ways, particularly by displacing animals from an area.

MacArthur et al. (1982) reported elevated heart rates and flight among mountain sheep approached by humans. Withdrawal distances and elevated heart rates were greatest when hikers approached from over a ridge or when they were accompanied by dogs. Reactions to hikers were greater than reactions to road traffic, helicopters, or fixed-wing aircraft.

A variety of studies have reported on disturbances of trumpeter swans (Henson and Grant 1991), common sandpipers (Yalden 1992), golden plovers (Yalden and Yalden 1990), eider ducklings (Keller 1991), and snow geese (Bélanger and Bédard 1989) by various pedestrian recreational activities. Most of the documented responses were behavioral and short-lived. Yalden (1992) reported that although disturbance reduced the size of the breeding population, apparently through redistribution, it had no effect on breeding success of the individuals remaining in the population.

Peak levels of hiking and skiing displaced chamois from nutritionally important habitats for prolonged periods (Hamr 1988). Orienteering activities in Denmark displaced roe and red deer from their home ranges; however, the animals eventually returned to these areas after disturbances ceased (Jeppesen 1987a, 1987b).

Cassirer et al. (1992) documented elk responses to cross-country skiers in Yellowstone National Park. In areas free from human disturbance, and when skiers approached within 400 meters, elk moved an average of 1,765 meters to steep slopes nearer trees and often into another drainage. Ferguson and Keith (1982), working in Alberta, found that moose numbers were negatively associated with cross-country ski trails, although elk numbers were unaffected.

ROCK CLIMBING/MOUNTAINEERING/SPELUNKING

Cliffs and mountains are important landscape features for a variety of wildlife (Maser et al. 1979). In recent years, the popularity of rock climbing and mountaineering has increased substantially, evidenced by the proliferation of guidebooks to climbing areas (e.g., Vogel 1987). Rock climbing and mountaineering have the potential to disrupt wildlife because they overlap both spatially and temporally with wildlife use of cliffs. Rock climbers choose routes that follow cracks, features that are commonly used for breeding, foraging, and roosting by wildlife. Additionally, ledges are used by climbers as sites to belay other climbers. These same ledges may also be used by wildlife as nest or perch sites. Ledges and cracks often support what little vegetation is found on cliffs, and this vegetation provides wildlife perching, feeding, and nesting sites. Climbers often remove vegetation because it hinders their climbing abilities. Temporal overlap also occurs because the most popular time to climb mountains and cliffs coincides with the peak of the breeding season for many wildlife species. For example, most birds nest during spring and summer. This is also the ideal time for rock climbing; snow and ice are off the cliff faces and the weather is favorable for climbing.

Cave exploration has increased in popularity and has been implicated in the decline of sensitive wildlife populations. Many species of bats use caves as roosting and maternity sites. Declines in gray bat numbers are attributable to disturbance by recreationists (Tuttle 1979; Rabinowitz and Tuttle 1980).

PETS

Pets are viewed as undesirable in wildlands because they both chase and kill wildlife (e.g., George 1974; Lowry and McArthur 1978). MacArthur et al. (1982) noted that the greatest increase in heart rates of bighorn sheep occurred when they were approached by humans with a dog. Prairie chickens showed a stronger fear response to domestic dogs than to native predators such as foxes. Of 41 visits by foxes to lekking areas, only 3 caused all the birds to flush; whereas, of 25 dog visits, 13 caused total flushes (Hamerstrom et al. 1965).

BATHING/SWIMMING/BEACHES

The recreational use of beaches may displace wildlife populations as well as alter wildlife communities (see Chapter 17). Eighty-five percent of Virginia's beaches are "protected" under the ownership of conservation organizations, whereas 75% of oceanfront in New Jersey allows unrestricted recreation (e.g., bathing). Over 80% of four beach-nesting seabirds (common tern, least tern, black skimmer, and herring gull) in coastal Virginia use natural barrier island beaches, while in New Jersey most of these species have been displaced from beaches to dredge-composition islands and natural marsh islands (Erwin 1980).

BOATING

During any time of the year, water-based recreation can deprive animals of roosting or feeding habitats whereas, in the breeding season, boating disturbances can cause reduced reproductive success or may otherwise render potential breeding areas unsuitable.

A number of studies have documented effects of both motorized and nonmotorized boating, particularly on waterfowl, wading birds, and raptors. Canvasback ducks disturbed by boaters had additional flight times of 34 to 44 minutes per day in one study (Kahl 1991) and one hour per day in another (Korschgen et al. 1985). At Keokuk Pool on the Mississippi River, boating disturbance of diving ducks was frequent (14.1 disturbances per day in autumn), and diving ducks have declined (Havera et al. 1992). Canoeing was negatively associated with green-backed heron abundance along the Ozark National Scenic Riverways in Missouri and human disturbance appeared to disrupt heron feeding activity (Kaiser and Fritzell 1984). Canoeing, however, was judged to have little impact on the distribution or daily activity of ruddy shelducks in Nepal (Hulbert 1990).

Total number of boats and people can be an inappropriate measure of recreational intensity because the presence of a single boat might be just as disturbing as that of many (Tuite et al. 1983; Knight and Knight 1984). Likewise, not all types of boats are equally disruptive to wildlife. Motorboats have the greatest disturbance potential because they involve both movement and noise, whereas sailing and canoeing are less disruptive as they involve only movement (Tuite et al. 1983). Speedboats precipitated nest failures in groundnesting osprey; incubating birds suddenly flushed from their nests, dragging or breaking eggs in the process (Ames and Mersereau 1964). Skagen (1980) reported a significant decrease in the proportion of bald eagles feeding at a site when motorized boating had occurred within 200 meters of that area during the preceding 30 minutes.

VEHICLES/SNOWMOBILES/MOTORCYCLES/
ORVS/ROADS AND HIGHWAYS

Vehicles, whether on or off highways, can potentially impact wildlife. For example, Yarmoloy et al. (1988) disturbed radio-collared mule deer with an all-terrain vehicle and noted that harassed deer altered feeding and spatial-use patterns, while undisturbed animals maintained normal usage. Additionally, only disturbed deer experienced decreased reproduction the following year. Between 1973 and 1983, in Denali National Park, Alaska, there was a 50% increase in daily vehicular traffic using the main park road (Singer and Beattie 1986). This elevated volume correlated with a 72% decrease in moose sightings per trip and a 32% decrease in grizzly-bear sightings. Sightings of Dall sheep and caribou were unaffected during this time interval.

A number of biologists have examined the effects of snowmobiles on white-tailed deer. These studies have taken place in harsh winter environments where abiotic factors (e.g., cold, snowfall) might be expected to play an important role in wildlife survival. Dorrance et al. (1975) studied two areas in Minnesota, one where snowmobile use averaged from 10 to 195 vehicles per day, and another in which snowmobile use was prohibited. In the first site, traffic displaced deer from areas immediately adjacent to snowmobile trails. At the other site, snowmobile activity caused increased animal home-range sizes and increased movement and displacement of deer from areas along trails. In contrast, Eckstein et al. (1979) found no changes in home-range size or habitat use by white-tailed deer in two Wisconsin deer yards where snowmobile activity was experimentally introduced. Snowmobiling did, however, cause some deer to leave the immediate vicinity of snowmobile trails.

Wolves in the Great Lakes region of North America do not occur in areas with road (passable by two-wheel-drive vehicles) densities greater than 0.61 kilometers per square kilometer (Thiel 1985; Jensen et al. 1986; Mech et al. 1988). It is not that roads themselves keep wolves away (but see Thurber et al. 1994); rather, the roads provide access to humans who deliberately, accidentally, or incidentally kill wolves by shooting, snaring, trapping, or through vehicular impact (DeVos 1949; Mech et al. 1988).

AIRCRAFT

Wildlife responses to aircraft are influenced by many variables including aircraft size, color, proximity (both height above ground and lateral distance), flight profile, engine noise, and sonic booms (Smith et al. 1988). Because rotary-winged aircraft and smaller fixed-winged aircraft fly more slowly and at lower altitudes than large commercial and military aircraft, they may be more likely to disturb wildlife. Within these types of aircraft, however, wildlife show markedly different responses (e.g., Watson 1993). Large, fixed-winged planes used by commercial and military organizations have less disturbance potential, except during takeoff and landing and when creating sonic booms.

Chamois in Austria did not respond to high-flying, large passenger aircraft; however, helicopters, glider planes, and hang-gliders evoked strong responses, causing panic in groups of females with kids (Hamr 1988). Bighorn sheep showed no changes in heart rate in response to rotary-winged or fixed-winged aircraft at distances beyond 400 meters. Heart rates of sheep increased two to three times, however, when helicopters passed directly overhead at heights of 90 to 250 meters (MacArthur et al. 1982). In Alaska, aircraft disturbed brant at greater distances than other disturbance types and affected more geese over a larger area than other stimuli. Helicopters caused the greatest response duration of all aircraft types (Ward et al. 1994).

Dunnet (1977) found no evidence that either fixed-wing or rotary-winged aircraft flying at elevations of about 100 meters above a mixed seabird colony (mainly kittiwakes and guillemots) affected the abundance of birds. Kushlan (1979) found that fixed-winged and rotary-winged flights as low as 60 meters over a mixed colony of mainly heron species in Florida did not adversely affect the birds. Kushlan attributed this observation to the fact that fixed-winged aircraft had been used for 20 years in southern Florida, and suggested that the colony had habituated to this form of disturbance. Alternatively, any plane below 500 meters and up to 1.5 kilometers away would put flocks of wintering brant geese to flight (Owens 1977). Slow, noisy aircraft were especially disruptive, while helicopters caused widespread panic. Owens suggested that the strong response of geese to aircraft, and their reluctance to habituate to planes, was partly due to the visual resemblance of aircraft to large predatory birds. More empirical evidence on the effects of aircraft on waterfowl is found in Chapter 14. In addition, Bowles (see Chapter 8) provides an excellent discussion of how noise may impact wildlife.

RECREATIONAL DEVELOPMENT

Bird communities at six lakes in southern Ontario with different levels of cottage and recreational development were studied by Robertson and Flood (1980). Developed areas had significantly more birds; however, they tended to have lower species diversity than undeveloped sites. Species richness remained fairly constant in both areas, but species evenness was significantly lower in the disturbed areas. Nesting success of common loons and eastern kingbirds was lower in the developed areas.

Campgrounds, resorts, and cottage developments along lakes in Alberta were negatively associated with numbers of breeding pairs of common loons (Vermeer 1973). Heimberger et al. (1983) found that hatching success of common loons declined with increasing numbers of cottages along the shorelines of northern lakes, presumably as a result of increased levels of human disturbance. Likewise, Lehtonen (1970) attributed the decline of Arctic loons in southwestern Finland to increasing numbers of summer cottages and boating activity.

Although bald eagles nest in close association with riparian areas, nests were consistently farther from shorelines with recreational development than from areas without cottages and homes (Fraser et al. 1985). This relationship was stronger for clusters of houses than for single homes. Fraser et al. surmised that clustered homes were more likely to be occupied year-round, and that single homes were hunting cabins used primarily in the autumn.

Watson (1979) reported the impacts on wildlife of a ski area in Scotland. Populations of native mountain hares and birds, mostly scavengers (e.g., crows and gulls), increased at the ski area.

Knowledge Gaps

That recreational activities disturb wildlife is well appreciated but poorly understood. Most popular forms of recreation in wildlands have yet to receive even one detailed study. For example, we were able to find only three published studies that looked at the effects of cross-country skiing on wildlife; we found none pertaining to rock climbing.

To develop effective management plans that will ensure viable wildlife populations and communities, we must develop a better understanding of both direct and indirect impacts of recreational disturbance on wildlife (Fig. 4.1). Given the present state of our understanding of wildlife/recreation impacts, there are gaps in our knowledge that can be resolved only through well-designed experimental studies (Gutzwiller 1991). At present, most studies are deficient in any number of ways: they may be too short in duration (Wiens 1984), not have adequate controls or replications (Hurlbert 1984), be anecdotal in nature, or have too many potentially confounding variables (Cooke 1980; van der Zande and Vos 1984; Bell and Austin 1985; Anderson 1988; Madsen 1988).

Perhaps because behavioral data (e.g., flushing responses, flushing distances) are easily obtained, most studies have focused on overt behavioral responses. Studies need to be placed in the context of both short- and long-term, and direct and indirect effects. Short- and long-term effects include both proximate and ultimate impacts on populations or communities, as well as the twin concepts of learning and change in gene frequencies. Direct effects are usually an animal's immediate response to disturbance; indirect effects involve potential additive or synergistic impacts.

An area of research in need of increased attention deals with the energetic effects of disturbance. Can individuals compensate for lost energy intake (i.e., foraging) and increased energy expenditure (i.e., fleeing from disturbance) due to recreational disturbance? The energetic consequences of disturbance are frequently alluded to in studies, but are seldom documented (but see Stalmaster 1983). The potential additive or synergistic aspects of this topic also require study.

Assuming that disturbance causes animals to flee, and that suitable habitat elsewhere is occupied, what happens to animals that abandon an area because of disturbance? Is redistribution after disturbance adaptive? If an animal leaves an area, is it able to resume its normal activities (e.g., foraging, breeding) elsewhere? Within a population of individuals, is there a continuum of tolerance levels to disturbance, so that individuals are able to distribute themselves according to levels of human activity and individual tolerance? Such redistribution may have negative repercussions, such as lowering access to food for less tolerant individuals, or contributing to overcrowding of individuals in some areas, which in turn may lead to increased aggression and lower fitness. Studies on redistribution should ideally be conducted on a regional basis, since it is important to understand the population responses when individuals are displaced (Batten 1977). If recreational disturbance was heavy, but only in a limited area, then perhaps wildlife populations would be able to show adaptive redistribution. If disturbance was intense but evenly distributed throughout a region, then redistribution would not be possible.

Recreational activities must not be viewed in isolation. There may be synergisms or interactions when more than one recreational activity is occurring simultaneously. For example, at a reservoir in South Wales, sailing was not viewed as detrimental to waterfowl because it occurred in deep waters while waterfowl preferred the shallows. However, when bank fishing occurred, waterfowl retreated to the deeper central waters where they encountered sailboats. Angling and sailing, therefore, resulted in birds being deprived of any part of the reservoir (Bell and Austin 1985).

What are the implications of disturbance on the persistence of wildlife populations, both short- and long-term? Scientists have suggested that disturbance may be an important reason for the decline of populations (e.g., Flemming et al. 1988), but what is the relationship between disturbance and inclusive fitness? If animals are denied access to areas that are essential for reproduction and survival, then that population will decline. Likewise, if animals are disturbed while performing essential behaviors, such as foraging or breeding, that population will also likely decline. The importance of examining disturbance impacts on animal populations over extended time periods is illustrated by van der Zande and Verstrael (1984). They studied the effects of recreational disturbance upon kestrel populations in the Netherlands, and found that disturbance had its greatest impact during years when vole populations were low. They speculated that disturbance during years of diminished prey base resulted in lost foraging time which, in turn, caused female kestrels to leave an area or not breed at all. This could cause kestrel population fluctuations that would only be detectable in a long-term study.

If recreational disturbance does alter animal populations, then one must assume this response may alter the dynamics of a wildlife community (Wiens 1989). Accordingly, research on the role of recreational disturbance on community structuring may yield important findings regarding a region's biological diversity (e.g., Skagen et al. 1991).

Literature Cited

Ames, P.L. and G.S. Mersereau. 1964. Some factors in the decline of the osprey in Connecticut. *Auk* 81:173–185.

Anderson, D.W. 1988. Dose-response relationship between human disturbance and brown pelican breeding success. *Wildlife Society Bulletin* 16:339–345.

Bartmann, R.M., G.C. White, and L.H. Carpenter. 1992. Compensatory mortality in a Colorado mule deer population. Wildlife Monograph 121.

Batcheler, C.L. 1968. Compensatory response of artificially controlled mammal populations. *Proceedings of the New Zealand Ecological Society* 15:25–30.

Batten, L.A. 1977. Sailing on reservoirs and its effects on water birds. *Biological Conservation* 11:49–58.

Bélanger, L. and J. Bédard. 1989. Responses of staging greater snow geese to disturbance. *Journal of Wildlife Management* 53:713–719.

Bell, D.V. and L.W. Austin. 1985. The game-fishing season and its effects on overwintering wildfowl. *Biological Conservation* 33:65–80.

Boyle, S.A. and F.B. Samson. 1985. Effects of nonconsumptive recreation wildlife: a review. *Wildlife Society Bulletin* 13:110–116.

Cassirer, E.F., D.J. Freddy, and E.D. Ables. 1992. Elk responses to disturbance by cross-country skiers in Yellowstone National Park. *Wildlife Society Bulletin* 20:375–381.

Cole D.N. and R.L. Knight. 1990. Impacts of recreation on biodiversity in wilderness. In *Proceedings of a Symposium on Wilderness Areas: Their Impact*, 33–40. Logan, Utah: Utah State University.

Cooke, A.S. 1980. Observations on how close certain passerine species will tolerate an approaching human in rural and suburban areas. *Biological Conservation* 18:85–88.

De Vos, A. 1949. Timber wolves (*Canis lupus lycaon*) killed by cars on Ontario highways. *Journal of Mammalogy* 30:197.

Dorrance, M.J., P.J. Savage, and D.E. Huff. 1975. Effects of snowmobiles on white-tailed deer. *Journal of Wildlife Management* 39:563–569.

Douglas, M.J.W. 1971. Behaviour responses of red deer and chamois to cessation of hunting. *New Zealand Journal of Science* 14:507–518.

Dunnet, G.M. 1977. Observations on the effects of low-flying aircraft at seabird colonies on the coast of Aberdeenshire, Scotland. *Biological Conservation* 12:55–64.

Earl, J.P. 1950. Production of mallards on irrigated land in the Sacramento Valley, California. *Journal of Wildlife Management* 14:332–342.

Eckstein, R.G., T.F. O'Brien, O.J. Rongstad, and J.G. Bollinger. 1979. Snowmobile effects on movements of white-tailed deer: a case study. *Environmental Conservation* 6:45–51.

Erwin, R.M. 1980. Breeding habitat use by colonially nesting waterbirds in two mid-Atlantic U.S. regions under different regimes of human disturbance. *Biological Conservation* 18:39–51.

Ferguson, M.A.D. and L.B. Keith. 1982. Influence of nordic skiing on distribution of moose and elk in Elk Island National Park, Alberta. *Canadian Field-Naturalist* 96:69–78.

Flemming, S.P., R.D. Chiasson, P.C. Smith, P.J. Austin-Smith, and R.P. Bancroft. 1988. Piping plover status in Nova Scotia related to its reproductive and behavioral responses to human disturbance. *Journal of Field Ornithology* 59:321–330.

Fraser, J.D., L.D. Frenzel, and J.E. Mathisen. 1985. The impact of human activities on breeding bald eagles in north-central Minnesota. *Journal of Wildlife Management* 49:585–592.

Geist, V. 1978. Behavior. In *Big Game of North America: Ecology and Management*, eds., J.L. Schmidt and D.L. Gilbert, 283–296. Harrisburg, Pennsylvania: Stackpole Books. 494 pp.

George, W.G. 1974. Domestic cats as predators and factors in winter shortages of raptor prey. *Wilson Bulletin* 86:384–396.

Götmark, F. 1992. The effects of investigator disturbance on nesting birds. In *Current Ornithology*, ed., D.M. Power, 63–104, vol. 9. New York: Plenum Press.

Guth, R.W. 1978. Forest and campground bird communities of Peninsula State Park, Wisconsin. *Passenger Pigeon* 40:489–493.

Gutzwiller, K.J. 1991. Assessing recreational impacts on wildlife: the value and design of experiments. *North American Wildlife and Natural Resources Conference* 56:248–255.

Gutzwiller, K.J., R.T. Wiedenmann, K.L. Clements, and S.H. Anderson. 1994. Effects of human intrusion on song occurrence and singing. Consistency in subalpine birds. *Auk* 111:28–37.

Hamerstrom, F., D.D. Berger, and F.N. Hamerstrom Jr. 1965. The effect of mammals on prairie chickens on booming grounds. *Journal of Wildlife Management* 29:536–542.

Hammond, M.C. and W.R. Forward. 1956. Experiments on causes of duck nest predation. *Journal of Wildlife Management* 20:243–247.

Hamr, J. 1988. Disturbance behaviour of chamois in an alpine tourist area of Austria. *Mountain Research and Development* 8:65–73.

Havera, S.P., L.R. Boens, M.M. Georgi, and R.T. Shealy. 1992. Human disturbance of waterfowl on Keokuk Pool, Mississippi River. *Wildlife Society Bulletin* 20:290–298.

Heimberger, M., D. Euler, and J. Barr. 1983. The impact of cottage development on common loon reproductive success in central Ontario. *Wilson Bulletin* 95:431–439.

Henson, P. and T.A. Grant. 1991. The effects of human disturbance on trumpeter swan breeding behavior. *Wildlife Society Bulletin* 19:248–257.

Hulbert, I.A.R. 1990. The response of ruddy shelduck *Tadorna ferruginea* to tourist activity in the Royal Chitwan National Park of Nepal. *Biological Conservation* 52:113–123.

Hurlbert, S.H. 1984. Pseudoreplication and the design of ecological field experiments. *Ecological Monograph* 54:187–211.

Jensen, W.F., T.K. Fuller, and W.L. Robinson. 1986. Wolf (*Canis lupus*) distribution on the Ontario-Michigan border near Sault Ste. Marie. *Canadian Field-Naturalist* 100:363–366.

Jeppesen, J.L. 1987a. The disturbing effects of orienteering and hunting on roe deer (*Capreolus capreolus*). *Danish Review of Game Biology* 13:1–24.

Jeppesen, J.L. 1987b. Immediate reactions of red deer (*Cervus elaphus*) to orienteering and hunting in a Danish environment (in Danish with an English summary). *Danske Vildtundersogelser* 43:1–26.

Jeppesen, J.L. 1987c. Seasonal variation in group size, and sex and age composition in a Danish red deer (*Cervus elaphus*) population under heavy hunting pressure.Comm. No. 212, Vildtbiologisk Station, Dalo, Denmark.

Kahl, R. 1991. Boating disturbance of canvasbacks during migration at Lake Poygan, Wisconsin. *Wildlife Society Bulletin* 19:242–248.

Kaiser, M.S. and E.K. Fritzell. 1984. Effects of river recreationists on green-backed heron behavior. *Journal of Wildlife Management* 48:561–567.

Keller, V.E. 1991. Effects of human disturbance on eider ducklings *Somateria mollissima* in an estuarine habitat in Scotland. *Biological Conservation* 58:213–228.

King, M.M. and G.W. Workman. 1986. Response of desert bighorn sheep to human harassment: management implications. *Transactions of the North American Wildlife and Natural Resources Conference* 51:74–85.

Klein, D.R. 1971. Reaction of reindeer to obstructions and disturbance. *Science* 173:393–398.

⤙ Klein, M.L. 1993. Waterbird behavioral responses to human disturbances. *Wildlife Society Bulletin* 21:31–39.

Knight, R.L. and R.E. Fitzner. 1985. Human disturbance and nest site placement in black-billed magpies. *Journal of Field Ornithology* 56:153–157.

Knight, R.L. and S.K. Knight. 1984. Responses of wintering bald eagles to boating activity. *Journal of Wildlife Management* 48:999–1004.

Knight, R.L. and S. A. Temple. 1986a. Methodological problems in studies of avian nest defence. *Animal Behavior* 34:561–566.

Knight, R.L. and S.A. Temple. 1986b. Nest defense in the American goldfinch. *Animal Behavior* 34:887–897.

Knight, R.L. and S.A. Temple. 1986c. Why does intensity of avian nest defense increase during the nesting cycle? *Auk* 103:318–327

Knight, R.L., D.P. Anderson, and N.V. Marr. 1991. Responses of an avian scavenging guild to anglers. *Biological Conservation* 56:195–205

Korschgen, C.E., L.S. George, and W.L. Green. 1985. Disturbance of diving ducks by boaters on a migrational staging area. *Wildlife Society Bulletin* 13:290–296.

Kushlan, J.A. 1979. Effects of helicopter censuses on wading bird colonies. *Journal of Wildlife Management* 43:756–760.

Kuss, F.R., A.R. Graefe, and J.J. Vaske. 1990. *Visitor Impact Management: A Review of Research.* Washington, D.C.: National Parks and Conservation Association.

Lehtonen, L. 1970. Zur Biologie des Prachttauchers. *Gavia a. arctica* (L.). *Ann. Zool. Fennici* 7:25–60.

Lowry, D.A. and K.L. McArthur. 1978. Domestic dogs as predators on deer. *Wildlife Society Bulletin* 6:38–39.

MacArthur, R.A., V. Geist, and R.H. Johnston. 1982. Cardiac and behavioral responses of mountain sheep to human disturbance. *Journal of Wildlife Management* 46:351–358.

MacInnes, C.D. and R.K. Misra. 1972. Predation on Canada goose nests at McConnell River, Northwest Territories. *Journal of Wildlife Management* 36:414–422.

MacIvor, L.H., S.M. Melvin, and C.R. Griffin. 1990. Effects of research activity on piping plover nest predation. *Journal of Wildlife Management* 54:443–447.

Madsen, J. 1988. Autumn feeding ecology of herbivorous wildfowl in the Danish Wadden Sea, and impact of food supplies and shooting on movements. Comm. No. 217, Vildtbiologisk Station, Dalo, Denmark.

Marzluff, J.M. 1988. Do pinyon jays alter nest placement based on prior experience? *Animal Behavior* 36:1–10.

Maser, C., J.M. Geist, D.M. Concannon, R. Anderson, and B. Lovell. 1979. Wildlife habitats in managed rangelands—the Great Basin of southeastern Oregon: geomorphic and edaphic habitats. U.S. Forest Service General Technical Report PNW-99.

Mech, L.D., S.H. Fritts, G.L. Radde, and W.J. Paul. 1988. Wolf distribution and road density in Minnesota. *Wildlife Society Bulletin* 16:85–87.

Newton, I. and C.R.G. Campbell. 1973. Feeding of geese on farmland in east-central Scotland. *Journal of Applied Ecology* 10:781–802.

Owens, N.W. 1977. Responses of wintering brent geese to human disturbance. *Wildfowl* 28:5–14.

Pomerantz, G.A., D.J. Decker, G.R. Goff, and K.G. Purdy. 1988. Assessing impact of recreation on wildlife: a classification scheme. *Wildlife Society Bulletin* 16:58–62.

Rabinowitz, A. and M.D. Tuttle. 1980. Status of summer colonies of the endangered gray bat in Kentucky. *Journal of Wildlife Management* 44:955–960.

Robertson, R.J. and N.J. Flood. 1980. Effects of recreational use of shorelines on breeding bird populations. *Canadian Field-Naturalist* 94:131–138.

Singer, F.J. and J.B. Beattie. 1986. The controlled traffic system and associated wildlife responses in Denali National Park. *Arctic* 39:195–203.

Skagen, S.K. 1980. Behavioral responses of wintering bald eagles to human activity on the Skagit River, Washington. In *Proceedings of the Washington Bald Eagle Symposium*, eds., R.L. Knight et al., 231–241. Seattle, Washington: The Nature Conservancy.

Skagen, S.K., R.L. Knight, and G.H. Orians. 1991. Human disturbance of an avian scavenging guild. *Ecological Applications* 1(2):215–225.

Small, R.J., J.C. Holzwart, and D.H. Rusch. 1991. Predation and hunting mortality of ruffed grouse in central Wisconsin. *Journal of Wildlife Management* 55:512–520.

Smith, D.G., D.H. Ellis, and T.H. Johnson. 1988. Raptors and aircraft. In *Proceedings of the Southwest Raptor Management Symposium and Workshop*, eds., R.L. Glinski et al., 360–367. Washington, D.C.: National Wildlife Federation.

Snyder, H.A. and N.F.R. Snyder. 1974. Increased mortality of Cooper's hawks accustomed to man. *Condor* 76:215–216.

Squibb, R.C., J.F. Kimball Jr., and D.R. Anderson. 1986. Bimodal distribution of estimated conception dates in Rocky Mountain elk. *Journal of Wildlife Management* 50:118–122.

Stalmaster, M.V. 1983. An energetics simulation model for managing wintering bald eagles. *Journal of Wildlife Management* 47:349–359.

Strang, C.A. 1980. Incidence of avian predators near people searching for waterfowl nests. *Journal of Wildlife Management* 44:220–222.

Thiel, R.P. 1985. Relationship between road densities and wolf habitat suitability in Wisconsin. *American Midland Naturalist* 113:404–407.

Thornburg, D.D. 1973. Diving duck movements on Keokuk Pool, Mississippi River. *Journal of Wildlife Management* 37:382–389.

Thurber, J.M., R.D. Peterson, T.D. Drummer, and S.A. Thomasma. 1994. Gray wolf response to refuge boundaries in Alaska. *Wildlife Society Bulletin* 22:61–68.

Tuite, C.H., M. Owen, and D. Paynter. 1983. Interaction between wildfowl and recreation at Llangorse Lake and Talybont Reservoir, South Wales. *Wildfowl* 34:48–63.

Tuttle, M.D. 1979. Status, causes of decline, and management of endangered gray bats. *Journal of Wildlife Management* 43:1–17.

van der Zande, A.N. and T.J. Verstrael. 1984. Impacts of outdoor recreation upon nest-site choice and breeding success of the kestrel. *Ardea* 73:90–99.

van der Zande, A.N. and P. Vos. 1984. Impact of a semi-experimental increase in recreation intensity on the densities of birds in groves and hedges on a lake shore in The Netherlands. *Biological Conservation* 30:237–259.

Vermeer, K. 1973. Some aspects of the nesting requirements of common loons in Alberta. *Wilson Bulletin* 85:429–435.

Vogel, R. 1987. The American guidebook. *Climbing* 101:98–101.

Ward, D.H., R.A. Stehn, and D.V. Derksen. 1994. Response of staging brant to disturbance at the Izembek Lagoon, Alaska. *Wildlife Society Bulletin* 22:220–228.

Watson, A. 1979. Bird and mammal numbers in relation to human impact at ski lifts on Scottish hills. *Journal of Applied Ecology* 16:753–764.

Watson, J.W. 1993. Responses of nesting bald eagles to helicopter surveys. *Wildlife Society Bulletin* 21:171–178.

Wiens, J.A. 1984. The place of long-term studies in ornithology. *Auk* 101:202–203.

Wiens, J.A. 1989. *The Ecology of Bird Communities,* vol. 2. New York: Cambridge University Press. 316 pp.

Wood, A.K. 1993. Parallels between old-growth forest and wildlife population management. *Wildlife Society Bulletin* 21:91–95.

Yalden, D.W. 1992. The influence of recreation disturbance on common sandpipers *Actitis hypoleucos* breeding by an upland reservoir in England. *Biological Conservation* 61:41–49.

Yalden, P.E. and D.W. Yalden. 1990. Recreational disturbance of breeding golden plovers *Pluvialis apricarius*. *Biological Conservation* 51:243–262.

Yarmoloy, C., M. Bayer, and V. Geist. 1988. Behavior responses and reproduction of mule deer, *Odocoileus hemionus*, does following experimental harassment with an all-terrain vehicle. *Canadian Field-Naturalist* 102:425–429.

CHAPTER 5

Factors That Influence Wildlife Responses to Recreationists

Richard L. Knight and David N. Cole

A number of factors influence the nature and severity of recreational impacts on wildlife. Some activities may have serious consequences, while others have little or no effect. It is critical that these factors be understood and managed to mitigate recreational impacts. Impacts are the result of interactions between recreational disturbances and animals. Therefore, the two broad categories of factors we discuss in this chapter are: (1) characteristics of the recreational disturbance, and (2) characteristics of the affected animals.

Recreational Influences

Characteristics of the Disturbance

There are at least six distinct factors of recreational disturbances that shape wildlife responses. Each of these variables is capable of pronounced impacts on wildlife; in addition, the potential for synergisms is substantial.

TYPE OF ACTIVITY

Chapter 4 provides a detailed description of wildlife responses to a variety of recreation types. Important distinctions can be drawn between motorized and nonmotorized activities. For example, motorized boating in Minnesota caused nest desertion by common loons, whereas the presence of canoe travelers did not (Titus and VanDruff 1981). Other obvious distinctions can be drawn between land-, water-, and snow-based activities; air- versus ground-based disturbances; and between activities that have localized effects (e.g., downhill skiing) and those with more widespread impacts (e.g., cross-country skiing).

Recreationist's Behavior

Although virtually unstudied, the behavior of recreationists can have a profound influence on wildlife responses (Klein 1993). For example, depending on context, speed can influence wildlife responses. Rapid movement directly toward wildlife frightens them, while movement away from or at an oblique angle to the animal is less disturbing. Snowmobiles moving at high speeds alarmed white-tailed deer more easily than those at low speeds (<16 km/h), but when people stopped to view deer, they invariably caused the deer to flush (Richens and Lavigne 1978). Slow-moving disturbances in any spatial context appear to elicit a milder response from wildlife. Humans that slowly approached roosting waterbirds flushed fewer birds than did humans moving rapidly (Burger 1981).

Because recreationist behavior is important, natural resource managers have developed visitor education programs, while other groups and organizations have developed codes of ethical conduct (e.g., Glinski 1976, American Birding Association). Klein (1993) found that visitors to a Florida wildlife refuge who encountered roving refuge personnel were less likely to disturb wildlife than visitors who did not.

Predictability

Predictability of a given activity shapes wildlife response to it. When animals perceive a disturbance as frequent enough to be "expected" and nonthreatening, they show little overt response. During periods of peak recreational use in the Sheep River Wildlife Sanctuary, Alberta, bighorn sheep encountered 25–30 vehicles per hour. Behavioral reactions of sheep to passing vehicles were minimal; less than 1% evoked withdrawal responses. Likewise, less than 9% of these encounters caused an increased heart rate among sheep (MacArthur et al. 1982).

If wildlife perceive disturbance as predictable and *threatening* (e.g., active persecution), they react quite differently. Feeding bald eagles were more vigilant and fed less in areas where active persecution occurred than in sites where birds were not harmed (Knight and Knight 1986).

Frequency and Magnitude

If disturbance negatively affects wildlife, then perhaps the degree of impact will depend on the disturbance's frequency and magnitude of the disturbance. A number of studies have compared reproductive success of birds with frequently visited nests to those with infrequently visited nests. In general, nests visited more often exhibited lower reproductive success (reviewed in Götmark 1992).

There appear to be thresholds of disturbance frequencies above which substantial impacts to wildlife can occur. For example, when human activity levels during Missouri's hunting season exceeded 0.45 hours per hectare, white-tailed deer movement increased (Root et al. 1988). Bird densities in the Netherlands decreased between recreational intensity levels of 8 and 37 visitors per hectare (maximum number of people present at the same time)(van der Zande and Vos 1984). Four species of waterbirds in South Wales virtually abandoned areas when levels of recreation exceeded 8 to 10 boats on a lake at any one time (Tuite et al. 1983). Pink-footed geese avoided using fields adjacent to roads with traffic of 20 to 50 cars per day; even highways with 10 cars per day had a depressing effect on use of fields (Madsen 1985).

TIMING

Recreational disturbance has traditionally been viewed as most detrimental to wildlife during the breeding season. Recently, it has become apparent that disturbance outside of the animal's breeding season may have equally severe effects (e.g., Hobbs 1989; Skagen et al. 1991). The consequences of disturbance during these two periods in an animal's annual cycle are quite different, however. Disturbance during the breeding season may affect an individual's productivity; disturbance outside of the breeding season may affect the individual's energy balance and, therefore, its survival.

Wildlife may respond to disturbance during the breeding season by abandoning their nests or young, leading to total reproductive failure. Human activity can also alter parental attentiveness, increasing the risks of the young being preyed upon, disrupting feeding patterns, or exposing the young to adverse environmental stress. Bird studies dominate research in this area; little is known for other taxa.

The impacts of disturbance are not consistent throughout the breeding season; the period of greatest sensitivity to disturbance occurs during nest-building and incubation (Götmark:102). In addition, if parents are disturbed from their nests, and are reluctant to return, predators may visit nests and consume eggs or young (e.g., Choate 1967).

Outside of the breeding season, animals are not restrained to a nest or den site, and young depend less on their parents. Wildlife, however, can still be disturbed in a way that potentially reduces energy acquisition (i.e., foraging) or increases energy expenditure (i.e., fleeing)(Owens 1977). For example, black bears entered winter dens 31 days earlier in areas with high levels of outdoor recreation; bears were also more likely to abandon dens when disturbed (Goodrich and Berger 1994).

During winter, processes influencing energy intake, rather than energy expenditure, have a much greater impact on energy balance of ungulates (Hobbs 1989), suggesting that disruption of wildlife while feeding is of greater concern than causing wildlife to flee. Mammals show a weaker response to humans during the winter months than at other times of the year. Hamr (1988) reported that chamois were least sensitive to recreationists when snow was deep, forage was inaccessible, and energy conservation was decisive to survival.

LOCATION

The relative locations of wildlife and disturbance can influence animals' responses. Wildlife often show a more pronounced response to activities from above, apparently because they perceive them as a greater threat to their safety and ability to escape. Hikers approaching bighorn sheep from above elicited a stronger reaction than those approaching downslope from the sheep (Hicks and Elder 1979). Though nesting peregrine falcons were disturbed by recreationists at the base of their nesting cliffs, any approach from the cliff top triggered, by comparison, a more immediate and more intense alarm (Herbert and Herbert 1965).

In certain situations, wildlife feel more secure when they have greater open distance between themselves and potential threats. Wintering avian scavengers preferred to feed on gravel bars in the middle of rivers and away from vegetative cover, which may have screened predators (Skagen et al. 1991). Geese in Scotland fed in large rather than small fields, regardless of whether those fields contained a more plentiful food source (Newton and Campbell 1973). Pink-footed geese in Denmark avoided areas where vegetation or topography hindered their views (Madsen 1985). Wildlife vary greatly in their reaction to noise (see Chapter 8), and this response is shaped by the context in which the noise occurs. Wintering bald eagles disregarded most noises when they were visually shielded from the disturbing activity (Stalmaster and Newman 1978). Likewise, mountain goats ignored the sounds of visitors and trains passing unseen (Singer 1978).

Characteristics of Wildlife

TYPE OF ANIMAL

Animals with different life-history traits and evolutionary strategies (e.g., longevity, parental care, reproductive effort) vary in their reactions to recreational disturbance. Species with specialized food and shelter requirements are more vulnerable to disturbance than species with generalized requirements. Likewise, species that live in relatively stable environments have not evolved mechanisms to respond to rapid changes, whereas species in more fluctuating environments are better able to adjust to stochastic events. With the former group, there has been selection for longevity and specialization; with the latter group there has been selection for rapid reproduction (i.e., r- and K-selection) (MacArthur and Wilson 1967; Pianka 1970).

Preliminary evidence shows that body size is important in determining a species' response to disturbance. Comparison studies indicate that larger species flush at greater distances than smaller species (Cooke 1980; Skagen et al. 1991; Holmes et al. 1993). Presently, there are two explanations for this pattern: (1) Because of greater human persecution, larger, more conspicuous species have increased reason to be wary (Cooke 1980); (2) These variations in flushing are due to differing energetics of size (Holmes et al. 1993). Smaller species have greater ratios of surface area to body weight, so they expend relatively more energy than larger species (Hayes and Gessaman 1980; Koplin et al. 1980; Wasser 1986). Disturbance produces increased energy expenditure because of unnecessary avoidance flights, and decreased energy intake due to shortened foraging and feeding times (Stalmaster 1983). Small species are consequently more energetically stressed if they are repeatedly forced to take flight. These individuals, therefore, may show a greater tolerance for human activity in order to minimize energy expenditure.

GROUP SIZE

Animals feeding in groups respond to approaching threats at greater distances and are less vulnerable to attack than solitary individuals (Rubenstein 1978; Morse 1980; Pulliam and Caraco 1984). In general, the time devoted to vigilance by feeding individuals decreases as flock or herd size increases. In addition, wildlife show a positive correlation between distance to human intruders at which they take flight and group size (e.g., Owens 1977; Batten 1977; Greig-Smith 1981; Madsen 1985; but see Richens and Lavigne 1978; Knight and Knight 1984). This variation in flight distance is probably due to different tolerances among flock members. Larger flocks have an increased likelihood of containing individuals who are more sensitized to humans and will flush at a greater distance, thereby causing other group members to follow suit.

AGE AND SEX

Age and sex composition of groups may also shape wildlife responses to humans. Cow and calf groups of caribou were more likely to flee than all-cow groups; bull groups were the least likely to flee (Singer and Beattie 1986). Stalmaster and Newman (1978) observed that adult bald eagles flushed at greater distances than immatures and attributed this difference to experience; older birds had learned that humans were to be avoided.

Knowledge Gaps

In this chapter we provide a review of wildlife responses to recreational activities. Intentionally, we broke this topic into two categories: (1) characteristics of the recreationists, and (2) characteristics of the wildlife being affected.

Although there are several generalizations to be made about the latter, there is virtually no work on the former (see Klein 1993 for an exception). Understanding the role recreationists play in affecting wildlife is particularly critical, because natural resource managers may be more capable of changing recreationist's behavior than the characteristics of wildlife that predispose them to impacts (see Chapters 2 and 3).

The impacts of frequency and magnitude of disturbance on wildlife are little known. Does the passing of a single boat along a river have the same effect as several boats together? At what frequency, interval, and duration of disturbance will wildlife alter their use of a critical area? If studies fail to find significant influences of recreation on animals, is it because the disturbance is unimportant or simply too subtle to be easily detected (van der Zande and Vos 1984; Anderson 1988)? Likewise, sex, age, and group size of wildlife affect their response to disturbance, yet we do not adequately understand how the location or composition of wildlife groups shape wildlife response to recreationists.

As with any field of study, our understanding of the factors that influence wildlife responses to recreationists is replete with dogma (Romesburg 1981). Perhaps this is due to the methodological complexity of the problem (Van der Zande et al. 1984). Attributing cause-and-effect to observed patterns usually requires controlled experiments. Fraser et al. (1985) and others (e.g., Anderson et al. 1987) have decried the continued emphasis of conducting observational studies with retrospective analyses to understand wildlife responses to human activities. The lack of systematic studies of wildlife and recreationist interactions is surprising, since it is not difficult to experimentally simulate disturbance (e.g., Yarmoloy et al. 1988; Skagen et al. 1991).

Finally, conducting research of recreational influences on wildlife involves a close, cooperative working relationship between scientists and natural resource managers (MacNab 1983). Managers commonly manipulate habitats or populations to achieve some objective. These manipulations could be designed, not only as ways to achieve specific goals, but as treatments to increase our knowledge of how outdoor recreation affects wildlife. Controls and random assignments of treatments would enable the manager to act as researcher and make a contribution to our understanding of this topic.

Literature Cited

Anderson, D.R., K.P. Burnham, J.D. Nichols, and M.J. Conroy. 1987. The need for experiments to understand population dynamics of American black ducks. *Wildlife Society Bulletin* 15:282–284.

Anderson, D.W. 1988. Dose-response relationship between human disturbance and brown pelican breeding success. *Wildlife Society Bulletin* 16:339–345.

Batten, L.A. 1977. Sailing on reservoirs and its effects on water birds. *Biological Conservation* 11:49–58.

Burger, J. 1981. The effect of human activity on birds at a coastal bay. *Biological Conservation* 21:231–241.

Choate, J.S. 1967. Factors influencing nesting success of eiders in Penobscot Bay, Maine. *Journal of Wildlife Management* 31:769–777.

Cooke, A.S. 1980. Observations on how close certain passerine species will tolerate an approaching human in rural and suburban areas. *Biological Conservation* 18:85–88.

Fraser, J.D., L.D. Frenzel, and J.E. Mathisen. 1985. The impact of human activities on breeding bald eagles in north-central Minnesota. *Journal of Wildlife Management* 49:585–592.

Glinski, R.L. 1976. Birdwatching etiquette: the need for a developing philosophy. *American Birds* 30:655–657.

Goodrich, J.M., and J. Berger. 1994. Winter recreation and hibernating black bears *Ursus americanus*. *Biological Conservation* 67:105–110.

Götmark, F. 1992. The effects of investigator disturbance on nesting birds. In *Current Ornithology*, ed., D.M. Power, 63–104, vol. 9. New York: Plenum Press.

Greig-Smith, P.W. 1981. Responses to disturbance in relation to flock size in foraging groups of barred ground doves *Geopelia striata*. *Ibis* 123:103–106.

Hamr, J. 1988. Disturbance behaviour of chamois in an alpine tourist area of Austria. *Mountain Research and Development* 8:65–73.

Hayes, S.R. and J.A. Gessaman. 1980. The combined effects of air temperature, wind and radiation on the resting metabolism of avian raptors. *Journal Therm. Biology* 5:119–125.

Herbert, R.A. and K.G.S. Herbert. 1965. Behavior of peregrine falcons in the New York City region. *Auk* 82:62–94.

Hicks, L.L. and J.M. Elder. 1979. Human disturbance of Sierra Nevada bighorn sheep. *Journal of Wildlife Management* 43:909–915.

Hobbs, N.T. 1989. Linking energy balance to survival in mule deer: development and test of a simulation model. Wildlife Monograph 101. 39 pp.

Holmes, T.L., R.L. Knight, L. Stegall, and G.R. Craig. 1993. Responses of wintering grassland raptors to human disturbance. *Wildlife Society Bulletin* 21:461–468.

Klein, M.L. 1993. Waterbird behavioral responses to human disturbances. *Wildlife Society Bulletin* 21:31–39.

Knight, R.L. and S.K. Knight. 1984. Responses of wintering bald eagles to boating activity. *Journal of Wildlife Management* 48:999–1004.

Knight, S.K. and R.L. Knight. 1986. Vigilance patterns of bald eagles feeding in groups. *Auk* 103:263–272.

Koplin, J.R., M.W. Collopy, A.R. Bammann, and H. Levenson. 1980. Energetics of two wintering raptors. *Auk* 97:795–806.

MacArthur, R.A., V. Geist, and R.H. Johnston. 1982. Cardiac and behavioral responses of mountain sheep to human disturbance. *Journal of Wildlife Management* 46:351–358.

MacArthur, R.H. and E.O. Wilson. 1967. *The Theory of Island Biogeography.* Princeton, New Jersey: Princeton University Press.

MacNab, J. 1983. Wildlife management as scientific experimentation. *Wildlife Society Bulletin* 11:397–401.

Madsen, J. 1985. Impact of disturbance on field utilization of pink-footed geese in West Jutland, Denmark. *Biological Conservation* 33:53–63.

Morse, D.H. 1980. *Behavioral Mechanisms in Ecology.* Cambridge, Massachusetts: Harvard University Press. 383 pp.

Newton, I. and C.R.G. Campbell. 1973. Feeding of geese on farmland in east-central Scotland. *Journal of Applied Ecology* 10:781–802.

Owens, N.W. 1977. Responses of wintering brent geese to human disturbance. *Wildfowl* 28:5–14.

Pianka, E.R. 1970. On r- and K-selection. *American Naturalist* 104:592–597.

Pulliam, H.R. and T. Caraco. 1984. Living in groups: is there an optimal group size? In *Behavioral Ecology: An Evolutionary Approach,* ed., J.R. Krebs and N.B. Davies, 122–147, 2nd ed. Oxford, England: Blackwell Scientific Publishing.

Richens, V.B. and G.R. Lavigne. 1978. Response of white-tailed deer to snowmobiles and snowmobile trails in Maine. *Canadian Field-Naturalist* 92:334–344.

Romesburg, H.C. 1981. Wildlife science: gaining reliable knowledge. *Journal of Wildlife Management* 45:293–313.

Root, B.G., E.K. Fritzell, and N F. Giessman. 1988. Effects of intensive hunting on white-tailed deer movement. *Wildlife Society Bulletin* 16:145–151.

Rubenstein, D.I. 1978. On predation, competition and the advantages of group living. In *Perspectives in Ethology,* eds., P.P.G. Bateson and P.H. Klopfer, 205–231, vol. 3. New York: Plenum Press.

Singer, F.J. 1978. Behavior of mountain goats in relation to U.S. Highway 2, Glacier National Park, Montana. *Journal of Wildlife Management* 42:591–597.

Singer, F.J. and J.B. Beattie. 1986. The controlled traffic system and associated wildlife responses in Denali National Park. *Arctic* 39:195–203.

Skagen, S.K., R.L. Knight, and G.H. Orians. 1991. Human disturbance of an avian scavenging guild. *Ecological Applications* 1:215–225.

Stalmaster, M.V. 1983. An energetics simulation model for managing wintering bald eagles. *Journal of Wildlife Management* 47:349–359.

Stalmaster, M.V. and J.R. Newman. 1978. Behavioral responses of wintering bald eagles to human activity. *Journal of Wildlife Management* 42:506–513.

Titus, J.R. and L.W. VanDruff. 1981. Response of the common loon to recreational pressure in the Boundary Waters Canoe Area, northeastern Minnesota. Wildlife Monograph 79:1–58.

Tuite, C.H., M. Owen, and D. Paynter. 1983. Interaction between wildfowl and recreation at Llangorse Lake and Talybont Reservoir, South Wales. *Wildfowl* 34:48–63.

van der Zande, A.N. and P. Vos. 1984. Impact of a semi-experimental increase in recreation intensity on the densities of birds in groves and hedges on a lake shore in The Netherlands. *Biological Conservation* 30:237–259.

van der Zande, A.N., J.C. Berkhuizen, H.C. van Latesteijn, W.J. ter Keurs, and A.J. Poppelaars. 1984. Impact of outdoor recreation on the density of a number of breeding bird species in woods adjacent to urban residential areas. *Biological Conservation* 30:1–39.

Wasser, J.S. 1986. The relationship of energetics of falconiform birds to body mass and climate. *Condor* 88:57–62.

Yarmoloy, C., M. Bayer, and V. Geist. 1988. Behavior responses and reproduction of mule deer, *Odocoileus hemionus*, does following experimental harassment with an all-terrain vehicle. *Canadian Field-Naturalist* 102: 425–429.

CHAPTER 6

Origin of Wildlife Responses to Recreationists

Richard L. Knight and Stanley A. Temple

Recreational Influences

One of the greatest challenges facing natural resource managers attempting to ensure the coexistence of wildlife and recreationists is dealing with the variety of wildlife responses to disturbance. For example, individuals of a population of peregrine falcons in New Mexico showed differences by a factor of 22 in the distances at which they responded to disturbance (Johnson 1988). Flight distances among bald eagles differ within and between sites, as well as seasonally (Knight and Knight 1984; Fraser et al. 1985).

There is also much between-species variation in response to disturbance. In the Netherlands, recreation activity negatively influenced eight species of passerines, while five others were unaffected (van der Zande et al. 1984). Moose in Denali National Park were more alert to vehicle traffic than were caribou (Singer and Beattie 1986).

This wide range in intra- and inter-specific variation has both learned and innate components. To the degree that wildlife managers understand the causes of this variation, they can design and implement management strategies to minimize harmful effects of recreationists on wildlife.

Learned Responses of Wildlife

There are three learned responses wildlife may show to recreationists: (1) habituation, (2) attraction, and (3) avoidance (Knight and Cole 1991). Outdoor recreationists may have a variety of impacts on wildlife ranging from injury, which might result in avoidance behavior, to feeding wildlife, which could cause attraction behavior. Other activities may neither positively nor negatively affect wildlife; such instances could result in wildlife habituating to humans.

This variety of wildlife responses, based on the outcome of wildlife-human interactions, can create challenging management dilemmas. A particular wildlife population could receive different stimuli from different types of outdoor recreationists; the upshot of these mixed stimuli and outcomes would create a mixed response from this population. It is possible, for example, for these types of interactions to occur at certain times of the year, resulting in seasonally different responses to humans. A population of black bear that is hunted in the fall may show avoidance behavior during fall or winter and attraction behavior during spring or summer when hikers and campers feed animals.

When different user groups have different expectations of their interactions with wildlife, a management impasse may arise. In the scenario above, wildlife viewers may expect wildlife to show attraction behavior to allow photography and nature study. Sportsmen, on the other hand, may desire wild behavior to ensure a quality hunt. A land-management agency such as the National Park Service, however, may want bear to show neither attraction nor avoidance behavior to people.

Geist (1978) has suggested that if people are to mix in wildlands with wildlife, then wildlife need to habituate to humans. This requires that wildlife learn to ignore humans. If wildlife, however, associate either rewards or punishments with people, then neither desired outcome will occur. For example, in areas open to hunting (e.g., national forests), black bears avoid roads and people, while in protected sites (e.g., national parks), bears are attracted to roads because of human food wastes (Reiffenberger 1974; Hamilton 1978; Brown 1980; Villarrubia 1982; Brody and Pelton 1989).

There are no easy answers to these ever-increasing management challenges. To develop strategies that enable humans and wildlife to coexist, resource managers will need to appreciate the complexity of learned wildlife responses to humans.

HABITUATION

Habituation is defined as a waning of a response to a repeated stimulus that is not associated with either a positive or negative reward (Eibl-Eibesfeldt 1970). Although there is abundant anecdotal evidence that animals do habituate to human beings, little experimental data on wildlife populations exist (Knight and Skagen 1988).

As an indication of the difficulties associated with verifying habituation, Knight and Knight (1984) attempted to determine whether bald eagles habituated to people (Stalmaster and Newman 1978; Russell 1980; Skagen 1980). Flushing responses of wintering eagles were measured on two adjacent rivers, one with recreational boating and the other with none. On the heavily traveled river, eagles were much less likely to flush from boaters than on the river with no boating activity, suggesting that eagles had indeed habituated to humans. A possible alternative explanation for the observed difference between rivers, however, was that there was less food available on the river with high human activity. If this was the case, then perhaps eagles were less likely to fly because they were energetically stressed (e.g., Owen 1972). Indeed, when food availability (i.e., spawned salmon carcasses) on the two rivers was compared, the river with heavy recreational traffic (and where eagles were less likely to flush) had nine times less food than the undisturbed river (Knight and Knight 1984). Accordingly, the reluctance for eagles to fly on the river with recreational activity could be explained by either: (1) an increased level of human activity resulting in habituation or (2) decreased food abundance resulting in food-deprived birds.

Another field study designed to look at habituation examined whether American crows responded differently to humans in cities, where they were protected, than to humans in rural areas, where crows were persecuted. It was hypothesized that in the absence of persecution, but in the presence of high human activity, crows would habituate to humans in order to complete their daily activities. Urban crows largely ignored humans, suggesting they had indeed habituated to nonthreatening activities. In contrast, rural crows showed elevated avoidance behavior to humans (Knight et al. 1987).

Ungulates may habituate to predictable events such as highway traffic, which they learn is not dangerous (Yarmoloy et al. 1988). Nonetheless, dogs and humans away from roads and trails are unpredictable, and ungulates did not habituate to these disturbances (Geist 1978; Geist et al. 1985). Birds also habituate to stimuli that are predictable and nonthreatening. Brant geese habituated to routine sounds, but unexpected sounds (e.g., gun shots) quickly put geese to flight (Owens 1977). Vos et al. (1985) reported that great blue herons habituated to repeated, nonthreatening activities such as fishermen boating past a heronry. Unexpected disturbances, however, put the herons to flight. In areas of high levels of human activity, nesting osprey habituated to a variety of nonthreatening activities, but in more remote sites where human presence was abrupt and sporadic, osprey did not habituate (Swenson 1979; Poole 1981).

ATTRACTION

We define attraction as the strengthening of an animal's behavior because of rewards or reinforcement. Examples of attraction are evident in most wildlands where recreationists may occur. Whether it be a chipmunk or gray jay, travelers are accustomed to encountering wildlife that approach them for some form of food reward. An interesting example of this form of learning is when caribou follow the sound of chain saws used in logging operations to feed on the lichens of downed trees (Klein 1971; Bergerud 1974).

This attraction of wildlife to humans can be harmful to both humans and wildlife. In the extreme case, where attraction brings humans into contact with potentially dangerous animals, it results in the "problem" animals being killed or humans injured. The occurrence of black bear incidents at backcountry sites was associated with high numbers of visitor nights (Singer and Bratton 1980). In other cases, it may alter some important aspect of the animal's behavior that, in the absence of the food reward, could affect the animal's survival. Wolves frequented a weather station on Ellesmere Island because human food wastes were left at a garbage dump (Grace 1976). Grace speculated that if a widespread dump-foraging habit arose among wolves, it might impair their health as well as decrease their effectiveness as natural predators.

Perhaps the best documented example of a species that became sensitized to human beings are the grizzly bears in Yellowstone National Park. Up until the early 1970s, a portion of the park's grizzly-bear population subsisted, to varying degrees, on human food wastes at garbage dumps within the park. Following the sudden closures of the dumps by the National Park Service, there were expansions in the size of bear home ranges and decreases in body size, reproductive rate, and average litter size (Despain et al. 1986). The sudden change in nutrition from human food wastes to natural foods may explain a number of these life-history differences. In addition, bears had to relearn skills required to obtain live prey and carrion. "In a garbage dump situation, the most successful feeder might be the bear that can best defend a pile of potatoes from other bears; in a hunting situation, the most successful feeder might be the bear that can best chase and catch elk" (Despain et al. 1986).

At the extreme end of this spectrum of sensitization would be where a keystone species could even alter an ecosystem. Clark's nutcrackers are attracted to scenic turnouts in Rocky Mountain National Park where they are fed by tourists. Nutcrackers are an important seed-dispersal species for limber pine, a tree of the subalpine ecosystem of the Front Range of Colorado. Tomback and Taylor (1986) speculated that if tourist activities discouraged normal nutcracker seed harvesting and storing activities, a decline in afforestation rates of limber pine was possible.

It is well appreciated that attraction behavior by wildlife has created management problems for natural resource agencies. Wildlife potentially harmful to human beings (e.g., grizzly bear), and those that seek out humans for food, can create dangerous situations for recreationists (Herrero 1976). For example, of 107 personal injuries to humans by black bears in Great Smokey Mountains National Park, 35 occurred while people were either feeding or petting bears (Singer and Bratton 1980). These, and situations with less grave consequences to humans, have encouraged natural resource agencies to attempt to alter recreationists' behavior so they do not reward wildlife for close approaches.

Avoidance

When given positive rewards (e.g., food), wildlife might be expected to be attracted to humans, so wildlife should learn to avoid humans when these interactions are associated with pain or punishment. One illustration of this is recreational hunting, where pronounced behavioral shifts in wildlife, including alterations in foraging ecology and habitat use (see Chapter 4), have been reported. Grizzly bears in Glacier National Park moved immediately away from people only 5% of the time; in a nearby area where they were persecuted, bears always fled from people, in most occasions 1 km or farther (McLellan and Shackleton 1989).

The hypothesis that persecution causes avoidance behavior in wildlife was examined for common ravens, American crows, and black-billed magpies (Knight 1984; Knight et al. 1987; Kenney and Knight 1992). For each species, birds nesting in areas of high persecution were more wary, and showed strong avoidance behavior and reduced nest defense than individuals in areas without persecution.

Wildlife may show avoidance behavior in ways other than simply moving away from humans. Black-billed magpies placed their nests in less accessible sites each year after humans climbed to their nests (Knight and Fitzner 1985). When constrained by nesting substrate, the number of nesting magpies declined in subsequent years.

Origin of Learned Responses

The learned component in wildlife responses to humans has been attributed to the *number and outcome* of interactions between individuals, and human stimuli over the individual's lifetime (Newton 1979; Poole 1981; Buitron 1983; Fraser 1984; Knight and Temple 1986a). Desert bighorn sheep increased their avoidance responses to human beings with an increasing number of negative encounters (King and Workman 1986). Red-winged blackbirds, American robins, and American goldfinches whose nests were repeatedly visited by researchers became much more aggressive over time (Knight and Temple 1986b, 1986c). Parent birds at nests visited only once, but at equivalent time periods during the nesting season, did not show elevated levels of aggressiveness. With repeated visits, bald eagles flew at increasing distances to observers approaching their nests (Fraser et al. 1985). Owens (1977) and Madsen (1988) found that when brant geese were frequently bothered, they became more easily disturbed on subsequent occasions. Pink-footed geese also showed increased wariness with elevated harassment (Madsen 1985).

These findings, and the studies listed under the three types of learning described earlier, all support the concept that learning is the result of the number and outcome of interactions between an individual and its environment over the individual's lifetime. Additional variables must certainly also shape an animal's response, including other individuals with which it is associating, its motivational state, and the degree and intensity of disturbance (see Chapter 5).

Genetic Responses of Wildlife

An animal's behavior is the product of both innate and learned components. Unlike learning, the innate responses of a species will be performed in a uniform and stereotyped fashion. Hailman (1967) used the concept of "learning of an instinct" to explain the interrelationship of genetic and learned responses. Animals are genetically predisposed to certain behaviors that are in turn mediated by environmental factors.

Bighorn sheep and mountain goats withdrew to cliffs when they heard sudden, loud noises, apparently an innate response to avalanches and rockfalls (Geist 1971, 1978). This genetically determined behavior can be subsequently reinforced through learning by the discharge of firearms in a hunted population. Hamr (1988) believed that the alarm response shown by chamois to airplanes was due to an innate fear the species had of golden eagles.

Newton (1979) hypothesized that intraspecific differences in nest-defense behavior of Falconiformes were due to past levels of human persecution. If shooting disproportionately eliminated aggressive birds, then nest-defense aggressiveness would vary with the persecution history of an area. Newton implied that natural selection was the mechanism that modified a species' behavior. Although there have been laboratory experiments on the innate responses of animals to stimuli, there has been little work on this topic in wild populations. One attempt to examine this hypothesis compared nest-defense behavior in seven widely separated populations of red-tailed hawks in North America that differed in the number of years since European settlement (range: 75–215 years). Length of European settlement was assumed to correlate positively with the duration of persecution. Pre-European persecution by native North Americans was assumed to be similar among areas, and was limited until the acquisition of firearms from Europeans. There was a highly negative correlation between the number of years since European settlement and the aggressive behavior of the hawks to humans, with the most aggressive birds occurring in the areas of most recent Anglo settlement (Knight et al. 1989).

Knowledge Gaps

The role of learning in shaping wildlife responses is replete with dogma, yet there are few studies that have critically demonstrated any of these three types of learning responses in wild animal populations. None have attempted to demonstrate all three experimentally.

Any study on learning would be incomplete without an understanding of the motivation or the decision-making process preceding the animal's response. Animals are capable of making complex contextual decisions (e.g., Orians 1981) and, in a given situation, they should be expected to weigh the many factors involved, their internal state, and the external ecological circumstances. In general, one should expect animals to act in a way that maximizes benefit and minimizes cost to themselves. Assuming that learning plays an important part in individual responses to disturbance, one can ask whether increased habituation, attraction, or avoidance responses are adaptive or maladaptive. If wildlife tolerate humans, how will mortality be affected if persecution patterns change? Likewise, if animals seek out humans because of some reward (e.g., food), how will this affect the health of natural populations?

Although learning apparently plays a disproportionate role in influencing wildlife responses to disturbance, there is also a genetic component. Persecution over an extended period of time might result in altered gene frequencies that could predispose animals to react differently. Although studies of this nature are difficult, and presently not available, it would be useful to address learning issues in the greater context of a species' evolutionary history.

Literature Cited

Bergerud, A.T. 1974. The role of the environment in the aggregation, movement and disturbance behavior of caribou. In *The Behavior of Ungulates and its Relation to Management,* eds., V. Geist and F. Walther, 552–584, vol. 2. Morges, Switzerland: IUCN New Serial Publication 24.

Brody, A.J. and M.R. Pelton. 1989. Effects of roads on black bear movements in western North Carolina. *Wildlife Society Bulletin* 17:5–10.

Brown, W.S. 1980. Black bear movements and activities in Pocahontas and Randolph counties, West Virginia. Masters thesis. Morgantown, Virginia: West Virginia University. 91 pp.

Buitron, D. 1983. Variability in the responses of black-billed magpies to natural predators. *Behaviour* 78:209–236.

Despain, D., D. Houston, M. Meagher, and P. Schullery. 1986. *Wildlife in Transition: Man and Nature on Yellowstone's Northern Range.* Boulder, Colorado: Roberts Rinehart. 142 pp.

Eibl-Eibesfeldt, I. 1970. *Ethology: The Biology of Behavior.* New York: Holt, Rinehart, and Winston. 530 pp.

Fraser, J.D. 1984. The impact of human activities on bald eagle populations— a review. In *The Bald Eagle in Canada,* eds., J.M. Gerard and T.M. Ingram, 68–94. Winnipeg, Canada: Proceedings of Bald Eagle Days.

Fraser, J.D., L.D. Frenzel, and J.E. Mathisen. 1985. The impact of human activities on breeding bald eagles in north-central Minnesota. *Journal of Wildlife Management* 49:585–592.

Geist, V. 1971. *Mountain Sheep: A Study in Behavior and Evolution.* Chicago, Illinois: University of Chicago Press. 383 pp.

Geist, V. 1978. Behavior. In *Big Game of North America, Ecology and Management,* eds., J.L. Schmidt and D.L. Gilbert, 283–296. Harrisburg, Pennsylvania: Stackpole Books. 494 pp.

Geist, V., R.E. Stemp, and R.H. Johnson. 1985. Heart-rate telemetry of bighorn sheep as a means to investigate disturbances. In *The Ecological Impact of Outdoor Recreation on Mountain Areas in Europe and North America,* eds., N.G. Bayfield and G.C. Barrow, 92–99. Recreational Ecology Research Group Report No. 9. Wye, England: Wye College.

Grace, E.S. 1976. Interactions between men and wolves at an Arctic outpost on Ellesmere Island. *Canadian Field-Naturalist* 90:149–156.

Hailman, J.P. 1967. The ontogeny of an instinct. *Behaviour* (Suppl.) 15:1–159.

Hamilton, R.J. 1978. Ecology of the black bear in southeastern North Carolina. Masters thesis. Athens, Georgia: University of Georgia. 214 pp.

Hamr, J. 1988. Disturbance behaviour of chamois in an alpine tourist area of Austria. *Mountain Research and Development* 8:65–73.

Herrero, S. 1976. Conflicts between man and grizzly bears in the national parks of North America. In *Bears—Their Biology and Management,* eds., M.R. Pelton, J.W. Lentfer, and G.E. Folk, 121–145. Binghamton, New York: Third International Conference on Bear Resource Management.

Johnson, T.H. 1988. Responses of breeding peregrine falcons to human stimuli. In *Proceedings of the Southwest Raptor Management Symposium and Workshop,* eds., Glinski et al., 301–305. Washington, D.C.: National Wildlife Federation.

Kenney, S.A. and R. L. Knight. 1992. Flight distances of black-billed magpies in different regimes of human density and persecution. *Condor* 94:545–547.

King, M.M. and G.W. Workman. 1986. Response of desert bighorn sheep to human harassment: management implications. *Transactions of the North American Wildlife and Natural Resources Conference* 51:74–85.

Klein, D.R. 1971. Reaction of reindeer to obstructions and disturbance. *Science* 173:393–398.

Knight, R.L. 1984. Responses of nesting ravens to people in areas of different human densities. *Condor* 86:345–346.

Knight, R.L. and D.N. Cole. 1991. Effects of recreational activity on wildlife in wildlands. *Transactions of the North American Wildlife and Natural Resources Conference* 56:238–247.

Knight, R.L. and R.E. Fitzner. 1985. Human disturbance and nest site placement in black-billed magpies. *Journal of Field Ornithology* 56:153–157.

Knight, R.L. and S.K. Knight. 1984. Responses of wintering bald eagles to boating activity. *Journal of Wildlife Management* 48:999–1004.

Knight, R.L. and S.K. Skagen. 1988. Effects of recreational disturbance on birds of prey: a review. In *Proceedings of the Southwest Raptor Management Symposium and Workshop,* eds., L. Glinski, et al., 355–359. Washington, D.C.: National Wildlife Federation.

Knight, R.L. and S.A. Temple. 1986a. Methodological problems in studies of avian nest defence. *Animal Behaviour* 34:561–566.

Knight, R.L. and S.A. Temple. 1986b. Nest defence in the American goldfinch. *Animal Behaviour* 34:887–897.

Knight, R.L. and S.A. Temple. 1986c. Why does intensity of avian nest defense increase during the nesting cycle? *Auk* 103:318–327.

Knight, R.L., D.J. Grout, and S.A. Temple. 1987. Nest defense behavior of the American crow in urban and rural areas. *Condor* 89:175–177.

Knight, R.L., D.E. Andersen, M.J. Bechard, and N.V. Marr. 1989. Geographic variation in nest-defence behaviour of the red-tailed hawk *Buteo jamaicensis. Ibis* 131:22–26.

Madsen, J. 1985. Impact of disturbance on field utilization of pink-footed geese in West Jutland, Denmark. *Biological Conservation* 33:53–63.

Madsen, J. 1988. Autumn feeding ecology of herbivorous wildfowl in the Danish Wadden Sea, and impact of food supplies and shooting on movements. Comm. No. 217, Vildtbiologisk Station, Dalso, Denmark.

McLellan, B.N. and D.M. Shackleton. 1989. Immediate reactions of grizzly bears to human activities. *Wildlife Society Bulletin* 17:269–274.

Newton, I. 1979. *Population Ecology of Raptors.* Vermillion, South Dakota: Buteo Books. 399 pp.

Orians, G.H. 1981. Foraging behavior and the evolution of discriminatory abilities. In *Foraging Behavior: Ecological, Ethological, and Psychological Approaches,* eds., A.C. Kamil and T.D. Sargent, 389–406. New York: Garland Press. 534 pp.

Owen, M. 1972. Some factors affecting food intake and selection in white-fronted geese. *Journal of Animal Ecology* 41:79–92.

Owens, N.W. 1977. Responses of wintering brent geese to human disturbance. *Wildfowl* 28:5–14.

Poole, A. 1981. The effects of human disturbance on osprey reproductive success. *Colonial Waterbirds* 4:20–27.

Reiffenberger, J.C. 1974. Range and movements of West Virginia black bear during summer and autumn 1973. *Proceedings of the Eastern Workshop on Black Bear Management and Research* 2:139–142.

Russell, D. 1980. Occurrence and human disturbance sensitivity of wintering bald eagles on the Sauk and Suiattle rivers, Washington. In *Proceedings of the Washington Bald Eagle Symposium*, eds., R.L. Knight, G.T. Allen, M.V. Stalmaster, and C.W. Servheen, 165–174. Seattle, Washington: The Nature Conservancy.

Singer, F.J. and J.B. Beattie. 1986. The controlled traffic system and associated wildlife responses in Denali National Park. *Arctic* 39:195–203.

Singer, F.J. and S.P. Bratton. 1980. Black bear/human conflicts in the Great Smoky Mountains National Park. In *Bears—Their Biology and Management*, eds., C.J. Martinka and K.L. McArthur, 137–139. IUCN Publ. New Ser. 40.

Skagen, S.K. 1980. Behavioral responses of wintering bald eagles to human activity on the Skagit River, Washington. In *Proceedings of the Washington Bald Eagle Symposium*, eds., R.L. Knight, G.T. Allen, M.V. Stalmaster, and C.W. Servheen, 231–241. Seattle, Washington: The Nature Conservancy.

Stalmaster, M.V. and J.R. Newman. 1978. Behavioral responses of wintering bald eagles to human activity. *Journal of Wildlife Management* 42:506–513.

Swenson, J.E. 1979. Factors affecting status and reproduction of ospreys in Yellowstone National Park. *Journal of Wildlife Management* 43:595–601.

Tomback, D.F. and C.L. Taylor. 1986. Tourist impact on Clark's nutcracker foraging activities in Rocky Mountain National Park. In *Proceedings of the Fourth Triennial Conference on Research in the National Parks and Equivalent Reserves*, ed., F.J. Singer, 158–172. Fort Collins, Colorado: Colorado State University.

van der Zande, A.N., J.C. Berkhuizen, H.C. van Latesteijn, W.J. ter Keurs, and A.J. Poppelaars. 1984. Impact of outdoor recreation on the density of a number of breeding bird species in woods adjacent to urban residential areas. *Biological Conservation* 30:1–39.

Villarrubia, C.R. 1982. Movement ecology and habitat utilization of black bears in Cherokee National Forest, Tennessee. Masters thesis. Knoxville, Tennessee: University of Tennessee. 159 pp.

Vos, D.K., R.A. Ryder, and W.D. Graul. 1985. Response of breeding great blue herons to human disturbance in north-central Colorado. *Colonial Water-birds* 8:13–22.

Yarmoloy, C., M. Bayer, and V. Geist. 1988. Behavior responses and reproduction of mule deer, *Odocoileus hemionus*, does following experimental harassment with an all-terrain vehicle. *Canadian Field-Naturalist.* 10:425–429.

PART II

Specific Issues

CHAPTER 7

Physiological Responses of Wildlife to Disturbance

Geir Wing Gabrielsen and E. Norbert Smith

Walter Cannon, at Harvard Medical School, was the first to systematically investigate the physiological responses of animals to disturbance. As early as 1929, he showed that disturbance produced dramatic physiological changes that helped animals survive during an emergency. In the laboratory, when an animal was confronted with situations that evoked pain, fear, or rage, a well-defined set of physiological reactions that prepare it to meet the threat was activated. Cannon used the term "fight or flight response" to describe this event (Cannon 1929), a concept every student of physiology is familiar with.

The complexity of physiological and behavioral responses to disturbance has been studied by other investigators (Folkow and Neil 1971; Mayes 1979). A more descriptive term for the fight or flight response is the "active defense response." Some of the physiological adjustments described for birds and mammals include increased heart rate and respiration, increased blood flow to skeletal muscle, increased body temperature, elevation of blood sugar, and reduced blood flow to the skin and digestive organs (Fig. 7.1). A detailed list of physiological adjustments associated with the active defense response is shown in Table 7.1. Most of these adjustments are controlled by the sympathetic portion of the autonomic nervous system and involve release of adrenaline. Each of these alterations improves the chances of survival under conditions where prolonged strenuous activity might be necessary, as in fighting or fleeing.

One example is a rabbit disturbed by a predator. If the rabbit runs and is pursued by the predator, the rabbit is then running for its life, and its survival depends on speed, agility, and knowledge of the area. The rabbit makes sudden turns for cover to try to elude the predator. The physiological adjustments made by the rabbit include increased sympathetic activity, which

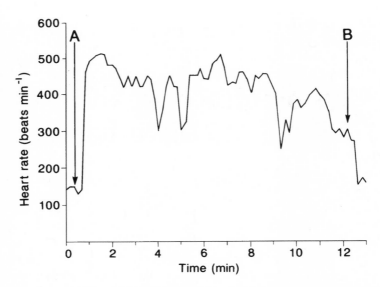

Figure 7.1 Heart rate response in a wild, nonincubating ptarmigan hen when approached by a human, showing the "fight or flight response." The disturbance begins at point A, and the human leaves the area at point B. From Steen et al. (1988).

Table 7.1

Physiological Changes Associated
with the Active Defense Response

The Following Parameters Increase:

Behavioral activity	Respiration rate
Heart rate	Respiration depth
Metabolism	Oxygen consumption
Blood sugar	Brain blood flow
Body temperature	Heart blood flow
Skeletal muscle blood flow	

The Following Parameters Decrease:

Blood flow to the gut
Gut motility
Digestive secretions
Blood flow to the skin?

increases heart rate and cardiac output. Blood sugar is increased to support prolonged activity. Blood flow to skeletal muscle is increased to enable greater speed, agility, and endurance. Increased blood flow to the brain and sense organs heightens perception and reduces reaction time.

If an animal is cornered or caught by the predator, fear turns to anger or defensive aggression. Many animals will make threatening sounds or assume threatening postures ready to fight and defend themselves or their young. Even the most loving lap-cat arches its back, spits, and becomes a formidable combatant when approached by a strange dog. Again, the sympathetically dominated active defense response is involved. Many of the same physiological responses that enabled increased speed, agility, and sensory acuity while running prepare an animal for fighting.

Assume our hypothetical rabbit is feeding, grooming, or sleeping. Before it can run or hide, it must be aware of the danger. Any mild sensory stimuli will alert the rabbit. It may be sight, sound, smell, or some other sensory modality such as ground vibration from a person walking. Most mammals will look around, sniff the air, and move their ears. This behavioral response was called the "What is it?" response by Pavlov in 1927. Today it is called the "orienting response," and it usually precedes either the active or passive defense response (Fig. 7.2). The orienting response involves head and eye movements in which the animal orients the sense organs toward the stimulus, apparently in an attempt to identify it. If the stimulus is neutral, such as the distant call of a crow, the rabbit will continue its original behavior. If instead it announces the approach of a potential predator or a human, the animal will apply either the active or passive defense response.

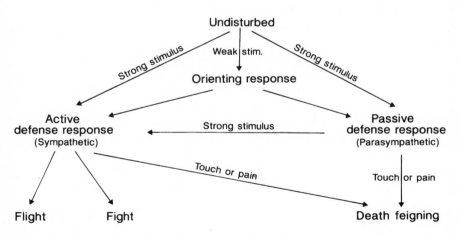

Figure 7.2   A schematic showing how disturbance affects wildlife.

The outcome depends on the age of the prey, the amount of cover, distance to the predator or human, and several other factors. Jacobsen (1979), for example, showed that the near approach of humans will cause newborn white-tailed deer to drop to the ground and hide. After the fawns are two-weeks old, the same stimulus will cause them to run.

The passive defense response involves profound physiological adjustments. Some of the major physiological adjustments for animals exhibiting the response include inhibition of activity, decreased blood flow to skeletal muscle, reduced blood flow to the digestive system, reduced heart and respiratory rate, and a reduction of body temperature. A detailed list of physiological adjustments associated with the passive defense response is shown in Table 7.2. Most of these alterations are controlled by the parasympathetic portion of the autonomic nervous system, and are opposite to the effects of the active defense response.

Two examples will illustrate the typical passive defense response. Perhaps the most insightful study of this response was done in Canada, with deer mice. Rosenmann and Morrison (1974) found that the response could easily be elicited by moving the shadow of a fan across the cage. In response to this stimulus, the mice would stop all activity for the duration of the stimulus. Respiration rate and depth decreased. Heart rate slowed suddenly and oxygen consumption was reduced. Body temperature dropped, especially when the ambient temperature was low, probably as a result of the reduced circulation to the muscles and skin area. Their study demonstrated the profound and far-reaching physiological alterations associated with the passive defense response.

The second example is from studies of willow grouse at Karlsøy, an island off the coast of northern Norway (Gabrielsen et al. 1977, 1985; Steen et al. 1988). Radio telemetry measurements of the heart rate of wild, incubating

Table 7.2
Physiological Changes Associated
with the Passive Defense Response

The Following Parameters Decrease:	
Behavioral activity	Respiration rate
Heart rate	Respiration depth
Metabolism	Oxygen consumption
Blood sugar	Brain blood flow
Body temperature	Heart blood flow
Skeletal muscle blood flow	
Digestive blood flow?	
Skin blood flow?	

willow grouse hens varied from 120 to 140 beats per minute (bpm). Upon approach to a distance of two to four meters by a human or dog, the bird would become motionless, and the heart rate would drop to less than 30 bpm. On one occasion, the heart rate remained at 40 bpm throughout a five-minute period of provocation. The heart rate returned to pre-stimulus values when the human or dog retreated (Fig. 7.3). In both of these examples, the passive defense response was elicited by the apparent danger, and the animals remained motionless. It would appear from these and other studies that the response may help the animal remain hidden in the presence of perceived danger.

Hiding is not the only expression of the passive defense response. Under certain conditions many animals appear to faint or feign death. From antiquity, hunters no doubt observed prey species freezing or "playing dead" when approached. Death-feigning in the American opossum is so well known that the term *playing possum* has become widely used in American culture. Experimentally induced death-feigning was reported over 300 years ago by Kircher

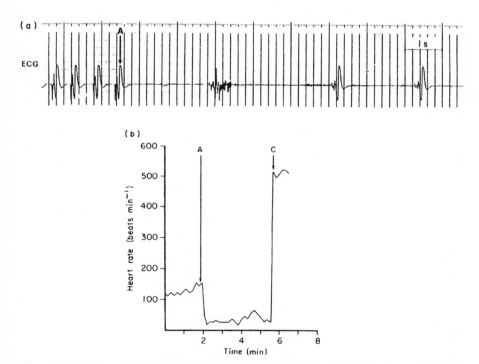

Figure 7.3 Provocation of a wild incubating willow ptarmigan hen. (a) ECG records from a wild incubating ptarmigan hen approached by one person. (b) Heart rate response in a wild incubating ptarmigan hen when approached by one person. The provocation starts at point A, and the hen flies away from the nest at point C. From Steen et al. 1988.

(1646). He found that certain manipulations of chickens resulted in "entrancement."

The passive defense response has been described for a wide array of vertebrates (Table 7.3). Fish stop swimming in response to ground or sound vibrations. Turtles become motionless when disturbed while free diving. Chickens become motionless and drop their heart rate at the sight of a hawk. Deer fawns become motionless, drop their heart rate, and stop breathing for minutes when alarmed by approaching humans. Regrettably, there are nearly as many

Table 7.3
Animals Showing the Passive Defense Response with Documented Bradycardia

Fish	Bluegill sunfish, *Lepomis marcochirus rafinesque*
	Chum salmon, *Oncorhynchus keta*
	Eel, *Anguilla anquilla*
	Ganoids, *Acipencer sturio*
	Atlantic cod, *Gadus morhua*
	Atlantic salmon, *Salamo salar*
Reptiles	Ornate box turtle, *Terepene ornata*
	Spectacled caiman, *Caiman crocodilus*
	American alligator, *Alligator mississippiensis*
Birds	Willow grouse, *Lagopus lagopus*
	Svalbard ptarmigan, (*L. mutus hyperboreus*)
	Pochards, *Aythya ferina*
	Tufted ducks, *Aythya fuligula*
	Common eider, *Somateria mollissima*
	Crested cormorants, *Phalcrocorax auritus*
	Canadian goose, *Branta canadensis*
Mammals	White-tailed deer, *Odocoileus virginianus*
	Red deer, *Cervus elaphus*
	Deer mouse, *Peromysus maniculatus*
	Swamp rabbit, *Sylvilagus aquaticus*
	Cotton tail rabbit, *Sylvilagus floridanus*
	Eastern chipmunk, *Tamias striatus*
	Ground squirrels, *Citellas armatus*
	Woodchuck, *Marmota monax*
	Fox squirrel, *Sciurus niger*
	Grey squirrel, *Sciurus carolinesis*
	Brazilian opossum, *Didelphis albiventris*
	American opossum, *Didelphis marsupialis*
	Harp seal, *Phoca groenlandica*
	Manatee, *Trichus manatus*
	Man, *Homo sapiens*

names given to the passive defense response as there have been workers investigating it. Kircher (1646) first used *entrancement*, but the term is too anthro-
pomorphic to be accepted today. Throughout most of the nineteenth century,
it was thought to be a form of hypnosis. At the turn of the century, Darwin
(1900) introduced the term "death-feint." It is now generally accepted that this
is one aspect of the passive defense response.

Recreational Influences

Over the last twenty years, we have telemetered a variety of wild animals, and
investigated their behavioral and heart responses to humans or predators
(Gabrielsen et al. 1977; Smith and Sweet 1980; Smith and Woodruff 1980;
Causby and Smith 1981; Smith et al. 1981; Smith and Johnson 1984;
Gabrielsen and Smith 1985; Gabrielsen 1985; Gabrielsen et al. 1985; Steen et
al. 1988). In each case, if the animals had cover or a safe hiding place, they exhibited the passive defense response when threatened.

Woodchucks were telemetered and released on a small island on an Oklahoma lake (Smith and Woodruff 1980). When disturbed by a human or dog,
they retreated toward their burrows and showed the active defense response,
including increased heart rate. Upon reaching their burrows they immediately
responded with the passive defense response, including reduced respiration
and heart rate. The amount the heart slowed was stimulus dependent. When a
dog or human approached the burrow and began digging, the heart rate would
again slow down. Similar results were obtained with eastern chipmunks.

The next series of animals included fox squirrels, grey squirrels, swamp
rabbits, and cottontail rabbits (Smith and Sweet 1980; Causby and Smith
1981; Smith and Johnson 1984). These animals were released on the same island. After an adjustment period ranging from a few days to a few weeks, their
heart rate was measured by radio-telemetry with and without disturbance.
Nondisturbance values were used to establish normal values. Measurements
were made from a blind while observing the animals behavior. Animals were
then approached by a human or a dog. If the animals were away from cover,
they would first retreat to cover with the active defense response and increased
heart rate, and then engage in the passive defense response. When the squirrels were disturbed, they would simply hide by slipping around the tree out of
sight. Heart rate would often drop by 60%. If the human or dog retreated, the
heart rate would return to predisturbance values. If the human or dog approached too closely, the animal would flee with increased heart rate and the
active defense response. Heart rate was observed to increase from less than 200
bpm to over 450 bpm within a single heart-beat interval.

Similar results were obtained with rabbits, which would often hide by remaining motionless, lower their profile, and decrease their heart rate by about 60% to 80% (Causby and Smith 1981). If the intruder backed off, their behavior and heart rate would return to pre-stimulus values. If instead, the intruder approached the rabbit to within two to three meters, it would immediately switch over to the active defense response. Heart rate often increased from less than 50 bpm to over 200 bpm within a single heart beat.

Wildlife display either of the two fundamental strategies available to prey species when confronted with humans. Depending on the situation, either the fight-and-flight or the freezing/playing-dead response may be initiated. The physiological capacity for both options is available. Newborn deer or a ptarmigan hen on the nest will freeze. The deer's chance of being killed if moving and the hen's chance of losing her eggs if she is flushed are much greater than if they remain motionless. Both animals rely on cryptic coloration for camouflage. That this strategy is appropriate is attested to by the observation that neither a trained dog nor a keen researcher's eyes are of much use in finding incubating ptarmigan hens.

The passive defense response in squirrels, rabbits, opossums, deer fawns, and ptarmigan hens differ both behaviorally and physiologically. The deer fawn and opossum play dead; they will not move even if touched and can only be awakened by removing the provocation. Foxes hunting ground-breeding ducks (Sargeant and Eberhardt 1975) appeared to be fooled by this response, and cats would not bite a mouse as long as it remained still. In contrast, rabbits, eiders, and ptarmigan hens on the nest are fully awake and will switch to flight reaction once the disturbance becomes an immediate threat to their life.

A review of the literature suggests that the passive defense response is present in all animals: invertebrates, fish, amphibians, reptiles, birds, and mammals, when subjected to strong fear, pain, anger, sudden shock, or situations with few possibilities for escape. The passive defense response is related to age, and is well developed in newborn mammals. Passive defense may be as widespread as active fight and flight, and may be an important alternative behavioral and physiological response to disturbance. We suggest that both freezing or feigned death are developments of the more general passive defense response. We recognize that important differences exist between these responses and agree that more data from a variety of animals under natural conditions are needed.

Management Options for Coexistence

Behavioral and physiological research on the effect of human disturbance on wildlife has shown that several species are very tolerant of the noise from aircraft and from engines of cars, motorcycles, and snowsmobiles at a distance of one to two km. (MacArthur et al. 1979, 1982; Olsson and Gabrielsen 1990; Tyler 1991; Langvatn and Andersen 1991). At shorter distances, the active defense response may be activated when the vehicles head directly toward the victim. However, the most dramatic physiological defense response is observed when wildlife are provoked by humans (also see MacArthur et al. 1979, 1982). This is probably because mechanical disturbance is most often very brief, while humans walking take more time to cover the same distance, and thus have a much more profound effect. The magnitude of the response also depends on the distance, the movement pattern of the provoker, and the animal's access to cover. Most animals seem to tolerate disturbance better in woodland than in open terrain. They also seem to have a greater defense response to humans moving unpredictably in the terrain than to humans following a distinct path. To reduce the effects of human disturbance, permanent paths should be used or traffic should be restricted or reduced to certain times ⟧ Mgmt of the year in sensitive areas.

Studies of the effect of human disturbance on wildlife have revealed that there are two critical periods for many species of birds and mammals. The immediate postnatal period in mammals and the breeding period in birds are the most vulnerable. During these times, human disturbance should be prevented or reduced to a minimum. While most animals show active fear when disturbed and may respond by running, jumping, swimming, or flying away rapidly, there are some animals that, during a certain stage of the breeding cycle, show the passive defense response. Both the behavioral and the physiological responses are profound and are important to the individual's fitness.

This can be illustrated by incubating common eiders, which were studied in the high Arctic (Gabrielsen 1985; Gabrielsen et al. 1991). Female eiders do not eat for 25 days during incubation. To avoid predation and to maintain a constant egg temperature, they must rely on stored body reserves, mainly body fat, during the incubation period. At this time they lose 40% of their body weight. By seldom leaving the nest, and by using as little energy as possible, the eiders lose only 20 to 25 grams of body weight per day. Disturbance, including provocation by humans, and repeated reheating of eggs on their return to the nest, results in increased energy requirements, meaning further loss of essential energy needed later to raise their young. An increase in activity level of 10% per day would result in an extra daily weight loss of four to five grams. In time, this would result in a weight loss that would cause the bird to abandon incubation or to give up its chicks at hatching in order to save itself.

In the spring, people may find newborn deer freezing or playing dead. People should avoid touching these animals, because the parents may then abandon them. Human touch has a profound effect on animals. This can be illustrated using an example from studies of chinstrap penguins in Antarctica (Rory Wilson, pers. comm.). A breeding bird was provoked and touched for 30 seconds, but not removed from the nest. Such an event increased the stomach temperature by 2°C and it took two to three hours to return to normal (37.5° to 38°C). This clearly shows a great physiological effect of human disturbance on the penguin. Such an increase in body temperature is accompanied by an increase in energy expenditure. Depending on the bird's body resources and the number of provocations made during the breeding period, it would only be a question of time before such disturbances caused a reduction in the bird's breeding success.

Disturbance by humans also affects an animal's ability to habituate. Upon repeated stimulation, most behavioral and physiological concomitants decrease in intensity and gradually disappear. This is referred to as habituation, and it is probably why some species are better adapted to human activities and noises than others. Several studies of adult reindeer, deer, or moose that have habituated to mechanical sounds indicate that they have shorter flight distances than animals in remote areas not exposed to disturbances (Freddy et al. 1986; Tyler 1991). Similarly, birds nesting close to humans tolerate more disturbance than birds nesting in remote areas (Gabrielsen et al. 1985). However, both birds and mammals habituate more rapidly to mechanical noise than to human presence. Mammals and birds nesting close to human settlements seem to have built up a higher tolerance threshold toward vehicles and humans as a result of habituation than animals living in remote areas. Based on these preliminary results, less strict regulations may be needed to protect wildlife in areas close to human settlements.

Knowledge Gaps

Many questions related to the physiological effect of human disturbance on wildlife still need to be investigated in order to support good management of recreation and conservation. The use of physiological data from telemetered free-ranging animals will enable us to get much better data both from disturbed and undisturbed animals. Such data will be more precise and will enable us to quantify the impact of human disturbance on many wild species.

Physiological and behavioral research on the impact of human disturbance on wildlife is still lacking in many areas. There are, for example, few data on the effects of human disturbance on wildlife during migration, during the winter, and when animals experience drought or lack of food. In polar regions many animals must rely on stored body reserves and on maintaining low levels of activity to survive winter. Increased human activity in these areas due to increased tourism or industry activity, for example, will certainly affect their behavior and physiology. The energetic cost of disturbance by humans and vehicles for both mammals and birds should also be investigated. Another interesting field for future research includes studies of the degree of habituation among wildlife. For example, do birds that breed close to human settlements, or near paths regularly used by humans, have lower energetic investments and higher breeding success than animals breeding in remote areas? There is also a need for studies of the effects of human disturbance on marine mammals breeding in coastal areas. Whether or not certain species of seals stop to breed in coastal areas as a result of increased human activity and the habituation of seal pups to the presence of humans are also in need of study.

Physiological and behavioral data gathered from a variety of wild animals strongly indicate that wildlife and humans can coexist in most situations. However, both birds and mammals have several critical periods that need to be identified. During these periods traffic and human disturbance should be regulated in order to prevent a reduction in the animals' reproductive success. Future research efforts should be directed towards such critical periods both in birds and mammals.

Literature Cited

Cannon, W.B. 1929. *Bodily Changes in Pain, Hunger, Fear, and Range.* New York: Appleton Press.

Causby, L.A. and E.N. Smith. 1981. Control of fear bradycardia in swamp rabbit (*Sylvilagus aquaticus*). *Comparative Biochemical Physiology* 69C:367–370.

Darwin, C. 1900. *A Posthumous Essay on Instinct, Mental Evolution in Animals*, ed., G.S. Romanes. New York: Appleton Press.

Folkow, B. and E. Neil. 1971. *Circulation.* London, England: Oxford University Press.

Freddy, D.J., W.B. Bronaugh, and M.C. Fowler. 1986. Response of mule deer to disturbance by persons at foot and snowmobiles. *Wildlife Society Bulletin* 14:63–68.

Gabrielsen, G.W. 1985. Do not disturb nesting eiders! In *Norsk Polarinstitutt Årbok 1984*, 21–24.

Gabrielsen, G.W. and E.N. Smith. 1985. Physiological responses associated with feigned death in the American opossum. *Acta Physiologica Scandinavia* 123:393–398.

Gabrielsen, G.W., J. Kanwisher, and J.B. Steen. 1977. Emotional bradycardia: a telemetry study on incubating willow grouse (*Lagopus lagopus*). *Acta Physiologica Scandinavia* 100:255–257.

Gabrielsen, G.W., A.S. Blix, and H. Ursin. 1985. Orienting and freezing responses in incubating ptarmigan hens. *Physiology and Behavior* 34:925–934.

Gabrielsen, G.W., F. Mehlum, H.E. Karlsen, O. Andresen, and H. Parker. 1991. Energy cost during incubation and thermoregulation in female common eiders (*Somateria mollissima*). *Norsk Polarinstitutt Skrifter* 195:51–62.

Jacobsen, N.K. 1979. Alarm bradycardia in white-tailed deer fawns. *Journal of Mammalogy* 60:343–349.

Kircher, A. 1646. Experimentum mirabile. De imaginatione gallinae, *Ars Magna Lucis et Umbrae*, 154–155, Rome.

Langvatn, R. and R. Andersen. 1991. Støy og forstyrrelser, metodikk til registrering av hjortedyrs reaksjon på militær aktivitet. *NINA Oppdragsmelding* 98:48 pp.

MacArthur, R.A., R.H. Johnston, and V. Geist. 1979. Factors influencing heart rate in free-ranging bighorn sheep: a physiological approach to the study of wildlife harassment. *Canadian Journal of Zoology* 57:2010–2021.

MacArthur, R.A., V. Geist, and R.H. Johnston. 1982. Cardiac and behavioral responses of mountain sheep to human disturbance. *Journal of Wildlife Management* 46:351–358.

Mayes, A. 1979. The physiology of fear and anxiety. In *Fear in Animals and Man*, ed., W. Sluckin, 24–55. New York: Van Nostrand Reinhold Company.

Olsson, O. and G.W. Gabrielsen. 1990. Effects of helicopters on a large and remote colony of Brunnich's guillemots (*Uria lomvia*) in Svalbard. *Norask Polarinstitutt Rapport-serie* 64:36 pp.

Pavlov, J.P. 1927. *Conditioned Reflexes*. London, England: Oxford University Press.

Rosenmann, M. and P. Morrison. 1974. Physiological characteristics of the alarm reaction in the deer mouse. *Physiologica Zoologica* 47:230–241.

Sargeant, A.B. and L.E. Eberhardt. 1975. Death feigning by ducks in response to predation by red foxes (*Vulpes vulpes*). *American Midland Naturalist* 94:108–119.

Smith, E.N. and C. Johnson. 1984. Fear bradycardia in the eastern fox squirrel (*Sciurus niger*) and eastern gray squirrel (*S. carolinensis*). *Comparative Biochemical Physiology* 78A:409–411.

Smith, E.N. and D.J. Sweet. 1980. Effect of atropine on the onset of fear bradycardia in eastern cottontail rabbits (*Sylvilagus floridanus*). *Comparative Biochemical Physiology* 66C:239–241.

Smith, E.N. and R.A. Woodruff. 1980. Fear bradycardia in free-ranging woodchucks (*Marmota monax*). *Journal of Mammalogy* 61:750–753.

Smith, E.N., C. Johnson, and K.J. Martin. 1981. Fear bradycardia in captive eastern chipmunk (*Tamius striatus*). *Comparative Biochemical Physiology* 70A:529–532.

Steen, J.B., G.W. Gabrielsen, and J. Kanwisher. 1988. Physiological aspects of freezing behavior in willow ptarmigan hens. *Acta Physiologica Scandinavia* 134:299–304.

Tyler, N.J.C. 1991. Short-term behavioral responses of Svalbard reindeer (*Rangifer tarandus platyrhynchus*) to direct provocation by snowmobiles. *Biological Conservation* 56:179–194.

CHAPTER 8

Responses of Wildlife to Noise

Ann E. Bowles

Recreational Influences

Many species of animals depend on *meaningful sounds* to communicate, navigate, avoid dangers, and find food; they detect these sounds against a background of *noise*. Humans produce both meaningful sounds and noise, but wildlife managers cannot always distinguish the two. For the purposes of this review, human-made noise will include both meaningful sounds and meaningless noise, by definition "any human-made sound that alters the behavior of animals or interferes with their normal functioning." From a legal point of view, such noise *takes* animals (e.g., Endangered Species Act of 1973; Marine Mammal Protection Act of 1972).

The most important "takes" are those that *damage* animals by harming their health or altering reproduction, survivorship, habitat use, distribution, abundance, or genetic composition. However, noise can also harass animals by *disturbing* them (causing detectable change in behavior). From a legal point of view, harassment includes behaviors that indicate an animal has heard a sound, as well as behaviors that indicate aversion. However, from a common sense point of view, harassment refers to a disturbance that is threatening or uncomfortable, especially one that is deliberately directed at an individual. When harassed in this narrower sense, animals become sensitized, whereas they often learn to ignore disturbances that are not directed at them. To make this distinction, the term *harassment* will refer strictly to disturbances that threaten or cause discomfort.

Noise Descriptors

Human-made noise has a number of acoustic properties that can change the probability of behavioral and physiological responses. *Dose-response models* describing the relation between these properties and animal responses are still in the earliest stages of development. What follows is a brief description of the noise descriptors that predict responses in humans and laboratory animals, a good starting point for studies of wild animals.

Sound is a physical disturbance in a medium that can be detected by the ear. Most often, it is measured as sound pressure and expressed in relative logarithmic units, called decibels (dB). On this scale, human perception of loudness is roughly linear. Decibels are defined by $20 \log(P/P_o)$, where P is the pressure of the sound and P_o is the reference level. In air, the reference level is the threshold of human hearing, approximately 20 µPa. In water, the reference level is 1 µPa. Table 8.1 gives the levels of various sounds on this scale.

Sounds differ in frequency, perceived as pitch. Frequency is measured in cycles per second (or Hertz [Hz]). Both humans and animals have an *auditory threshold function* that describes their frequency-specific sensitivity, or auditory "filter." For example, humans and many terrestrial animals cannot hear sounds like sonic booms that have their *peak frequency* (the frequency with the most energy) at around 10 Hz, which is below the range of hearing. They detect such sounds indirectly, by hearing the rattle of windows or feeling vibrations induced in the ground or in their muscles and lungs. The *spectrum* (plot of amplitude vs. frequency) of a sound must be weighted by the auditory threshold function of an animal to characterize its audibility. The most commonly used weighting function, *A-weighting*, emphasizes frequencies that humans hear best.

Sounds also differ in duration. For sounds lasting less than a second (s), instantaneous level integrated over time determines loudness. Constant tones lasting longer than 1 s are perceived as having a constant level, no matter how long they last. Acousticians commonly use fast time-averaging (125 ms) to characterize short (*impulsive or transient*) sounds. They use sound exposure level (SEL) to characterize longer discrete sounds, and to compare discrete sounds and impulses. SEL is defined as sound exposure integrated over 1 s.

Any measurement of noise must specify, at minimum, the type of measure (e.g., sound pressure level, sound exposure level), any frequency-weighting functions that were applied (e.g., A-weighting), and any time-averaging (e.g., fast response). Without this information, a given level cannot be compared to others. Most authors of studies on terrestrial animals have reported A-weighted sound pressure levels (*sound levels*) measured with readily available, hand-held *sound level meters*. Unfortunately, taking sound measurements is not simply a matter of holding out a meter and reading a number. Choice of frequency-weighting, time averaging, and type of measure greatly affect the usefulness of noise measurements, especially in the real world, where wind noise makes low-frequency noise hard to measure, and where the auditory

Table 8.1

Sound Pressure Levels of Various Familiar Sounds in Air and Water [a]

Example source	In air		In water	
	SPL in air (dB)	Intensity (watts/m²)	SPL in water (dB)	Example source
Sound just audible to nocturnal carnivore	−20 dB	1×10^{-14}	42	Sound just audible to bottlenose dolphin
Sound just audible to humans	0	9.5×10^{-13}	61	Quiet ambient in small bodies of water
Quiet desert	20	9.5×10^{-11}	81	Ocean ambient, no wind
Nighttime home	40	9.5×10^{-9}	101	Sound just audible to salmonids, tuna
Normal speech	60	9.5×10^{-7}	121	50% of mysticetes respond to human-made noise
Safe limit continuous noise	70	9.5×10^{-6}	131	
	90	9.5×10^{-4}	151	Safe limit for small fish (20–30 min exposures)
Startle reflex stops habituating	100	9.5×10^{-3}	161	Noise made by vessels
Threshold of auditory pain in humans	120	9.5×10^{-1}	181	Sounds causing discomfort in seals, divers
Jet aircraft at 50 m	130	9.5	191	Maximum level of whale calls
Rocket noise at 500 m	160	9.5×10^{3}	221	Intense engine noise underwater
One-quarter stick of dynamite at 1 m	180	9.5×10^{5}	241	
	200	9.5×10^{7}	261	Intense seismic survey impulses

[a] Levels in the two media can be compared by calculating sound pressure level of a sound with equivalent intensity in each medium (center column).

characteristics of the species receiving the noise can vary greatly. Instruments are not always readily available to collect the measures of interest, for example, measures of cumulative noise level underwater or measures of sound energy passing through a unit area, called *intensity* (watts/cm^2).

Community noise monitors are important instruments for measuring cumulative noise exposure in air. Among the measures they collect is L_{eq}, the *equivalent continuous sound level*, defined as L_{exi} + 10 log N_i + 10 log 1/T, where T is a specified duration, such as an hour, N_i is the ith occurrence of a noise, and L_{exi} is the level of the noise. L_{eq} is usually A-weighted. It correlates well with human annoyance in response to most constant noises, but it is less predictive of responses to transients and impulses, such as aircraft overflights (Fidell et al. 1985). Most agencies define exposure limits using a modified version of this measure, the A-weighted day-night sound level (DNL or L_{dn}). DNL is the 24-hour L_{eq} with nighttime sound exposure weighted by an added 10 dB.

Noise duration, SEL, and L_{eq} are the measures that best predict effects on humans, but other measures are important too. These include intermittency or duty cycle, the proportion of time that a sound is on; onset rate (dB/s), which determines the potential to startle; bandwidth, which determines the potential to mask meaningful sounds; modulation rate in amplitude and frequency, which predicts potential for annoyance; variability, which affects arousal; and timbre or quality, which also predicts annoyance. The ear determines meaning using fine scale time-varying temporal and spectral properties.

Sound Transmission in Natural Environments

Sound attenuates (fades) as it travels along a path from a source to a receiver. It is transmitted through a medium (air, land, or water) and it is filtered by the auditory system before it is received. In an ideal world, sound in air attenuates away from a point source as a function of the square of the distance, but attenuation is also dependent on a host of environmental factors such as altitude of the source, temperature, humidity, terrain, wind conditions, and vegetation. Reception is dependent on background noise, called ambient noise. Because transmission loss is complex, it is best to measure noise exposure at the receiver. Unfortunately, most studies up to now have had to depend on estimates of received levels from models of transmission loss or empirical measurements at various ranges from a source. Currently, several groups are working on devices to measure exposure that can be mounted on animals (Kugler and Barber 1993).

At high frequencies, sound is increasingly absorbed by air and water, increasing attenuation. At low frequencies (below 100 Hz), absorption is not very important. Therefore, low frequency noise is transmitted most efficiently.

Low frequency noise from massive noise sources such as storms and rockets is detectable hundreds of miles away (Griffin and Hopkins 1974). Most terrestrial mammals cannot detect these sounds, but some birds, marine mammals, and fish respond to low-frequency sound (Kreithen and Quine 1979; Ketten 1992; Enger et al. 1993).

Sound travels five times faster underwater than it does in air, a consequence of the greater density of water. Underwater, intense low-frequency sources remain detectable halfway around the world under the right conditions (Munk and Forbes 1989). This is why shipping traffic and other large vessels have elevated ambient levels by 15–30 dB at low frequencies over much of the northern hemisphere. Similar increases in ambient noise could affect fish in the Great Lakes and other deep bodies of water. In the right substrate, low frequency sound can also travel through the ground. Birds can detect low-frequency man-made noise transmitted through the ground before it arrives in air (Higgins 1974).

Sound generated in air does not transmit well into denser media, including water. Young (1973) and Urick (1972) have shown that the sound pressure level of a noise generated by a low-flying aircraft in air has an equivalent sound pressure level just under the surface and directly below the source, but that it attenuates very rapidly both with depth and with angle away from vertical.

Potential Effects of Noise on Animals

Figure 8.1 shows the ranges at which noise sources can have particular kinds of acute effects. This hypothetical example shows that the range of a response is dependent on the rate of transmission loss and the ambient level. Noisy vehicles like snowmobiles will be detectable at ranges of several km in a typical, quiet wilderness environment. If animals respond as soon as they detect a sound, noisy vehicles will affect them at much greater ranges than humans. However, if they are habituated to vehicle noise at levels that are not aversive, humans laughing and yelling can arouse responses at greater ranges than snowmobiles. This illustrates the importance of knowing both the received level of a noise and its meaning to an animal.

Studies of acute effects on wild animals have been conducted since the 1950s. Unfortunately, much of this work is buried in government reports and other so-called "gray literature." The work often remained unpublished when the results were not significant evidence of impact. While negative results are difficult to interpret from a scientific point of view, they are important to managers, who need a reasonable list of concerns to use as guidelines for preparing management plans and environmental impact statements. The list of potential effects that follows emphasizes effects supported by empirical evidence, even if incomplete, and downplays effects that were not supported by

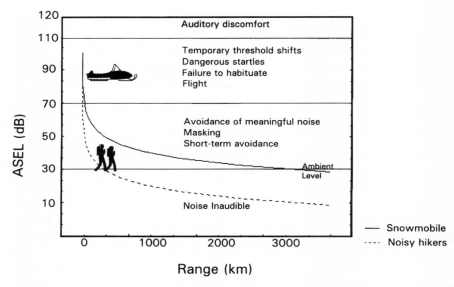

Figure 8.1   Schematic showing the ranges at which acute effects of noisy disturbances might occur under ideal conditions in air (attenuation as a function of r²), assuming a quiet wilderness area with a background ambient of 30 dB. The limits shown are based on standards used for humans.

experiments. It also downplays unsupported speculations and circumstantial evidence.

ANNOYANCE IN HUMANS AND LABORATORY ANIMALS

In humans, any change in sound level that is just detectable, a change as little as 1 dB, can annoy a sensitive individual. Such sensitivities are most likely when a sound has negative meaning or when it interferes with a difficult listening task. In animals, the response corresponding to human annoyance is aversion, usually measured by avoidance responses. There is much literature from the laboratory on annoyance in humans and aversion in animals, from which the following useful generalizations can be gleaned.

Noise is an environmental stressor, like heat and cold (Committee on Pain and Distress in Laboratory Animals 1992). Intense noise can be frightening, especially to naive individuals (Gray 1987). With repeated exposure, all vertebrates habituate or adapt behaviorally and physiologically (Peeke and Herz 1973; Borg 1981). One form of adaptation is sensitization (an increase in responsiveness) resulting from negative experiences associated with noise; vertebrates from fish to mammals can learn to avoid noise associated with danger (Soria et al. 1993). This kind of avoidance is very different from avoidance occasioned by noise per se. Cats and laboratory rodents learn complex tasks to avoid intense noise unassociated with other negative consequences, but not as reliably as they learn after exposure to other stressors, such as pain and handling. This means noise is not as aversive as pain and handling. In the laboratory, noise is rarely used in negative conditioning experiments by itself because it is such a mild stressor. Aversion to intense noise may result because an animal experiences auditory pain, but more often it is the result of the unpleasant sensation of startling or fright. Motivations such as hunger that keep animals from paying attention to noise lessen its aversiveness. Aversion is also lessened if animals can control or predict exposures.

Aversive noise stimulates mood changes. Humans and laboratory rodents become irritable when exposed chronically to high levels of noise (Anthony and Ackerman 1955). This irritability leads to increases in agonistic behavior (aggression and defense) and suppressed food intake (Sackler et al. 1959). Intense impulses or variable noise can interfere with performance in tasks that require intense concentration, although both humans and laboratory rodents learn to correct for this interference when performing familiar tasks.

HEARING LOSS

Hearing loss is the most important and best-understood health effect of noise in humans. Guidelines and standards for human exposure prevent two kinds of hearing loss, acute and chronic. Acute exposures are brief and occasional. The acceptable limits for acute exposure vary with duration, frequency, and sound level. Acute exposures damage hearing at levels over 140–150 dB (peak overpressure) in the frequency range heard best by humans. Chronic human exposure means daily exposures over a typical working lifetime (up to 40 years). DNL values of 70 dB(A) are used by most agencies as a limit for long-term exposure in air, and L_{eq} values of 80–100 dB(A) for shorter exposures (a few hours to a few minutes). Above about 110 dB(A), even short-term exposure to continuous noise is uncomfortable.

Chronic exposures are rarely relevant to wildlife because high levels of human-made noise are rare in the wilderness. For example, Bowles et al. (1993) measured hourly and daily L_{eq} in an area exposed to as many as 167 low-altitude jet overflights per day with sound levels frequently in excess of 100 dB ASEL. They reported hourly values of around 58–67 dB when aircraft were present and daily L_{eq} values of around 48 dB.

Guidelines that protect human hearing apply to many terrestrial mammals because they are based on studies of laboratory animals (rats, chinchillas, gerbils, hamsters, and cats). Susceptibility does vary somewhat with species, but the models currently in use are conservative. These limits will not apply to mammals with very different hearing, such as odontocete and mysticete cetaceans or bats.

SPEECH AND SLEEP INTERFERENCE

Noise interferes with human speech understanding and sleep, so it can doubt-less affect communication and sleep in animals as well. Predictive models for humans may not apply to wild animals, but a few of the observations on humans are important. Noise annoys humans at any sound level that masks speech or other meaningful sounds. Humans and laboratory mammals tol-erate very high noise levels during sleep after they have adapted behaviorally and physiologically, a process that can take several months (Suter 1992). After adaptation, detectable physiological changes persist, such as changes in REM sleep, but the subject is not aware of wakening to noise. The sounds that in-terfere most successfully with sleep are meaningful sounds and transients with high onset rate. After adaptation, subjects continue to awaken in re-sponse to meaningful sounds, such as a baby crying.

EFFECTS ON WILD ANIMALS

Noise is suspected of causing stress-related illness in both humans and ani-mals, but the causal link has been difficult to prove (NRC 1981; FICON 1992). A number of effects can occur in wild animals that would not be important to humans or laboratory animals. In wild animals the question of noise meaning becomes crucial. Wild animals view humans as potential predators, so they often respond to human-made noise at levels that would never arouse the in-terest of unstressed laboratory animals (e.g., quiet speech). Wild animals con-tinue to respond to disturbances even when they have had a chance to adapt because they are sensitized by predator harassment or other dangers. Wild an-imals can abandon favored habitat in response to disturbances or incur ener-getic expenses after reacting repeatedly when they cannot escape. Masking and hearing loss represent a life-threatening hazard in predator-prey interactions (e.g., Webster and Webster 1972). Meaningful human-made noise might also confuse animals into behaving in maladaptive ways (Brattstrom and Bondello 1983). Aversive levels of noise might cause wild animals to become irritable, affecting feed intake, social interactions, or parenting. All these effects might eventually result in population declines. Even if populations were unaffected, genetically determined differences in susceptibility might exert subtle selec-tion that eventually could affect fitness. Each of these potential effects will be considered.

Effects of Noise on Auditory Physiology and Sensory Perception

The auditory systems of animals are undoubtedly susceptible to noise. Audi-tory effects on wild animals have received little attention because human-made noise is not generally intense enough or constant enough in wilderness areas to cause problems with hearing. What follows is a brief summary of the

auditory abilities of animals, to put the information on hearing effects in perspective. Fay (1988) has provided an excellent compilation of all the data available on vertebrate hearing for those that need more detail.

AUDITORY ABILITIES OF ANIMALS

Mammal hearing varies, although abilities are fairly consistent within families
and sub-families. Taken together, mammals can hear in the bandwidth from
below 10 Hz to over 150 kHz, with sensitivity down to −20 dB. Rodents, small
odontocete cetaceans, and bats hear best at high frequencies, while mysticete
cetaceans appear to hear best at low-frequencies (Ketten 1992). Nocturnal
mammals (cats and nocturnal canids) have the most sensitive hearing among
terrestrial vertebrates.

The mammalian ear is most susceptible to noise within its best range, the
bandwidth it hears best. Hearing in this range is also damaged if individuals
are exposed at frequencies an octave below (half the frequency) the best range.
An animal's best sensitivity (the faintest sound it can hear) is a good predictor
of susceptibility to hearing loss. For example, cats are 15–20 dB more sensitive
than humans and laboratory rodents at their respective best frequency (the
frequency they hear best) and are likely to lose their hearing to noise at levels
18 dB lower than lab animals (Nixon and von Gierke 1966).

Birds have more uniform hearing capabilities than mammals (Dooling
1980). All birds hear well in the range from 100 Hz to around 8–10 kHz, and
most have best sensitivities around 0–10 dB. There are a few exceptions to the
rule; owls are 15–20 dB more sensitive in their best range than other birds.

Some birds may be able to detect very low-frequency noise (Kreithen and
Quine 1979). Birds can localize sounds at lower frequencies than might be expected based on their head size because they have an air-filled space passing all
the way through the head between the ears, which acts as an extremely sensitive detector of pressure differentials. This adaptation may also help them to
detect very low frequencies. Kreithen and Quine's work is considered somewhat controversial because it has not been replicated in other laboratories, but
there is nothing in their work that is inconsistent with studies of higher frequency sound. Birds are not sensitive at these very low frequencies, but they
could detect intense low frequency sources, such as storms and rocket
launches.

Reptile hearing is poorly-studied (Fay 1988). Reptiles as a group have more
limited range than birds, from 50 Hz to at most 2 kHz. Testudinates (turtles
and tortoises), both terrestrial and marine, hear poorly, with best sensitivities
of 40–50 dB or worse and dramatic rolloff above 1 kHz. They respond to
meaningful sounds that they can hear, of course. Snakes hear as poorly as testudinates (Wever and Vernon 1960). Lizards hear better, with best sensitivities

of 10–20 dB (measured electrophysiologically). The leopard lizard has good sensitivity at low frequencies (Wever et al. 1966).

Birds and reptiles are also highly sensitive to vibration (e.g., Shen 1983), which low-frequency noise can induce in an animal or the substrate. Vibration sensitivity is an important source of information about approaching predators and prey. Because they have relatively insensitive hearing, reptiles may detect noise using induced vibrations.

Amphibians have variable hearing capacities specialized for the perception of social and other meaningful signals. Overall, their bandwidth lies between 100 Hz and 2 kHz. Their best sensitivities range widely from 10 dB to 60 dB. However, they have exquisite sensitivity to vibration (Lewis and Narins 1985). There are species that use low-frequency acoustic cues detected via ground vibrations to communicate, to time their emergence from burrows (Dimmitt and Ruibal 1980), and to avoid predators. In amphibians, induced vibrations are important predictors of effect.

SUSCEPTIBILITY OF HEARING TO NOISE

Sound is transduced by hair cells in the inner ear. Noise-induced hearing loss usually results from hair-cell loss. Intense noise can also damage the underlying membranes, supportive tissue, and nervous tissue, but ruptures and breaks in the bones surrounding the ear are always the result of blast and impact injuries (explosions, hitting the head). "Bleeding ears" result from ruptures of the tympanic membrane, the result of abrupt changes in static pressure. Animals do not develop bleeding ears from noise exposure (Suter 1992). Bleeding into the inner ear results from exposure to intense noise (Reinis 1976; Brattstrom and Bondello 1983). Exposures in these cases are intense enough to cause acute hearing loss regardless of bleeding.

Hair-cell loss in mammals is typically permanent. The relationship between hair-cell loss and hearing loss is not simple, however. Individuals can have damaged hair cells without detectable deficits in hearing, or, alternately, can experience obvious hearing losses with apparently minor hair-cell damage. Hearing deficits should always be measured directly, either as temporary shifts in threshold (TTS) or permanent hearing loss (PTS). TTS of greater than around 25 dB is a good indicator of PTS in mammals, but smaller losses are not (Harris 1991). Recovery time from TTS is proportional to the logarithm of exposure time.

In birds, hearing loss is harder to characterize. Birds regenerate hair cells even after substantial losses (Corwin and Cotanche 1988; Ryals and Rubel 1988), a process that can take around two months, depending on the degree of damage. Thus, neither TTS nor PTS measured over short periods is a good predictor of permanent hearing loss in birds.

Hearing losses are poorly studied in reptiles. Reptile hearing can be damaged by exposure at close range by impulsive noise from ATV's (Bondello 1976). It is likely that reptiles regenerate hair cells like other poikilotherms, but this has yet to be proven.

Amphibians experience hair-cell growth throughout life (Corwin and Warchol 1991), but hearing damage has not been studied extensively.

MASKING

Masking occurs when noise interferes with the perception of a sound of interest. The ear can be viewed as a series of band-pass filters about ⅓ octave wide, called critical bands. Masking occurs when the energy of a signal and of noise within a critical band are equal. Most research on masking has been done with continuous wide-band noise; the problem becomes much more complicated with varying real-world noise.

Human-made noise can certainly mask meaningful sounds. However, animals are not helpless in the presence of noise, particularly the sort of intermittent, fluctuating, and impulsive noise that is typical of recreational activities. Trained humans can hear over 20 dB into broad-band masking noise. This is an extreme case, but it is not unusual for both humans and animals to be able to hear 5–10 dB into noise by taking advantage of binaural cues, internal auditory templates of the signal of interest, and frequency and amplitude tracking.

Masking will affect predator avoidance most because detecting predators is a time-critical activity. Webster and Webster (1972) showed that elimination of the auditory bulla, which controls low-frequency hearing in kangaroo rats, increased the risk of capture by rattlesnakes on dark nights.

If noise masks social signals for long periods, it can foreshorten the effective range of communication. Busy highways affect the effective range of birds that sing softly over distances of several hundred meters. Birds compensate for attenuation due to terrain and vegetation by choosing a spot to listen and sing that minimizes attenuation. They can control the masking effects of human-made noise through behavioral adaptation as well. So far, no one has provided evidence of any serious consequences of such behavioral changes.

Animals can also compensate for noise masking by avoiding it. They can move out of the insonified area, wait until the noise stops, or shift the frequency and level of their signals to avoid masking. These behaviors could affect energy expenditure. In particular, sound production is not an efficient process; the energy output of a call is a few percent (1%–2%) of the energy input (Ryan 1988). Compensating by increasing sound output could increase energetic expenditure significantly. Increases in energetic expenditure in the presence of human-made noise have not been measured, but energy consumption is probably a limiting factor in frogs trying to outperform calling conspecifics (Ryan 1988).

Masking may also affect the information in social signals. In one series of experiments, the Puerto Rican coqui stopped producing portions of its call in the presence of man-made masking noise (Narins 1982). The consequences of such changes are unknown.

STRESS AND DISTRESS IN WILD ANIMALS

Rodents experience adrenal hypertrophy after chronic exposure to noise (Anthony and Ackerman 1955; Chesser et al. 1975; Pritchett et al. 1978). However, hypertrophy did not affect population density of free-ranging individuals in the study by Chesser et al. Continuous, intense noise exposure can cause health effects in the laboratory, but extensive experiments with intermittent noise does not (Borg 1981), apparently because animals recover between successive exposures. Therefore, the sort of cumulative measures of exposure used to predict human annoyance may not reliably predict nonauditory health effects.

Some authors have measured heart rate, catecholamine levels, and corticosteroid levels after frightening animals briefly with a noisy disturbance (jets, simulated jet noise, sonic booms, and helicopters hovering). They found elevated heart rates for one to five minutes, elevated catecholamines for at most a half hour and elevated cortisol levels for two to four hours in domestic stock (Beyer 1983; Krausman et al. 1992; LeBlanc et al. 1991). Unfortunately, short-term increases in these measures do not correlate well with distress (Moberg 1985, 1987; Lefcourt 1986). In fact, responsive individuals may be better equipped to deal with stressors (Gray 1987). Recovery between exposures and psychologically mediated internal feedback mechanisms greatly affect the potential for cumulative effect.

Because the relation between short-term responses and distress is often tenuous, Moberg (1987) recommends using outcome measures to correlate exposure to stressors with distress. Outcome measures include reproductive success, rate of growth, incidence of disease, and changes in survivorship. These measures are especially important in wild animals because handling definitely causes distress (Kock et al. 1987). Careful controls would be needed before physiological measures collected from wild animals that had been handled would be convincing evidence of noise-induced distress.

EFFECTS ON REPRODUCTIVE PHYSIOLOGY

Reproduction is the outcome measure easiest to correlate with disturbances. In mammals, severe disturbances, such as intrusions, pursuit, handling, and transport can cause loss of fertility and abortions. Noise has been suspected of causing similar effects.

Conception and pregnancy   Spontaneous noise-induced abortions do not occur in well-established pregnancies. Disease and handling stress were the most parsimonious explanations for abortions in uncontrolled tests with pregnant domestic stock (Heicks 1985; Heuweiser 1982; Schriever 1985). The affected stock were infected with diseases known to cause abortions. In healthy herds, abortion rates did not increase, either compared with typical rates in the literature or with matched controls (LeBlanc et al. 1991; Erath 1984; Beyer 1983; Krüger 1982). Fright is not a recognized cause of abortions in clinical studies involving thousands of animals.

Theoretically, gametogenesis, conception, and the very early stages of development should be more sensitive to noise than an established pregnancy, because they are wholly controlled by the blood factors of the parent. The effects of noise on these early stages have not been studied, but Stoebel and Moberg (1982) reported interruptions of surges in luteinizing hormone after surprising electric shocks, a change that could affect fertilization.

Effects on hatchability   Effects on hatchability must result from parental neglect because repeated efforts to affect artificially incubated eggs with various kinds of intense noise have failed (Stadelman 1958; Heinemann and LeBrocq 1965; Cottereau 1972; Teer and Truett 1973; Cogger and Zegarra 1980; Bowles et al. 1991 and unpub.).

In the early 1970s there was a widespread belief that sonic booms could crack eggs (Bell 1972). This effect is physically impossible because one cannot generate sufficient sound pressure in air to crack eggs, and because sonic booms are unlikely to excite eggs at their resonant frequencies (chicken eggs resonate at frequencies of 400–800 Hz, whereas sonic booms have peak frequencies at 10 Hz; Bowles et al. unpub. data).

Effects on avian productivity   Domestic fowl sometimes experience declines in productivity after continuous exposure to noise at high levels (above 85 dB(A)) and intermittent noise at high-duty cycles (Belanovskii and Omel'yanenko 1982; Okamoto et al. 1963). Laying rates did not change in wild waterfowl after exposure to continuous noise from a compressor station, probably because levels never exceeded L_{eq} of 70 dB (Murphy and Anderson 1993).

Impulsive noise can break up broodiness in domestic fowl (Stadelman and Kosin 1957; Jeannotout and Adams 1961), but does not have similarly large effects on wild fowl (Lynch and Speake 1978). If noise were to reduce the motivation of wild parents to incubate or brood, small decreases in fledging success might be expected.

Persistent human disturbance or harassment by predators causes declines in productivity of colonies of birds (e.g., Anderson and Fortner 1988). Whether noisy disturbances can have the same effect is unclear. Figure 8.2 shows the distribution of differences in young fledged per nest between exposed nests and control nests in a series of studies on raptors. Individually, these studies did not find statistically significant effects on fledging success, but the sample sizes were always so small that the power of the tests was poor. Although the body of evidence in these studies is by no means a systematic sample, Figure 8.2 is suggestive of a small negative effect on fledging success resulting from humans approaching (59% of studies negative) and possibly from noisy disturbances (60% negative).

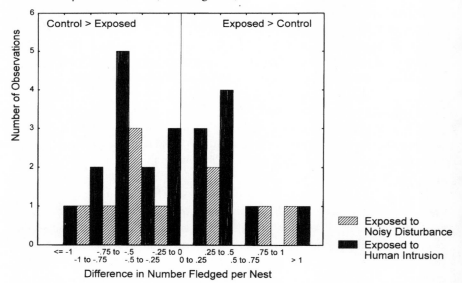

Figure 8.2   The frequency distribution of differences in numbers of young fledged per nest between groups of raptor nests monitored as controls and groups of nest exposed to disturbance. Dark shading indicates studies in which human approaches disturbed nests; cross-hatching indicates studies using noisy disturbances (mostly aircraft). A negative difference indicates that the control nests were more successful than the exposed nests (Holthuijzen et al. 1990; Platt 1977; Ellis 1981; Bednarz and Hayden 1988; Fraser and Mathisen 1985; Mathisen 1968; White et al. 1979; Poole 1981; Swenson 1979; Anderson 1989; Grier 1969; Windsor 1977).

EFFECTS ON ACTIVITY AND ENERGY CONSUMPTION

Noise and noisy disturbances may affect activity and energy consumption, but the effect is modulated by behavioral and physiological adaptation. Habituation keeps animals from expending energy and attention on harmless stimuli, whereas sensitization increases energy expenditure to help them avoid dangerous stimuli.

These responses are mediated by arousal, processing in the central nervous system, and the feedback mechanisms between brain and body. Noise can arouse animals, in the sense of getting their attention, waking them, and increasing their activity. Dangerous or unfamiliar noise is more likely to arouse them than harmless, familiar noise. Therefore, an animal must recognize a sound before it responds, requiring time-consuming processing in the central nervous system (CNS). There is a more rapid response, the startle response, which is a primitive reflex (Ekman et al. 1985) protecting all vertebrates from attack and accident. Individuals often do not know they have been startled until the response is well under way.

Startles are characterized by a rapid increase in heart rate and cardiac output, shutdown of the gut and other nonessential functions, and rapid mobilization of glucose reserves to supply the muscles. They habituate slowly, and, at high sound pressure levels, the cardiovascular component of the response does not habituate completely (Hoffman and Searle 1968). Animals can learn to control gross movements resulting from a startle, such as flight, alerting, and flinching. The feelings associated with the response are aversive.

In humans and laboratory animals, impulses at frequencies within the range of best hearing with onset rates better than around 20 dB/sec, and levels exceeding around 80 dB, trigger the reflex. At frequencies below 100 Hz, the response is considerably reduced. This means that impulses at these frequencies are less aversive.

Increases in heart rate after noisy disturbances are not by themselves evidence that an animal has been startled or frightened. Mild increases might mean that an animal has detected a sound, initiating transitory arousal. This is one of the pitfalls of using heart rate as an indicator of physiological effect. Meaningful noise, such as the sounds produced by biting insects, also causes increases in heart rate, sometimes as large or larger than increases caused by startles (Workman and Bunch 1991). In these cases, animals have clearly been harassed. Before heart rate can be a useful indicator of effect, arousal, startles, and responses to harassment have to be distinguishable.

Activity and energetic effects in mammals   If a noisy sound source arouses an animal, it has the potential to affect its metabolic rate by making it more active. Increased activity can, in turn, deplete energetic reserves.

A few studies have documented increases in activity after aircraft approaches. The overt behavioral component of these responses was fairly mild, such as starting a few steps or walking away slowly from the site of the disturbance (Harrington and Veitch 1991; Murphy et al. 1993). However, individuals tracked remotely moved greater distances in the 24 hours after exposure, on the order of 1–2 km.

Most tracking studies confirm these results. Wild ungulates and carnivores changed their movements in response to aircraft overflights, snowmobile approaches, traffic on logging roads, construction noise, military training activity, and walking park visitors (e.g., Krausman et al. 1986; Eckstein et al. 1979; Edge and Marcum 1985; Richens and Lavigne 1978). All these sounds were associated with approaches or human activity. Mammals exposed to constant noise unassociated with human activity have not been tracked.

One study reported possible effects on energetic reserves through changes in food intake. Stockwell et al. (1991) measured responses of bighorn sheep to helicopters in the Grand Canyon. They observed that sheep alerted more often while eating in the presence of helicopters than when undisturbed. The differences were only significant during seasons when the sheep were near the rim of the canyon, and hence within 500 m of the aircraft. By counting bites of food, the authors concluded that frequent alerting affected food intake.

Stockwell et al. (1991) studied a particularly interesting system because the bighorn sheep in their study were sought out and overflown by aircraft, sometimes at altitudes below 250 m. Close approaches by helicopters can harass animals, and the sheep responded with avoidance to the closest approaches. Unfortunately, the authors did not specify how often any group of sheep were approached this way under normal circumstances, nor did they show convincing evidence of changes that could have had long-term energetic consequences, such as a drop in food consumption over a 24-hour period, changes in 24-hour activity budget, or drops in weight relative to unexposed individuals. This kind of precision is important because bighorn sheep and other ungulates living in areas where they are exposed frequently to aircraft show little evidence of ill-health or reduced reproductive output (e.g., Krausman et al. 1993; Davis and Valkenburg 1985).

Activity and energetic effects in birds   The studies of energetic effects in birds have some of the same deficiencies already described, but the evidence for effects on migratory waterfowl is more convincing. Waterfowl are more overtly responsive to noise than other birds (Thiessen et al. 1957; Edwards et al. 1979), probably because they are hunted persistently by humans (i.e., loud guns) and because, as migrants, they are often unfamiliar with a given disturbance. They seem to be particularly responsive to aircraft (Owens 1977), possibly because they are frequently harassed by aerial predators. They can adjust to noisy disturbances eventually (Marsh et al. 1991), but the process is slow. Ward and Stehn (1989), Bélanger and Bédard (1990), and Davis and Wiseley (1974) showed that flight responses of migratory waterfowl exposed to overflights by light aircraft and helicopters did not habituate completely, and that changes in activity as a result of repeated exposure were extensive enough that energetic losses could be extrapolated.

One of the factors contributing to increased responsiveness was flock composition. Waterfowl flocks seemed to be as sensitive as the most responsive individual in the flock, so that larger flocks had a greater chance of responding than small ones. If so, it is possible that migrants in large flocks would never habituate to aircraft, particularly if their sensitivity was reinforced by aerial predators. Unfortunately, authors have not been able to equip individual waterfowl with instruments that document long-range movements and cumulative exposure. For example, Bélanger and Bédard suggested that snow geese did not increase foraging time at night to compensate for losses in foraging time induced by aircraft during the day, but they did not track individual birds over many days to show what they would do when consistently deprived of food.

Ward et al. (1986) and Miller (1991) provided some of the best evidence by monitoring individual Pacific black brants equipped with radio transmitters during multiple experimental exposures to Bell 206B helicopters. They found that the duration of responses was constant with repeated exposure, and that marked birds in flocks continued to respond. Unfortunately, they did not measure probability of flight with repeated exposure in a number of individuals. It is not surprising that the duration of responses did not change, as the birds were typically leaving the pond where they were exposed for other ponds; however, one might have expected a decline in probability of flight over periods of several months with repeated exposure.

Jensen (1990) and Ward and Stehn (1989) presented an analysis of the energetic losses of birds exposed to aircraft repeatedly. They admitted that their data might not represent real long-term energetic consequences, because the birds might have adjusted metabolically or behaviorally somehow over time. However, their extrapolated data predicted a significant reduction in energetic reserves if adaptation did not occur.

The brant geese in Ward's study exhibited flight responses at distances of up to 3 km, a range at which noise from helicopters would be just detectable. Therefore, the birds were sensitized in some way to the noise or the combination of noise and visual stimulus of the helicopter. The geese were harassed often by predatory bald eagles (Ward et al. 1986), which constituted a quarter of the disturbances observed.

EFFECTS ON IMMUNE FUNCTION AND GROWTH

Prolonged physiological stress and significant energetic expenditures would eventually compromise the health of animals by suppressing immune function, making them susceptible to infection and parasites, by altering growth, and by slowing recovery from food shortages.

These effects are theoretically possible but as yet unproven. Effects on growth have been studied in domestic species. Intense, prolonged exposure (levels ≥95 dB(A), eight hours per day) can affect growth (Fell et al. 1976), but studies using more typical exposures show little effect on growth. Food intake might be a useful measure of both increased activity (through increased food intake) and irritability (through decreased food intake) in both wild and domestic animals (Bradley et al. 1990).

EFFECTS ON SLEEP AND TORPOR

Sleep effects are little studied in animals, including laboratory animals. Brattstrom and Bondello (1983) stimulated estivating spadefoot toads with simulated ATV noise in the laboratory, causing them to emerge from their burrows. Normally, thunder and rain noise stimulates spadefoot toads to emerge (Dimmitt and Ruibal 1980) in time to take advantage of unpredictable desert rainstorms. Brattstrom found that the toads emerged repeatedly, but this response declined over the course of the experiment (temperature was held constant). Brattstrom expressed the concern that human-made impulses would cause toads to risk desiccation by emerging under inappropriate conditions, by forcing them to waste water, or by making them more vulnerable to predators.

Brattstrom did not systematically test for habituation, nor did he test the toads under natural conditions. In the wild, spadefoot toads respond differently depending on environmental conditions and their own internal physiological state. They do not emerge during winter rains, apparently because they ignore rain noise when hibernating. In the summer, they do not emerge during the day, and they often emerge at night in the absence of rain (Mayhew 1965). Future tests of this sort should consider the possibility that responses will differ under different conditions.

Sleeping, estivating, and hibernating mammals are difficult to arouse with noise, particularly meaningless noise. Tortoises and wild rodents hibernating in captivity enter and maintain torpor successfully, even in the presence of human activity. Torpor is easiest to interrupt during the early and late stages of hibernation. A study now underway is showing that populations of hibernating wild heteromyid rodents exposed to an average of 10 overflights/hr during the day at levels in excess of 80 dB do not differ significantly in abundance, reproductive output, residency, age and sex composition, or survivorship relative to unexposed populations (Bowles et al. 1993; McClenaghan pers. comm.).

Effects of Noise on Behavior

Most of the literature on noisy disturbances focuses on immediate aversive responses of animals because an Environmental Impact Analysis requires this information. There are actually three broad classes of responses to human-made noise—attraction, tolerance, and aversion—each with the potential for negative or positive effects. Birds, mammals, and fish respond overtly to noise and have been studied often.

In many ways, attraction can be the most dangerous response. Predators are often attracted to humans by an easy meal, after which they are attacked as dangers or competitors (Follman and Hechtel 1990; Stirling 1988; Anderson and Hawkins 1978). Other animals can be attracted as well. For example, the sound of snowmobiles and chain saws in the winter can attract deer to downed trees (Richens and Lavigne 1978). Raptors are attracted to military training activities that scare up prey (Andersen et al. 1986, 1990).

Motivation to find food can make animals tolerant of noise. Every review of the literature on pest control has concluded that one cannot drive animals away from attractive sites with noise over the long term (Shaughnessy et al. 1981; Bomford and O'Brien 1990; Marsh et al. 1991).

Tolerance to noise can have dangerous consequences. Animals that habituate to traffic noise are vulnerable to oncoming vehicles (automobiles, trucks, boats, and aircraft). Large ungulates collide with cars on roads, birds strike aircraft at airports (Burger 1983), and manatees are struck frequently by boats. These animals all hear well within the bandwidth of noise produced by the vehicles. They are hit because they are not sufficiently wary of the noise and because they cannot estimate a safe distance accurately. A number of commercial ventures have tried to improve the detectability of vehicles with noisemakers, but these devices are not designed to be aversive and they are uniformly ineffective in the long run. Unfortunately, the sort of signal that might work well would be very aversive to drivers.

Short-term behavioral responses, especially those that indicate aversion, have been studied often. The review of these responses is broken down by animal group, as follows.

SHORT-TERM RESPONSES

Short-term responses in mammals   The mildest responses of mammals cannot be distinguished from detection (ear movements, brief alerting). As intensity increases, animals alter their activity briefly or start mildly. Next, they exhibit wariness or mild aversion, walking slowly away, freezing, crouching, or making intention movements to run, engaging in mild aggression, and increasing social cohesion (e.g., mothers and calves approaching one another). Beyond this, responses are easily interpreted as aversion, including walking, trotting, running, or flying away, hiding in dens or burrows, and biting, kicking, or slapping. In the most intense responses, mammals panic, urinating or defecating and running at high speed. Some species exhibit species-specific agonistic and defensive behaviors that can indicate fear or aversion, for example the aerial behaviors of marine mammals, horses bucking, and the circling behavior of musk oxen.

Short-term responses in birds Birds have a similar continuum of responses. At the mildest level, they alert. Next, they exhibit mild aversion by flipping their wings (intention movements to fly), pecking at each other, and walking, swimming, or flying short distances. More intense aversion triggers longer movements, crouching on the nest, attacks on conspecifics or on the source of the disturbance (raptors, terns), and long interruptions of normal behavior. In the extreme case, individuals or flocks respond with panic flight or running.

Short-term responses in amphibians and reptiles Many species of reptiles and amphibians freeze in response to noise. Freezing is a behavioral response, and its intensity can be measured as latency to further activity if the reptiles are at all active to start with. It would be easiest to measure latencies in species that stop calling after they hear a noise, for example, chorusing frogs. Short-term behavioral responses can also result from confusion between a human-made sound and a meaningful natural stimulus (e.g., the responses of spadefoot toads to impulsive human-made noise).

CONSEQUENCES OF STARTLES AND FRIGHTS

Panic flight is the most dangerous of the responses to startling human-made noise. While accounts of unusual accidents are widely-quoted, systematic studies show that animals rarely respond with uncontrolled panic. Ungulates that run on flats do not injure themselves (Lenarz 1974). Ungulates and pinnipeds on precarious surfaces run less than those in the open (McCourt and Horstman 1974; Bowles and Stewart 1980). Raptor parents on nests tend to sit much more tightly than roosting parents (Table 8.2). Raptors that do fly from nests usually leave for a matter of a few minutes (Table 8.3). Although raptors can knock eggs from nests in panics (C. M. White pers. comm.), experiments using a total of 211 nests exposed to blasting, gunshots, and low-

Table 8.2

Responses of Perched and Brooding Raptors to Noisy Disturbances[a]

Previous activity	Response to disturbance				
	None	Alert	Move, no flight	Flight	TOTAL
Responses of perched individuals	42 16.09%	122 46.74%	37 14.18%	59 22.61%	260
Responses of incubating or brooding individuals	35 23.18%	85 56.29%	20 13.25%	10 6.62%	150
TOTAL	77 18.78%	207 50.48%	57 13.90%	69 16.83%	410

[a]Data on individual responses came from reports of experimental studies of seven raptor species exposed to close approaches by aircraft (Platt 1977; Windsor 1977; Ellis 1981; unpublished data from J. Hamber, Santa Barbara Museum of Natural History). Raptors incubating or brooding eggs or young were one-fourth as likely to fly.

altitude overflights found that no eggs or young were ever ejected. What follows is a review of the limited literature on losses in panics.

Panic-induced losses in birds   Many wildlife managers have been concerned about massive panic-induced losses in marine birds since the Sooty Tern Incident, in which most of the 50,000 nests of sooty terns on the Dry Tortugas failed. Both the initial abstract describing the Sooty Tern Incident and a longer account on file with the Van Tyne Library (Austin et al. 1972a,b) suggested that the sooty terns failed because they abandoned their eggs when exposed repeatedly to intense sonic booms. The authors arrived at this conclusion by eliminating other probable causes (e.g., pesticides, food shortages, unseasonable weather). The authors were not present on the rookery, however, when the losses occurred and did not examine it until over a month later, rendering their evidence entirely circumstantial.

Evidence internal to Austin et al.'s investigation and from other studies suggests that sonic booms are an unlikely cause for the massive losses. Brown noddies nesting on the islet fledged normal numbers of young, even though they were exposed to the same impulses. Experiments with real and simulated booms have failed to find any losses in other species (Ruddlesden 1971; Schreiber and Schreiber 1980). These experiments include observations of repeated exposures to simulated sonic booms that were more audible than real

Table 8.3

Durations of Raptor Flight Responses Following Aircraft Approaches

Species	Number of disturbances	Number of flights (%)[a]	Mean response duration in minutes[b]	Mean flight duration in minutes[b]	Median response duration in minutes (Range)	Median flight duration in minutes (Range)
California condor (J. Hamber, Santa Barbara Museum of Natural History, unpub.)	41	2 (5%)	3.6	62.5	0 (0–90)	— (35, 90)
Gyrfalcon (Platt and Tull 1975; Platt 1977)	25	10 (40%)	11.8	28.3	0 (0–120)	5.0 (0–120)
Peregrine falcon (Ellis et al. 1991; Platt 1977; Windsor 1977)	40	10 (25%)	7.2	26.7	0.2 (0–170)	2.75 (0.2–170)
Prairie falcon (Ellis et al. 1991)	102	10 (10%)	1.71	11.5	0.1 (0–75)	4.5 (0–75)
TOTAL	208	32 (15%)	4.3	24.71	0.10 (0–170)	4.0 (0–170)

[a] The proportion of flights in this table is biased, as most authors did not report the duration of brief responses such as alerts. Of the 490 responses recorded, 19% were flight responses, and only 1% were by incubating or brooding individuals.

[b] The distribution of response durations is skewed, so medians and ranges should be reported rather than means and standard deviations. Means are provided here for comparison with other studies.

sonic booms. Eyewitness accounts of other exposures to sonic booms on the Dry Tortugas indicated that sooty terns left their nests only briefly, in keeping with eyewitness accounts of responses in other species.

No one will ever know the truth of this incident because the examination of the colony came far too late to eliminate unknown, but natural, factors. In addition to weather and food shortages, the natural factors that should be considered after any future incidents are tick infestations or other parasites, and repeated attacks by terrestrial predators. Unfortunately, this incident often leads managers to presume that sonic booms are unusally dangerous to nesting birds, when in fact exposures to boats, low-flying aircraft, walking humans, and other directed approaches are much more likely to cause reproductive failures in colonial birds.

Panics induced by the approach of noisy disturbances can certainly cause egg loss in colonial birds, especially in naive birds that do not build nests (Bunnell et al. 1981; Hunt 1985) or colonies with predators resident on the rookery (Burger 1981). Bunnell et al. (1981) measured reproductive success of white pelicans exposed to low-flying aircraft. The pelicans were in a remote area with little previous exposure to air traffic and they were regularly attacked by coyotes. The authors did not provide any details about the overflights that caused losses, but reported that pelicans crushed 88% of the eggs on the rookery during one overflight incident shortly after laying. Overflights during brooding caused little loss in the same colony.

Habituation is the crucial determinant of success in the presence of noisy disturbances. Exposures of experienced birds produce no or minimal losses (Black et al. 1984). The habituation process can occur slowly, so it may not be detected by short-term experiments. In the long term, nesting birds become more tenacious and less responsive in the presence of human disturbance if they are not deliberately harassed (Burger and Gochfeld 1981; Knight et al. 1987). After habituation, loss rates, if present, are too low to be detected with the sample sizes typical of most field experiments.

Very low-altitude aircraft overflights can frighten flocking waterfowl into colliding with human-made structures such as power-lines (Blokpoel and Hatch 1976). In the few anecdotal accounts available, losses were the result of approaches within 100 m.

Panic-induced losses in mammals   There are anecdotal accounts of accidental injury to wild mammals when harassed by close pursuits from the air or with vehicles, but to the author's knowledge, deaths or traumatic injuries have not occurred during other noisy disturbances.

Cannibalism and parental neglect in carnivores   Wild and captive cats, dogs, mustelids, pigs, and raptors sometimes kill and consume their own young when severely startled or frightened (e.g., Gipson 1970), including anecdotal accounts of cannibalism in zoo animals (wild cats) that have been startled severely by aircraft. Experimental studies failed to induce the response in captive mink and pigs under controlled conditions (Travis et al. 1974; Winchester et al. 1959; Brach 1983). Future clinical reports of cannibalism should include (1) the proportion of individuals of each species affected; (2) age and condition of the young; (3) health of the parents; (4) conditions that might have sensitized individuals to disturbance (e.g., the presence of intruders); and (5) the typical rate of similar surprising noises in the area before the incident.

Mammals sometimes abandon newborn young when frightened by close approach of a noisy disturbance. Reports are rare, but such incidents are difficult to observe, especially in areas where mammals have little experience with humans. Johnson (1977) reported that harbor seals abandoned newborn pups after panics induced by various kinds of aircraft. In this case, the aircraft were within 150 m of the affected animals, and abandonments did not occur when mothers were already bonded to their young. There are no other similar reports in the literature, despite extensive studies of disturbances in pinniped rookeries (Johnson et al. 1989). Several studies (Harrington and Veitch 1992; Calef et al. 1976) have exposed Arctic ungulates with relatively young calves to approaches by aircraft, without detectable abandonment.

OTHER BEHAVIORS THAT COULD BE AFFECTED

The overwhelming majority of responses to intense noise are more subtle. These are:

Habitat use in birds and mammals   Large mammals alter their movements for periods of up to one to two days after exposure to noisy disturbances (Harrington and Veitch 1991; Malme et al. 1984; Davis et al. 1988; Gese et al. 1989; Klein 1973; Miller and Gunn 1979). Sometimes, this results in short-term changes in habitat use (Krausman et al. 1986), but mammals are clearly able to learn to adjust to these changes to a large degree. Aircraft, snowmobiles, military training activities, human voices, construction noise, blasting, predator calls, and artificial sounds designed to drive animals away have been tested.

The literature on ungulates is large enough to show the flexibility of mammalian habitat use in the face of disturbances. If noisy sources enter the habitat on a schedule (e.g., snowmobiles on weekends, construction noise), deer, sheep, and elk avoid areas when the noisy sources are present and return when they are not (Van Dyke et al. 1986; Dorrance et al. 1975; Edge and Marcum 1985; Leslie and Douglas 1980). If the exposure is brief or if mammals have good cover, differences in home-range size are not detectable (Eckstein et al.

1979; Edge et al. 1985). If mammals are exposed repeatedly to the same noisy stimulus without harassment, responses decline rapidly (Krausman et al. 1986; Valkenburg and Davis 1985). On the other hand, harassment by hunters (and probably predators) amplifies responses (King and Workman 1986).

Vertebrates from fish to mammals can track the direction of movement of disturbances (Burger and Gochfeld 1981, 1990), and they respond more to approaches than to tangential passes. Noise is a good source of information about the direction of movement of a disturbance, particularly underwater. If noise allows animals to track walking humans, movements are not as great (Kuck et al. 1985).

The few studies that have tracked bird movements in the presence of noisy disturbances show similar flexibility. Noisy human activity can cause raptors to expand their home ranges (Andersen et al. 1986, 1990), but the birds return to normal usage patterns when humans are not present. The most effective noisy disturbances are those that haze or harass, such as low-flying aircraft and boats that approach closely (Korschgen et al. 1985). The noisier the approaching vehicle, the greater the proportion of birds flying (Marsh et al. 1991). For example, noisy motorboats induce flight in waterfowl at ranges of 1–2 km, but quieter vessels do not stimulate flight outside 500 m (Owens 1977).

Birds familiar with aircraft can be almost completely insensitive to noise. Robbins (1966) reports on unsuccessful attempts to drive nesting Laysan albatrosses away from favored nesting sites with aircraft, intense tones, carbide cannons, gunfire, and various forms of direct harassment by humans (chasing and handling). The most intense noise sources (120 dB) repelled birds for short distances (up to 30 m), but not for long. Similar indifference to noisy disturbances is found on airfields elsewhere in a wide range of species (waterfowl, passerines, gulls, terns, and raptors). Hearing loss could be a factor when birds on runways adapt to noise at levels as high as 130 dB.

Courtship and mating   The effects of noise on courtship and mating have not been examined experimentally in any wild animal. Farm-raised mink exposed to aircraft overflights during courtship successfully formed pair bonds (Brach 1983) and bulls exposed to sonic booms during mating ignored the noise altogether (Cottereau 1972).

Predation and predator avoidance   Birds can lose eggs and young when predators attack nests after parents are startled into flight. One study suggests that mammals may be affected this way as well. Harrington and Veitch (1992) reported that caribou exposed to low-altitude overflights by jets lost more calves than unexposed caribou. The sample of caribou they tracked was small

(15 mothers), and the statistical comparison they used was invalid because they tested the same data repeatedly without correction. Nevertheless, their data were consistent with the hypothesis that mothers with calves exposed their young to wolves more often in movements stimulated by aircraft. The herds exposed in Harrington and Veitch's study were relatively inexperienced, and collared individuals were deliberately overflown by aircraft many times. In the absence of predators, success rates of ungulates exposed to noisy disturbances have been normal (e.g., Leslie and Douglas 1980; Weisenberger et al. in press).

Social communication   There are few studies documenting the effects of human-made noise on social communication. Wildlife stop acoustic communication to listen to an unusual noise, including human-made noise. Usually, this is a very transitory effect. It is most often described in marine mammals. High-latitude seals, humpback whales, sperm whales, and pilot whales become silent in the presence of industrial noise, narrow-band tomographic signals, and vessel noise (e.g., Terhune et al. 1979). Sperm whales are particularly prone to silence, sometimes for as long as the noise is present (Watkins and Schevill 1975; Watkins et al. 1985; Bowles et al. 1994).

Parental care   Birds minimize the time they spend off their nests. Median time off the nest after a noise-induced flight was less than five minutes in the few studies that have measured this parameter. Subtle changes in parenting behavior could occur, but attempts to measure them have been unsuccessful so far (Holthuijzen et al. 1990).

   Noisy disturbances have not affected mammal parenting for very long either. In caribou, nursing bouts were shorter during exposures to helicopters (Gunn et al. 1985) and for short periods afterwards. In musk oxen, nursing bouts were not affected when averaged over periods of several days (Miller et al. 1988), although immediate responses were observed. Mothers and offspring commonly approached each other during noisy disturbances (Gunn and Miller 1980; Miller et al. 1988; Kuck et al. 1985). Caribou were unresponsive when they had newborn calves, but several species of ungulates were most responsive in the period when calves were young (post-calving period; McCourt and Horstman 1974; Harrington and Veitch 1992; Lenarz 1974). In the extensive data-set provided by McCourt and Horstman, responsiveness during the post-calving period was probably the result of larger herd sizes. This was also the period when caribou were most vulnerable to wolves. Calf play was greater after overflights by helicopters, consistent with all the other evidence that activity levels increase after a disturbance.

Migration   Acute noise exposure does not affect the course of migrat
nificantly, although it can cause short detours or an increase in the ra~
travel. The best examples come from studies of marine mammals (Malme et
al. 1984), which respond by turning slightly to circumvent a noise source, and
caribou, which travel further along their migration route after exposure.

Intrusive noisy disturbances, such as heavy traffic on roads, are thought to
be capable of interrupting migrations. So far, the evidence indicates that
wildlife avoid traffic on roads, but not that roads interrupt migrations.
Whitten and Cameron (1983) studied the responses of radio-collared caribou
on roads built for the Trans-Alaska Pipeline (TAPS). They found a skew in the
sex ratio in favor of males of animals sighted along the roads, although this
skew was not present elsewhere. Studies on other species in wilderness areas
also suggest that large mammals avoid roads when there is traffic on them.
Edge and Marcum (1985) reported that elk leave a 500–1000 m buffer zone
around logging roads when traffic is high (at a rate of a few transits per day),
but not at other times. Similar observations have been made for deer (Dorrance et al. 1975; Singer and Beattie 1986), and coyotes (Gese et al. 1989). The
range at which animals avoided traffic was approximately the range at which
they could detect traffic noise, suggesting that traffic noise was meaningful
through association with human activity.

Effects on Populations

Even if proven significant, most of the effects of noisy disturbances are mild
enough that they may never be detectable as changes in population size or
population growth against the background of normal variation. Projections
using population models offer the best hope of determining population effects, but, to be valid, such models will require good estimates of population
parameters (age specific productivity and mortality, age at first reproduction,
etc.), and changes in these parameters with animal density, exposure, and
long-term adaptation to disturbance.

The small literature to date suggests that harassment is a crucial determinant of population effects in birds. Human occupation and activity are clearly
and directly correlated with declines in breeding populations of birds (Andersen 1988; Weseman and Rowe 1987; Burger 1984). However, the case for
declines after exposure to noisy disturbances is weaker. Bunnell et al. (1981)
projected a long-term decline in a small colony of white pelicans if exposures
to overflights during laying always caused egg losses of 88%. However, they
provided no details about the nature and frequency of overflights that could
cause such effects, or of pelican adaptation over time. The pelicans were naive
(resident in a remote area) and were frequently harassed by coyotes.

One pair of studies provided weak evidence that losses are caused by very close approaches. Lapland longspurs had lower reproductive success after exposure to helicopters at 16 m twice a day, measured by abandonments, addled eggs, and lowered fledging success (Gollop et al. 1974). Effects on population density were not found. A similar experiment using continuous gas compressor noise at 65–80 dB (C-weighted SPL) found no effects on nest density or reproductive success. The levels produced by the helicopters were not quantified. Both studies used three replicate sites that were very close to each other, so there could have been environmental differences among control and exposed nests. However, the results of these studies supported the hypothesis that intrusions were more damaging than elevated noise levels.

SPECIES AND POPULATION DIFFERENCES IN RESPONSES

Taxa with different predator avoidance strategies differ in susceptibility to disturbance. Genetic differences between populations of the same species and between similar species will play a role as well, but previous experience with human-made disturbance, motivational state, and recent exposure to predator pressure are very likely to explain large differences in sensitivity. These should always be considered before invoking genetic differences.

Having said this, some species do appear to be relatively responsive to disturbance, specifically the harbor seal, and migratory waterfowl, particularly snow, emperor, and brant geese. These species respond to human disturbances near the limits of detection (Allen et al. 1984; Salter and Davis 1974). Harbor seals are an interesting case study because their numbers are increasing in coastal areas where they are forced into contact with humans. Harbor seals that are exposed frequently have become much more tolerant of human activity (Brasseur 1993).

CHANGES IN GENETIC COMPOSITION OF POPULATIONS

If individuals with a particular genetic makeup were more susceptible to disturbance than others, subtle selection could be driven by noisy disturbances. For example, if flighty individuals exposed their young to predators more often, they would have lower reproductive success in the presence of frequent noise-induced disturbances. At present, this effect is a matter of speculation. Knight et al. (1987) reported an inverse correlation between the number of years of exposure to humans and response rates (diving and calling) in red-tailed hawks. This correlation is most parsimoniously viewed as the result of long-term behavioral adaptation, but genetic changes would be most likely to appear under such circumstances.

Management Options for Coexistence

Obviously, eliminating noisy disturbances would reduce the potential for noise-induced impact. When this is not an option, managers must control disturbances by limiting exposure. Most managers limit exposure by limiting approach distance, which Altman (1958) has recommended as a measure of reactivity of large animals. It is certainly the universal method for limiting approaches by aircraft and boats.

There are problems with setting approach limits using avoidance responses. Animals such as marine mammals whose responses are hard to detect at any distance may be approached as closely as 100 m, but ungulates and birds that fly visibly in response to aircraft can only be approached as closely as 300–500 m. These limits consistently protect animals from injury, but not from harassment. Preventing harassment often inconveniences recreationists, who want to use animal habitats and who like seeing animals close up. The following section discusses models that may help managers balance between these two conflicting needs.

Modeling Effects of Noise on Behavior

Figure 8.3 gives three examples of the relationship between approach distance and the proportion of a population running. The probability of flight does not vary linearly with distance, but increases slowly at long range and rapidly at close range. The exponential functions in Figure 8.3 are not necessarily the best models, but they have the advantage of being easy to compare with models of noise transmission loss. Although differences in sensitivity illustrated in Figure 8.3 might be explained by species differences, previous experience with disturbance is a more parsimonious explanation. The caribou in Miller and Gunn's study might never have seen humans, whereas the mountain sheep in Krausman and Hervert's study were exposed regularly.

The relationship between noise level and annoyance in humans is modeled by a logistic function that describes the relation between sound level and the percentage of the population highly annoyed. The proportions of animals responding to various levels of noise can be modeled by similar functions (Fig. 8.4), but these functions vary greatly with behavior. There is an inverse relation between the energy expended in a behavior and its likelihood, especially in animals that are somewhat habituated. The best example of this is given by Brown (1990), who used approaches by light aircraft to study responses of nesting crested terns. Brown found that alerting, a low-intensity response, had a threshold of around 65 dB(A) and involved nearly all the individuals in the colony. Flight responses, however, never involved more than

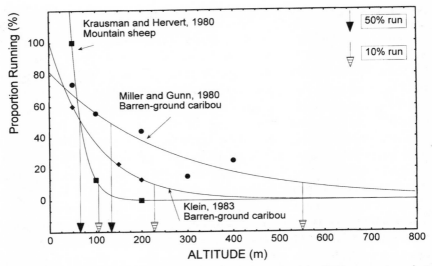

Figure 8.3   The relationship between aircraft approach distance and probability of response taken from three studies of ungulates. The data were fitted with an exponential function to estimate the range at which 50% (dark arrows) and 10% (hatched arrows) of the animals in each study area would run.

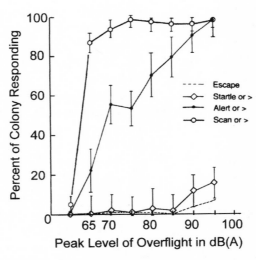

Figure 8.4   The relationship between peak sound level of light aircraft overflights and probability of four classes of behavioral responses in the crested tern. Each proportion is shown with 95% confidence intervals. The responses at the lowest sound levels were control measurements without the aircraft. The behaviors were graded in intensity, from scans (looking around) to escape flight. Data from Brown 1990.

20% of the colony and had a much higher threshold, around 90 dB(A). The terns were exposed frequently to aircraft, so the proportion of individuals responding with flight was probably smaller than it would have been in a naive colony.

The proportion of mammals and birds responding with flight varies greatly depending on previous experience, season, group size, age and sex composition, on-going activity, motivational state, reproductive condition, terrain, weather, temperament, and other natural factors. For example, the proportion of caribou running in response to aircraft at a standard distance varied with season from 5% to 40% (McCourt and Horstman 1974).

The limits that managers use now to prevent a particular behavior (e.g., flight) in a given proportion of the population (e.g., 10%, 50%) do not capture any of this complexity. In the future, managers will have to think more about predictive models, as humans place increasing pressure on wild populations. For example, if models show that the most dangerous exposures occur rarely, naive animals might best be protected by exposing them to man-made disturbances under controlled conditions so that they do not panic when exposed to uncontrolled activity. Figure 8.5 illustrates habituation as a result of relatively uncontrolled aircraft activity. Models would also help justify regulating total cumulative exposure to disturbance over the entirety of an

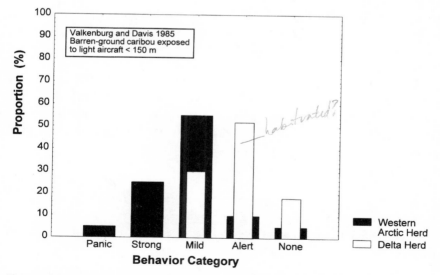

Figure 8.5 Response of barren-ground caribou from two herds to low-altitude overflights. The behaviors recorded in both herds were panic running, strong escapes (running), mild escapes (walking or trotting away), alerting, and no responses. The Delta herd, which was exposed frequently at levels below 150 m, never responded with dangerous running.

animal's range, despite the inevitable conflicts among private, industrial, military, subsistence, and recreational users of migratory corridors. Effects of disturbances on waterfowl may only be controllable this way because human and natural predation seem to preclude habituation.

Problems with Noise Standards for Animals

The approach limits now in use do not necessarily protect animals from hazardous noise exposure, so noise standards should be developed as well. At present, there are no well-established limits or standards for limiting noise exposure in animals. The Committee on Hearing, Biomechanics and Bioacoustics (CHABA) has published guidelines for evaluating noise exposure in Environmental Impact Assessments (CHABA 1977). These guidelines are derived from standards developed by bodies such as the American National Standards Institute (ANSI) and the Acoustical Society of America (ASA). CHABA guidelines incorporate frequency-weighting functions specific to human hearing and specify methods for quantifying noise exposure. They provide functions that describe the limits of tolerance in humans over a range of noise durations and frequencies. The CHABA document suggests using these guidelines for animals, but they are never used in practice.

Some agencies have informal working guidelines to limit noise exposure in animals. They are number limits expressed in dB without specifying integration or weighting. For the most part, these guidelines are not published, even as internal agency memoranda. The agencies that use informal guidelines are the U.S. Navy, the U.S. Fish and Wildlife Service, and the National Marine Fisheries Service.

The Animal Bioacoustics Technical Specialty Group of the ASA has recently begun to examine standards for animals. The CHABA guidelines are a good working model, but do not apply to many animals without modifying weighting functions, and do not apply at all to effects caused by meaningful noise.

Mitigation

What follows is a series of recommendations to limit noise effects in situations where noisy human activities and animals must coexist. Accidents are the easiest effects to prevent and have already been discussed at length in this document.

PROCEDURES TO LIMIT BEHAVIORAL EFFECTS

The most important behavioral responses are attraction and clear indications of harassment. Methods to limit important behavioral effects include:

(1) Keep noisy sources from approaching animals on a directed course. This method could allow approach limits to be reduced.

(2) Make noise sources predictable. For example, vehicles could be limited to roadways, boatways, and specified flight paths at predictable times. Aircraft could fly at constant slant distances without rapid changes in direction.

(3) Stop approaches if animals react with avoidance, defensive behaviors, or aggression.

(4) Gradually habituate animals to noise. In areas where animals will be exposed frequently, some active effort to habituate naive individuals could protect them from panics. This will not help animals made skittish by predators.

(5) Alter noise to make it less annoying. A meaningless masking noise might help mitigate effects of meaningful noise.

(6) Alter noise to make it less attractive. A particularly attractive noise could be removed altogether, or it could be masked by broadband noise.

PROCEDURES THAT PROTECT ANIMALS
FROM PHYSIOLOGICAL EFFECTS

The Committee on Pain and Distress in Laboratory Animals of the National Research Council (1992) defines distress as "an aversive state in which an animal is unable to adapt completely to stressors . . .," among which is noise (p. 4). To minimize the potential for distress, they recommend making noise more predictable and controllable. Their recommendations could apply to wild animals as follows:

(1) Limit cumulative exposure to noise to protect animal hearing. Safe limits will have to be specified by taxonomic group; use CHABA (1977) for most terrestrial animals in the interim.

(2) Limit the duty cycle and duration of noise to allow recovery between exposures.

(3) Eliminate or reduce meaningful noises. For example, silence boat motors in areas where hunters shoot birds from boats.

(4) Limit cumulative exposure to harassment. For example, restrict aircraft approaches during periods when animals are under greatest predator pressure.

(5) Provide safe cover in areas where noisy human activities are common. Cover will increase the sense of control over disturbance.

Knowledge Gaps

It is a tenet of modern behavioral ecology that animals weigh the costs and benefits of responding or not responding to disturbances against each other, either through behavioral adaptations or through genetic selection. For example, birds that abandon nests because they have detected a predator may benefit through decreased risk of predation, but will incur the costs of renesting. Good estimates of the costs and benefits of responses, whether behavioral or physiological, are needed. Here are some hypotheses that should be tested in support of such models:

(1) Animals adapt to noise-induced disturbances as well as they can to minimize the effect on their normal activities. This adaptation may take weeks or months.

(2) The probability of a response to a meaningless noise will be inversely related to the cost of response, energetic or otherwise. Alternately, the probability of response to a meaningful noise correlates directly with the cost of failure to respond.

(3) There is an inverse relation between the probability of interrupting an activity and the benefit of the activity (e.g., eating).

(4) Animals avoid noise that interferes with perception of meaningful sounds.

(5) Distress results most often when animals associate human-made noise with danger (e.g., gunshots).

(6) Habituation to low-level, meaningless noise is much more complete than habituation to noise associated with danger or noise with the potential to damage hearing.

(7) Over time, animals learn to distinguish human-made noise from meaningful natural sounds, although perhaps not perfectly.

Specific recommendations for research are the following.

(1) *Determine which features of noisy disturbances best predict behavioral responses.* Future studies should measure the relation between behavior and stimulus features other than noise level, such as duty cycle, onset rate, bandwidth, modulation rate, variability, and predictability. Future studies should also investigate the relationship between acoustic components of a noisy disturbance and other sensory modalities (vision, touch). For example, most authors believe helicopters to be more disturbing than

light aircraft even though they are less noisy. It would be easier to mitigate responses to helicopters if managers knew what features of helicopter overflights are most disturbing.

(2) *Study habituation to noisy disturbances in wild vertebrates.* All vertebrates are capable of habituation and behavioral adaptation, but the process is the most poorly understood component of responses to disturbance. Future studies should measure the responses of free ranging animals to repeated exposure, controlling for group size, predator pressure, and other factors thought to influence responsiveness.

(3) *Study hearing loss and natural exposure to noise.* No one knows how much noise exposure animals normally receive, or how they cope with it. Future studies should measure how much noise is normal and how it affects strategies for navigating, avoiding predators, finding prey, or communicating with conspecifics.

Hearing loss due to noise exposure merits further study, especially in taxa that have not received much attention, such as reptiles. In birds, fish, and mysticete whales, low-frequency hearing also merits further study, although the technical challenges of such studies will be great.

(4) *Determine the costs of coping with attacks.* Future studies should measure the costs and benefits of behavioral responses to predators, such as abandoning nests or territories, flying in response to approach by a predator, and maintaining a safe distance from a predator.

(5) *Measure noise-induced distress rather than noise-induced stress.* Several studies that have found evidence of physiological responses to noise might be extended to measure distress. For example:

- Chesser et al. (1975) found that small mammals under the flight path of an airport had adrenal hypertrophy. Further studies should show whether hypertrophy had negative consequences under difficult conditions, for example increased susceptibility to disease, declines in reproductive output, or poor growth.

- Davis and Wiseley (1974), Ward and Stehn (1989), and Bélanger and Bédard (1990) showed that migratory waterfowl expend more energy when exposed to repeated aircraft overflights, at least in the short term. Further studies should examine fat reserves, long-term energy consumption, survivorship, reproductive success, and so on in exposed and unexposed birds.

- Stockwell et al. (1991) and others have suggested that harassment by aircraft can reduce food intake. Proof of this effect will require better information on energy uptake, metabolic rate, and 24-hour activity budgets in disturbed and undisturbed individuals, preferably with continuous monitoring for weeks or months. Fat deposition might be measured directly as well.

(6) *Measure population parameters in the presence and absence of noisy disturbances.* Population effects will be difficult to measure. Studies must have multiple levels of controls to account for natural variability, which will usually exceed variation introduced by noise. Animals or groups can be their own controls over time (baseline vs. exposed). Exposed populations can be compared with matched, unexposed populations, as long as there are suitable (independent) replicates.

In any population study, sample sizes should be adequate to detect differences of less than 15% between exposed and unexposed groups. When adequate samples are not available, research on a similar but more abundant species should be considered.

(7) *Measure the consequences of adapting to noise.* Habituated animals may be more susceptible to losses through being less wary of humans. Exposure to noise certainly makes animals more tolerant of further noise exposure, and it might potentiate other kinds of tolerance. Approaches that might work:

• Measure responses to human intruders of wild animals that have been exposed to noisy disturbances (aircraft, vehicles on roads).

• Measure susceptibility of exposed and unexposed animals to collisions with vehicles, boats, or aircraft.

• Measure genetic differences between exposed and unexposed subsets of the same population and determine the fitness of exposed individuals.

Literature Cited

Allen, S., D.A. Ainley, G.W. Page, and C.A. Ribic. 1984.The effects of disturbance on harbor seal haul out patterns at Bolinas Lagoon, California. *Behavioural Processes* 82(3):61–68.

Altmann, M. 1958. The flight distance in free-ranging big game. *Journal of Wildlife Management* 22(2):207–209.

Andersen, D.E., O.J. Rongstad, and W.R. Mytton. 1986. The behavioral response of a red-tailed hawk to military training activity. *Raptor Research* 20:65–68.

Andersen, D.E., O.J. Rongstad, and W.R. Mytton. 1990. Home-range changes raptors exposed to increased human activity levels in southeastern Colorado. *Wildlife Society Bulletin* 18:134-142.

Anderson, D.J. and S. Fortner. 1988. Waved albatross egg neglect and associated mosquito ectoparasitism. *Condor* 90:727–729.

Anderson, D.W. 1988. In my experience . . . Dose–response relationship between human disturbance and brown pelican breeding success. *Wildlife Society Bulletin* 16:339–345.

Anderson, S.S. and A.D. Hawkins. 1978. Scaring seals by sound. *Mammal Review* 8(1,2):19–24.

Anthony, A. and E. Ackerman. 1955. Effects of noise on the blood eosinophil levels and adrenals of mice. *Journal of the Acoustical Society of America* 27(6):1144–1149.

Austin, O.L., Jr., W.B. Robertson, Jr., and G.E. Woolfenden. 1972a. Mass hatching failure in Dry Tortugas sooty terns (*Sterna fuscata*). In *Proceedings of the XVth International Ornithological Congress*, The Hague, Netherlands, 30 August–5 September, p. 627. Leiden, Netherlands: E.J. Brill.

Austin, O.L., Jr., W.B. Robertson, Jr., and G.E. Woolfenden. 1972b. Mass hatching failure in Dry Tortugas sooty terns ("*Sterna fuscata*"). Manuscript on file with the Van Tyne Library of Ornithology, University of Maryland.

Bélanger, L. and J. Bédard. 1990. Energetic cost of man-induced disturbance to staging snow geese. *Journal of Wildlife Management* 54(1):36–41.

Belanovskii, A.S. and V.A. Omel'yanenko. 1982. Acoustic stress in commercial poultry production. *Soviet Agricultural Science* 11:60–62.

Bell, W.B. 1972. Animal response to sonic booms. *Journal of the Acoustical Society of America* 51(2):758–765.

Beyer, D. 1983. [Study on the effects of low-flying aircraft on the endocrinological and physiological parameters in pregnant cows]. Translation by Translation Division, Foreign Technology Division, Wright-Patterson Air Force Base FTD-ID(RS) T-1052-89. Doctor Medicinae Veterinarie Thesis, Veterinary College of Hannover, München. 107 pp.

Black, B.B., M.W. Collopy, H.F. Percival, A.A. Tiller, and P.G. Bohall. 1984. Effects of low-level military training flights on wading bird colonies in Florida. Report by Florida Cooperative Fisheries and Wildlife Resources Unit. Gainesville, Florida: University of Florida, for U.S. Air Force. Technical Report Number 7. 190 pp.

Blokpoel, H. and D.R.M. Hatch. 1976. Snow geese, disturbed by aircraft, crash into power lines. *Canadian Field-Naturalist* 90:195.

Bomford, M. and P.H. O'Brien. 1990. Sonic deterrents in animal damage control: a review of device tests and effectiveness. *Wildlife Society Bulletin* 18:411–422.

Bondello, M.C. 1976. The effects of high-intensity motorcycle sounds on the acoustical sensitivity of the desert iguana, *Dipsosaurus dorsalis*. Masters thesis. Fullerton, California: California State University. 37 pp.

Borg, E. 1981. Physiological and pathogenic effects of sound. *Acta Oto-Laryngologica* (Supplement) 381:7–68.

Bowles, A.E. and B.S. Stewart. 1980. Disturbances to the pinnipeds and birds of San Miguel Island, 1979–1980. In *Potential Effects of Space Shuttle Sonic Booms on the Biota and Geology of the California Channel Islands: Research Reports*, eds., J.R. Jehl, Jr. and C.F. Cooper, 99–137. San Diego State University Foundation Technical Report Number 80-1.

Bowles, A.E., F.T. Awbrey, and J.R. Jehl, Jr. 1991. The effects of high-amplitude impulsive noise on hatching success: a reanalysis of the "sooty tern incident." Report by Hubbs-Sea World Research Institute for the U.S. Air Force Noise and Sonic Boom Impact Technology Program, Wright-Patterson Air Force Base. Technical Report Number HSD-TP-91-0006. 19 pp.

Bowles, A.E., L. McClenaghan, J.K. Francine, S. Wisely, R. Golightly, and R. Kull. 1993. Effects of aircraft noise on the predator-prey ecology of the kit fox (*Vulpes macrotis*) and its small mammal prey. *Proceedings of the 6th Annual International Congress on Noise as a Public Health Problem. Actes INRETS* 34(3): 462–469.

Bowles, A.E., M. Smultea, B. Würsig, D.P. DeMaster, and D. Palka. 1994. Relative abundance and behavior of marine mammals exposed to transmissions from the Heard Island Feasibility Test. *Journal of the Acoustical Society of America*, in press.

Brach, W. 1983. [Studies of the effects of aircraft noise on the peri-partal and post-partal losses in farm-raised mink (*Mustela vison*) f. dom.)]. Translation by Translation Division, Foreign Technology Division, Wright-Patterson Air Force Base FTD-0738-90. Doctor Medicinae Veterinarie Thesis, Veterinary College of Hannover, München. 162 pp.

Bradley, F., C. Book, and A.E. Bowles. 1990. Effects of low-altitude aircraft overflights on domestic turkey poults. Final report for period December 1989–June 1990. Report by BBN Laboratories, Inc. for Noise and Sonic Boom Impact Technology Program, Brooks Air Force Base. Technical Report No. HSD-TR-90-034. 127 pp.

Brasseur, S.M.J.M. 1993. Tolerance of harbor seals to human related disturbance sources during haulout. Presented at the Tenth Biennial Conference on the Biology of Marine Mammals, November 11–15, Galveston, Texas. Society for Marine Mammalogy.

Brattstrom, B.H. and M.C. Bondello. 1983. Effects of off-road vehicle noise on desert vertebrates. In *Environmental Effects of Off-road Vehicles: Impacts and Management in Arid Regions*, eds., R.H. Webb and H.G. Wilshire, 167–206. New York, New York: Springer-Verlag.

Brown, A.L. 1990. Measuring the effect of aircraft noise on sea birds. *Environment International* 16:587–592.

Bunnell, F.L., D. Dunbar, L. Koza, and G. Ryder. 1981. Effects of disturbance on the productivity and numbers of white pelicans in British Columbia— observations and models. *Colonial Waterbirds* 4:2–11.

Burger, J. 1981. Behavioral responses of herring gulls *Larus argentatus* to aircraft noise. *Environmental Pollution* 24:177–184.

Burger, J. 1983. Jet aircraft noise and bird strikes: why more birds are being hit. *Environmental Pollution* 30:143–152.

Burger, J. 1984. Colony stability in Least Terns. *Condor* 86:61–67.

Burger, J. and M. Gochfeld. 1981. Discrimination of the threat of direct versus tangential approach to the nest by incubating herring and great black-backed gulls. *Journal of Comparative and Physiological Psychology* (Series A) 95:676–684.

Burger, J. and M. Gochfeld. 1990. Risk discrimination of direct versus tangential approach by basking black iguanas (*Ctenosaura similis*): variation as a function of human exposure. *Journal of Comparative Psychology* (Series A) 104:388–394.

Calef, G.W., E.A. DeBrock, and G.M. Lortie. 1976. The reaction of barren-ground caribou to aircraft. *Arctic* 29:201–212.

Chesser, R.K., R.S. Caldwell, and M.J. Harvey. 1975. Effects of noise on feral populations of *Mus musculus*. *Physiological Zoology* 48(4):323–325.

Cogger, E.A. and E.G. Zegarra. 1980. Sonic booms and reproductive performance of marine birds: studies on domestic fowl as analogues. In *Potential Effects of Space Shuttle Sonic Booms on the Biota and Geology of the California Channel Islands: Research Reports*, eds., J.R. Jehl, Jr. and C.F. Cooper, 163–194. San Diego State University Foundation Technical Report No. 80-1.

Committee on Bioacoustics and Biomechanics (CHABA). 1977. *Guidelines for Preparing Environmental Impact Statements on Noise*. Report of CHABA Working Group Number 69. Washington, D.C.: National Academy of Sciences. 145 pp.

Committee on Pain and Distress in Laboratory Animals. 1992. *Recognition and Alleviation of Pain and Distress in Laboratory Animals*. Washington, D.C.: National Academy Press. 137 pp.

Corwin, J.T. and D.A. Cotanche. 1988. Regeneration of sensory hair cells after acoustic trauma. *Science* 240:1772–1774.

Corwin, J.T. and M.E. Warchol. 1991. Auditory hair cells: structure, function, development, and regeneration. *Annual Review of Neuroscience* 14:301–333.

Cottereau, P. 1972. Les incidences du "bang" des avions supersoniques sur les productions et la vie animales. *Revue de Medecine Veterinaire* 123(11): 1367–1409.

Cottereau, P. and M. Keller. 1993. [Report on experiments on the effects of the supersonic "bang" on fertilized hen's eggs]. Translation by Translation Division, Foreign Technology Division, Wright-Patterson Air Force Base. FASTC-ID(RS)T-0961-92. Report by Ministry of Agriculture, Lyon National Veterinary School.

Davis, J. L. and P. Valkenburg. 1985. Demography of the Delta Caribou Herd under varying rates of natural mortality and harvest by humans. Federal Aid in Wildlife Restoration—Final Report. 52 pp.

Davis, R.A. and A.L. Wiseley. 1974. Normal behaviour of snow geese on the Yukon-Alaska north slope and the effects of aircraft-induced disturbance on this behaviour, September, 1973. Arctic Gas Biological Report Series 27:1–85.

Davis, R.W., T.M. Williams, and F. Awbrey. 1988. Sea otter oil spill avoidance study. OCS Study. Report by Sea World Research Institute, Hubbs Marine Research Center for U.S. Department of the Interior, Minerals Management Service, Pacific Outer Continental Shelf Region. Technical Report No. MMS 88-0051. 65 pp.

Dimmitt, M.A. and R. Ruibal. 1980. Environmental correlates of emergence in spadefoot toads (*Scaphiopus*). *Journal of Herpetology* 14:21–29.

Dooling, R.J. 1980. Behavior and psychophysics of hearing in birds. In *Comparative Studies of Hearing in Vertebrates*, eds., A.N. Popper and R.R. Fay, 261–288. New York: Springer-Verlag.

Dorrance, M. J., P. J. Savage, and D. E. Huff. 1975. Effects of snowmobiles on white-tailed deer. *Journal of Wildlife Management* 39(3):563–569.

Eckstein, R.G., T.F. O'Brien, O.J. Rongstad, and J.G. Bollinger. 1979. Snowmobile effects on movements of white-tailed deer: a case study. *Environmental Conservation* 6(1):45–51.

Edge, W.D. and C.L. Marcum. 1985. Movements of elk in relation to logging disturbances. *Journal of Wildlife Management* 49(4):926–930.

Edge, W.D., C.L. Marcum, and S.L. Olson. 1985. Effects of logging activities on home-range fidelity of elk. *Journal of Wildlife Management* 49:741–744.

Edwards, R.G., A.B. Broderson, R.W. Barbour, D.F. McCoy, and C.W. Johnson. 1979. Assessment of the environmental compatibility of differing helicopter noise certification standards. Final Report. Washington, D.C.: Report for Department of Transportation, Federal Aviation Adminstration. Technical Report No. FAA-AEE-19-13.

Ekman, P., W.V. Friesen, and R.C. Simons. 1985. Is the startle reaction an emotion? *Journal of Personality and Social Psychology* 49(5):1416–1426.

Enger, P.S., H.E. Karlsen, F.R. Knudsen, and O. Sand. 1993. Detection and reaction of fish to infrasound. *ICES Marine Sciences Symposia* 196:108–112.

Erath, R. 1984. [Studies of the effects of aircraft noise and physiological parameters in pregnant mares kept in paddocks]. Translation by Translation Division, Foreign Technology Division, Wright-Patterson Air Force Base. FTD-ID(RS)T-1047-89. Doctor Medicinae Veterinarie Thesis, Veterinary College of Hannover, München. 121 pp.

Fay, R.R. 1988. *Hearing in Vertebrates: A Psychophysics Databook.* Winnetka, Illinois: Hill-Fay Associates. 621 pp.

Federal Interagency Committee on Noise. 1992. *Federal Agency Review of Selected Airport Noise Analysis Issues.* Washington, D.C.: FICON.

Fell, R.D., C.J. Ellis, and D.R. Griffith. 1976. Thyroid responses to acoustic stimulation. *Environmental Research* 12:208–213.

Follman, E.H. and J.L. Hechtel. 1990. Bears and pipeline construction in Alaska. *Arctic* 43(2):103–109.

Gese, E.M., O.J. Rongstad, and W.R. Mytton. 1989. Changes in coyote movements due to military activity. *Journal of Wildlife Management* 53(2):334–339.

Gipson, P.S. 1970. Abortion and consumption of fetuses by coyotes following abnormal stress. *Southwestern Naturalist* 21:558–559.

Gollop, M.A., R.A. Davis, J.P. Prevett, and B.E. Felske. 1974. Disturbance studies of terrestrial breeding bird populations: Firth River, Yukon Territory, June 1972. Arctic Gas Biological Report Series 14: 92–152.

Gray, J.A. 1987. *The Psychology of Fear and Stress.* 2nd ed. New York: Cambridge University Press.

Griffin, D.R. and C.D. Hopkins. 1974. Sounds audible to migrating birds. *Animal Behaviour* 22:672–678.

Gunn, A. and F. Miller. 1980. Responses of Peary caribou cow-calf pairs to helicopter harassment in the Canadian high Arctic. In *Proceedings of the 2nd International Reindeer/Caribou Symposium,* eds., E. Reimers, E. Gaare, and S. Skjenneberg, 479–507.

Gunn, A., F.L. Miller, R. Glahalt, and K. Jingfors. 1985. Behavioral responses of barren-ground caribou cows and calves to helicopters on the Beverly Herd calving ground, Northwest Territories. In *Caribou and Human Activity,* eds., A.M. Martell and D.E. Russell, 10–14. Proceedings of the 1st North American Caribou Workshop, Whitehorse, Yukon, 28–29 September, 1983.

Harrington, F.H. and A.M. Veitch. 1991. Short-term impacts of low-level jet fighter training on caribou in Labrador. *Arctic* 44(4):318–327.

Harrington, F.H. and A.M. Veitch. 1992. Calving success of woodland caribou exposed to low-level jet fighter overflights. *Arctic* 45(3):213–218.

Harris, C.M. 1991. *Handbook of Acoustical Measurements and Noise Control,* 3rd. ed. New York: McGraw-Hill.

Heicks, H. 1985. [Studies of the effect of aircraft noise on pregnant cows tied up in an open cattle shed]. Translation from Translation Division, Wright-Patterson Air Force Base. Doctor Medicinae Veterinarie Thesis, Veterinary College of Hannover, München. 135 pp.

Heinemann, J.M. and E.F. LeBrocq, Jr. 1965. Effects of sonic booms on the hatchability of chicken eggs. Rehl (K) Project No. 65-2. Report for Regional Environmental Health Laboratory (AFLC), U.S. Air Force. Technical Report No. SST 65-12.

Heuwieser, W. 1982. [The effect of flight noise on the pregnancy of cattle, with special consideration on endocrinological and physiological factors]. Translation by Translation Division, Foreign Technology Division, Wright-Patterson Air Force Base. Doctor Medicinae Veterinarie Thesis, Veterinary College of Hannover, München. 115 pp.

Higgins, T.H. 1974. The response of songbirds to the seismic compression waves preceding sonic booms. Report for Federal Aviation Administration. Technical Report No. FAA-RD-74-78. NTIS No. AD780050. 28 pp.

Hoffman, H.S. and J.L. Searle. 1968. Acoustic and temporal factors in the evocation of startle. *Journal of the Acoustical Society of America* 43(2):269–282.

Holthuijzen, A.M.A., W.G. Eastland, A.R. Ansell, M.N. Kochert, R.D. Williams, and L.S. Young. 1990. Effects of blasting on behavior and productivity of nesting prairie falcons. *Wildlife Society Bulletin* 18:270–281.

Hunt, G.L., Jr. 1985. Offshore oil development and seabirds: The present status of knowledge and long-term research needs. In *Long-term Environmental Effects of Offshore Oil and Gas Development*, eds., D.F. Boesch and N.N. Rabalais, 539–586. New York: Elsevier Applied Science.

Jeannoutot, D.W. and J.L. Adams. 1961. Progesterone versus treatment by high intensity sound as methods of controlling broodiness in broad breasted bronze turkeys. *Poultry Science* 40:512–521.

Jensen, K.C. 1990. Responses of molting Pacific black brant to experimental aircraft disturbance in the Tekshupek Lake special area, Alaska. Ph.D. dissertation in Wildlife and Fisheries Sciences. College Station, Texas: Texas A&M University. 72 pp.

Johnson, B. 1977. The effects of human disturbance on a population of harbor seals. In *Environmental Assessment of the Alaskan Continental Shelf*, 422–431. Annual Reports of Principal Investigators for the Year Ending March 1977, Vol. 1, Receptors-Mammals.

Johnson, S.R., J.J. Burns, C.I. Malme, and R.A. Davis. 1989. Synthesis of information on the effects of noise and disturbance on major haulout concentrations of Bering Sea pinnipeds. Report by LGL Alaska Research Associates, Inc. for U.S. Minerals Management Service, Alaska Outer Continental Shelf Region. Technical Report No. MMS 88-0092.

Ketten, D.R. 1992. The marine mammal ear: specializations for aquatic audition and echolocation. In *The Evolutionary Biology of Hearing*, eds., D.B. Webster, R.R. Fay, and A.N. Popper, 717–750. New York: Springer-Verlag. 859 pp.

King, M.M. and G.W. Workman. 1986. Response of desert bighorn sheep to human harassment: management implications. *Transactions of the 51st North American Wildlife & Natural Resources Conference* 51:74–85.

Klein, D.R. 1983. Reactions of reindeer to obstructions and disturbances. *Science* 173:393–398.

Knight, R.L., D.E. Andersen, M.J. Bechard, and N.V. Marr. 1987. Geographic variation in nest-defence behaviour of the red-tailed hawk (*Buteo jamaicensis*). *Ibis* 131:22–26.

Kock, M.D., D.A. Jessup, R.K. Clark, and C.E. Franti. 1987. Effects of capture on biological parameters in free-ranging bighorn sheep (*Ovis canadensis*): evaluation of drop-net, drive-net, chemical immobilization and the net-gun. *Journal of Wildlife Diseases* 23(4):641–651.

Korschgen, C.E., L.S. George, and W.L. Green. 1985. Disturbance of diving ducks by boaters on a migrational staging area. *Wildlife Society Bulletin* 13:290–296.

Krausman, P.R. and J.J. Hervert. 1983. Mountain sheep responses to aerial surveys. *Wildlife Society Bulletin* 11(4):372–375.

Krausman, P.R., B.D. Leopold, and D.L. Scarbrough. 1986. Desert mule deer response to aircraft. *Wildlife Society Bulletin* 14(1):68–70.

Krausman, P.R., M.C. Wallace, D.W. DeYoung, M.E. Weisenberger, and C.L. Hayes. 1993. The effects of low-altitude aircraft on desert ungulates. *Proceedings of the 6th International Congress on Noise as a Public Health Problem. Actes INRETS* 34(3):471–478.

Kreithen, M.L. and D.B. Quine. 1979. Infrasound detection by the homing pigeon: a behavioral audiogram. *Journal of Comparative Physiology* (Series A) 129(1):1–4.

Krüger, K.J. 1982. [The effects of aircraft noise on pregnancy in horses with special consideration of physiological and endocrinological factors]. Translation by Translation Division, Foreign Technology Division, Wright Patterson Air Force Base FTD-ID(RS)T-1048-89. Doctor Medicinae Veterinarie Thesis, Veterinary College of Hannover, München. 145 pp.

Kuck, L., G.L. Hompland, and E.H. Merrill. 1985. Elk calf response to simulated mine disturbance in southeast Idaho. *Journal of Wildlife Management* 49(3):751–757.

Kugler, B.A. and D.S. Barber. 1993. A method for measuring wildlife noise exposure in the field (abstract). *Journal of the Acoustical Society of America* 93(4):2378.

LeBlanc, M.M., C. Lombard, S. Lieb, E. Klapstein, and R. Massey. 1991. Physiological responses of horses to simulated aircraft noise. Final Report. Report by University of Florida for U.S. Air Force, Noise and Sonic Boom Impact Technology Program. 27 pp. + fig.

Lefcourt, A.M. 1986. Usage of the term "stress" as it applies to cattle. *Vlaams Diergeneeskundig Tijdschrift* 55(4):258–265.

Lenarz, M. 1974. The reaction of Dall sheep to an FH-1100 helicopter. Arctic Gas Biological Report Series 23:1–14.

Leslie, D.M., Jr. and C.L. Douglas. 1980. Human disturbance at water sources of desert bighorn sheep. *Wildlife Society Bulletin* 8(4):284–290.

Lewis, E.R. and P.M. Narins. 1985. Do frogs communicate with seismic signals? *Science* 227:187–189.

Lynch, T.E. and D.W. Speake. 1978. Eastern wild turkey behavioral responses induced by sonic boom. In *Effects of Noise on Wildlife*, eds., J.L. Fletcher and R.G. Busnel, 47–61. New York: Academic Press.

Malme, C.I., P.R. Miles, C.W. Clark, P. Tyack, and J.E. Bird. 1984. Investigations of the potential effects of underwater noise from petroleum industry activities on migrating gray whale behavior Phase II: January 1984 Migration. Report by Bolt Beranek and Newman Inc. for U.S. Department of the Interior. Technical Report No. 5586. 245 pp.

Marsh, R.E., W.A. Erickson, and T.P. Salmon. 1991. Bird hazing and frightening methods and techniques (with emphasis on containment ponds). Report by Department of Wildlife & Fisheries Biology for University of California at Davis, under contract with California Department of Water Resources. 233 pp.

Mayhew, W.W. 1965. Adaptations of the amphibian, *Scaphiopus couchi*, to desert conditions. *American Midland Naturalist* 74(1):95–109.

McCourt, K.H. and L.P. Horstman. 1974. The reaction of barren-ground caribou to aircraft. Arctic Gas Biological Report Series 23: 1–36.

Miller, F.L. and A. Gunn. 1979. Responses of Peary caribou and muskoxen to turbo-helicopter harassment, Prince of Wales Island, Northwest Territories, 1976–1977. Canadian Wildlife Service Occasional Paper 40. 88 pp.

Miller, F.L., A. Gunn, and S.J. Barry. 1988. Nursing by muskox calves before, during, and after helicopter overflights. *Arctic* 41(3):231–235.

Miller, M.W. 1991. A simulation model of the response of molting Pacific black brant to helicopter disturbance. Masters thesis. College Station, Texas: Texas A&M University. 87 pp.

Moberg, G. 1985. *Animal Stress*. Bethesda, Maryland: American Physiological Society.

Moberg, G.P. 1987. A model for assessing the impact of behavioral stress on domestic animals. *Journal of Animal Science* 65:1228–1235.

Munk, W.H. and A.M.G. Forbes. 1989. Global ocean warming: an acoustic measure? *Journal of Physical Oceanography* 19(11):1765–1778.

Murphy, S.M. and B.A. Anderson. 1993. Lisburne Terrestrial Monitoring Program. The effects of the Lisburne Development Project on geese and swans 1985–1989. Final Synthesis Report. Report by Alaska Biological Research, Inc. for ARCO Alaska, Inc. 202 pp.

Murphy, S.M., R.G. White, B.A. Kugler, J.A. Kitchens, M.D. Smith, and D.S. Barber. 1993. Behavioral effects of jet aircraft on caribou in Alaska. *Proceedings of the 6th International Congress on Noise as a Public Health Problem. Actes INRETS* 34(3):471–478.

Narins, P.M. 1982. Effects of masking noise on evoked calling in the Puerto Rican coqui (*Anura: Leptodactylidae*). *Journal of Comparative Physiology* (Series A) 147:439–446.

National Research Council. 1981. The effects on human health from long-term exposures to noise. Report for National Research Council, Committee on Hearing, Bioacoustics and Biomechanics. 16 pp.

Nixon, C.W. and H.E. von Gierke. 1966. Exposure to noise: mammals and roaches. In *Experimental Biology*, eds., P.L. Altman and D.S. Dittmer, 196–197. Bethesda, Maryland: Federation of American Societies for Experimental Biology.

Okamoto, S., I. Goto, and O. Koga. 1963. The effect of jet plane sounds on the economical performances in female chickens. *Science Bulletin of the Faculty of Agriculture, Kyushu University* 20(4):341–351.

Owens, N. 1977. Responses of wintering brent geese to human disturbance. *Wildfowl* 28:5–14.

Peeke, H.V.S. and M.J. Herz. eds. 1973. *Habituation*. Vol. 1. Behavioral Studies. New York: Academic Press, Inc. 290 pp.

Pritchett, J.F., M.L. Browder, R.S. Caldwell, and J.L. Sartin. 1978. Noise stress and in vitro adreno-cortical responsiveness to ACTH in wild cotton rats, *Sigmodon hispidus*. *Environmental Research* 16:29–37.

Reinis, S. 1976. Acute changes in animal inner ears due to simulated sonic booms. *Journal of the Acoustical Society of America* 60(1):133–138.

Richens, V.B. and G.R. Lavigne. 1978. Response of white-tailed deer to snowmobiles and snowmobile trails in Maine. *Canadian Field-Naturalist* 92(4):334–344.

Robbins, C.S. 1966. Birds and aircraft on Midway Islands—1959–63 investigations. Special Scientific Report—Wildlife No. 85. Report by U.S. Fish and Wildlife Service for Department of the Navy.

Ruddlesden, F. 1971. Some observations on the effect of bang type noises on laying birds. Report by Royal Aircraft Establishment for the Ministry of Defense, United Kingdom. Technical Report No. 71084. 24pp.

Ryals, B.M. and E.W. Rubel. 1988. Hair cell regeneration after acoustic trauma in adult coturnix quail. *Science* 240:1774–1776.

Ryan, M.J. 1988. Energy, calling, and selection. *American Zoologist* 28:885–898.

Sackler, A.M., A.S. Weltman, M. Bradshaw, and P.J. Jurtshuk, Jr. 1959. Endocrine changes due to auditory stress. *Acta Endocrinologica* 31:405–418.

Salter, R. and R.A. Davis. 1974. Snow geese disturbance by aircraft on the North Slope, September, 1972. Arctic Gas Biological Report Series 14:258–280.

Schreiber, E.A. and R.W. Schreiber. 1980. Effects of impulse noise on seabirds of the Channel Islands. In *Potential Effects of Space Shuttle Sonic Booms on the Biota and Geology of the California Channel Islands: Research Reports,* eds., J.R. Jehl, Jr. and C.F. Cooper, 163–194. San Diego State University Foundation Technical Report No. 80–1.

Schriever, K. 1985. [Studies of the effect of flight noise on physiological, endocrinological and ethological criteria on pregnant sows which are either running free or kept in stalls]. Translation by Translation Division, Foreign Technology Division, Wright-Patterson Air Force Base FTD-ID(RS)T-1053-89. Doctor Medicinae Veterinariae Thesis, Veterinary College of Hannover, München. Technical Report No. NP-7770114. 134 pp.

Shaughnessy, P.D., A. Semmelink, J. Cooper, and P.G.H. Frost. 1981. Attempts to develop acoustic methods of keeping fur seals (*Arctocephalus pusillus*) from fishing nets.*Biological Conservation* 21:141–158.

Shen, J.X. 1983. A behavioral study of vibrational sensitivity in the pigeon (*Columba livia*). *Journal of Comparative Physiology* 152(2):251–255.

Singer, F.J. and J.B. Beattie. 1986. The controlled traffic system and associated wildlife responses in Denali National Park. *Arctic* 39(3):195–203.

Soria, M., F. Gerlotto, and P. Fréon. 1993. Study of learning capabilities of tropical clupeoids using an artificial stimulus. *ICES Marine Sciences Symposia* 196:17–20.

Stadelman, W.J. 1958. Observations with growing chickens on the effects of sounds of varying intensities. *Poultry Science* 37:166–169.

Stadelman, W.J. and I.L. Kosin. 1957. The effects of sounds of varying intensities on chickens and hatching eggs. Wright Air Development Center Technical Report 57–87, AD110838. Report by Poultry Science Department, State College, Washington for Wright Air Development Center, U.S. Air Force, Wright-Patterson Air Force Base. 47 pp.

Stirling, I. 1988. Attraction of polar bears (*Ursus maritimus*) to offshore drilling sites in the eastern Beaufort Sea. *Polar Record* 24(148):1–8.

Stockwell, C A., G.C. Bateman, and J. Berger. 1991. Conflicts in national parks: a case study of helicopters and bighorn sheep time budgets at the Grand Canyon. *Biological Conservation* 56:317–328.

Stoebel, D.P. and G.P. Moberg. 1982. Repeated acute stress during the follicular phase and luteinizing hormone surge for dairy heifers. *Journal of Dairy Science* 65:92–96.

Suter, A.H. 1992. Noise sources and effects—a new look. *Sound and Vibration* January:18–38.

Teer, J.G. and J.C. Truett. 1973. Studies of the effects of sonic boom on birds. Report by James G. Teer and Company for the Department of Transportation, Federal Aviation Administration. Technical Report Number FFA-RD-73-148. NTIS Number AD-768-853/4. 81 pp.

Terhune, J.M., R.E.A. Stewart, and K. Ronald. 1979. Influence of vessel noises on underwater vocal activity of harp seals. *Canadian Journal of Zoology* 57:1337–1338.

Thiessen, G.J., E. A.G. Shaw, R.D. Harris, J.B. Gollop, and H.R. Webster. 1957. Acoustic irritation threshold of Peking ducks and other domestic and wild fowl. *Journal of the Acoustical Society of America* 29(12):1301–1306.

Travis, H.F., J. Bond, R.L. Wilson, J.R. Leekley, J.R. Menear, C.R. Curran. 1974. Effects of real and simulated sonic booms upon reproduction and kit survival of farm-raised mink (*Mustela vison*). In *Proceedings of the International Livestock Environment Symposium*, 157–172. Lincoln, Nebraska. American Society of Agricultural Engineers Special Publication SP-0174, April 17–19.

Urick, R.J. 1972. Noise signature from an aircraft in level flight over a hydrophone in the sea. *Journal of the Acoustical Society of America* 52:993–999.

Valkenburg, P. and J.L. Davis. 1985. The reaction of caribou to aircraft: a comparison of two herds. In *Caribou and Human Activity*, eds., A.H. Martell and D.E. Russell, 7–9. Proceedings of the 1st North American Caribou Workshop, Whitehorse, Yukon, 28–29 September, 1983.

Van Dyke, F.G., R.H. Brocke, H.G. Shaw, B.B. Ackerman, T.P. Hemker, and F.G. Lindzey. 1986. Reactions of mountain lions to logging and human activity. *Journal of Wildlife Management* 50:95–102.

Ward, D.H., E.J. Taylor, M.A. Wotawa, R.A. Stehn, D.V. Derksen, and C.J. Lensink. 1986. Behavior of the Pacific black brant and other geese in response to aircraft overflights and other disturbances at Izembek Lagoon, Alaska. 1986 Annual Report. Report by U.S. Fish and Wildlife Service, Alaska Fish and Wildlife Research Center.

Ward, D.H. and R.A. Stehn. 1989. Response of brant and other geese to aircraft disturbance at Izembek Lagoon, Alaska. Final Report by U.S. Fish and

Wildlife Service, Alaska Fish and Wildlife Research Center for the U.S. Minerals Management Service. 241 pp.

Watkins, W.A. and W.E. Schevill. 1975. Sperm whales (*Physeter catodon*) react to pingers. *Deep-Sea Research and Oceanographic Abstracts* 22:123–129.

Watkins, W.A., K.E. Moore, and P. Tyack. 1985. Sperm whale acoustic behaviors in the southern Caribbean. *Cetology* 49:1–15.

Webster, D.B. and M. Webster. 1972. Kangaroo rat auditory thresholds before and after middle ear reduction. *Brain, Behavior, and Evolution* 5:41–53.

Weisenberger, M.E., P.R. Krausman, M.C. Wallace, D.D. DeYoung, and O.E. Maughan. The effects of simulated low-level aircraft noise on heart rate and behavior of desert ungulates. *Journal of Wildlife Management,* in press.

Wesemann, T. and M. Rowe. 1987. Factors influencing the distribution and abundance of burrowing owls in Cape Coral, Florida. In *Integrating Man and Nature in the Metropolitan Environment,* eds., L.W. Adams and D.L. Leedy, 129–137. Chevy Chase, Maryland: Proceedings of a National Symposium on Urban Wildlife, 4–7 November, 1986.

Wever, E.G., M. Hepp-Reymond, and J.A. Vernon. 1966. Vocalization and hearing in the leopard lizard. *Proceedings of the National Academy of Sciences* 55:98–106.

Wever, E.G. and J.A. Vernon. 1960. The problem of hearing in snakes. *Journal of Auditory Research* (Supplement 5) 1:77–83.

Whitten, K. and R.D. Cameron. 1983. Movements of collared caribou, *Rangifer tarandus,* in relation to petroleum development on the Arctic Slope of Alaska. *Canadian Field-Naturalist* 97(2):143–146.

Winchester, C.F., L.E. Campbell, J. Bond, and J.C. Webb. 1959. Effects of aircraft sound on swine. Wright Air Development Center Technical Report 59-200. Delivery Order Number AF 33(616)-55-15, Project Number 7210. Report by U.S. Department of Agriculture for Aerospace Medical Laboratory, Wright Air Development Center, Wright-Patterson Air Force Base. 47 pp.

Workman, G.W. and T.D. Bunch. 1991. Sonic boom/animal stress project report on elk (*Cervus canadensis*). Report by Utah State University for U.S. Air Force. 87 pp.

Young, R.W. 1973. Sound pressure in water from a source in air and vice versa. *Journal of the Acoustical Society of America* 53:1708–1716.

CHAPTER 9

Recreational Disturbance and Wildlife Populations

Stanley H. Anderson

Dramatic changes in recreation have occurred during the past century (see Chapter 1). With more leisure time and more people living in cities, many have turned to fishing and wildlife activities. In 1993, the National Survey of Fishing, Hunting, and Wildlife Associated Recreation reported results from Americans surveyed about their wildlife related activity conducted in 1991. About 109 million people 16 years and older enjoyed some form of wildlife-related recreation including hunting, fishing, and nonconsumptive use (U.S. Fish and Wildlife Service 1993). People have an impact on wildlife habitat and all that depends on it, no matter what the activity. In Great Britain, for example, people involved in wildlife-related activities affect wildlife by destroying food, increasing pollution, and changing the environment in a variety of ways (Liddle and Scorgie 1980).

When trails and campsites are developed, habitat can be drastically altered. Discarded human food wastes provide different sources of food for animals, affecting their population structure. As people intrude into an area, the effects on animals can include altered behavior, increased stress, or changes in productivity and diet. The populations can change in size and distribution, and the species composition and interactions of whole communities can change (Knight and Cole 1991).

When suitable habitat is diminished, populations may decline. For example, rock climbing may make nest sites used by birds for years unsuitable. In the long term, if extensive habitat alteration occurs for animals that have a limited distribution, the population of a particular species may experience substantial declines.

In this chapter I will examine specific impacts on wildlife populations such as hunting, wildlife viewing, habitat encroachment, and stress. Then I will consider ways to maintain wildlife populations and allow outdoor recreational activities as well. Management options and research needs will also be addressed.

Recreational Influences

Many forms of recreation influence wildlife populations. Activities such as hunting can cause direct impacts while other forms of recreation have direct and indirect impacts, as they often bring people into wildlife areas. Habitat changes can occur, causing disruption of populations or actual stress on animals, both of which may affect reproduction and survival.

Hunting

Does hunting impact wildlife populations? State wildlife agencies attempt to regulate hunting so that the removal of animals will not decrease the population to a level at which it cannot sustain itself. In a study conducted in Ohio, the impact of hunting on squirrels was found to be negligible. The squirrel population fluctuated more in relation to seed productivity (Nixon et al. 1975). In another study, experiments were designed to determine the impact of hunting rock ptarmigan in the autumn and spring. Approximately 40% of the autumn population in a 10-kilometer square area was removed by shooting during three consecutive years. During May, 40% of the breeding population in a 7.5-kilometer square area was removed. Comparison of annual breeding density in experimental and control areas suggested that both the autumn and spring removal at 40% depressed the rock ptarmigan population. There was evidence that autumn removal may have contributed to a higher breeding density than the spring removal. In most cases, male ptarmigan taken in the spring were not replaced until the following year (McGowan 1975). Other studies have also shown that as long as hunting is controlled, it substitutes for mortality factors and therefore does not have an adverse impact on a population.

When hunting substitutes for other forms of mortality, it is said to be *compensatory*. When hunting compounds the total mortality, it is said to be *additive*. For some hunted populations such as waterfowl, there is a debate about how compensatory mortality or additive mortality occurs. Obviously, if hunting is additive then wildlife managers must reexamine the concept of hunting.

Hunting can alter predator-prey relationships. In the 1970s, the interrelationships of wolves, moose, caribou, and humans were studied in interior Alaska. Biologists found that mortality from severe winters, hunting, and wolf predation were largely additive on ungulate populations. For example, when hunting and severe weather took tolls on ungulates, wolves did not reduce their predation, causing the ungulates to decline to dangerously low levels for population viability. In effect, the typical feedback mechanism that operates in predator-prey relations did not work fast enough when hunting was included (Gasaway et al. 1983).

The hunting of ungulates brings up another concern. Often the largest animal with the biggest rack, or the one that is most conspicuous, is the animal targeted. If people continually hunt these larger animals, there may be impacts on the animals' population structure.

Other aspects of hunting can also affect populations of wildlife. The noise of shooting is known to cause animals to flee from an area. When gas compressors were used to simulate noise near flying snow geese and tundra swans, the birds broke their flight formations, flared, increased altitude, increased calling behavior, changed speed and/or landed. These disruptions caused the birds to alter their behavior and become disorganized (Wiseley 1974).

When feeding or loafing snow geese were frightened suddenly, entire flocks took off in near unison without normal pre-flight coordination of families. If flocks were large, birds rose in a confused, clamoring mass, and social groups frequently were broken in the disorder. Flocks of snow geese frightened by the same disturbance mixed while circling before landing again. Gunfire on refuge boundaries sometimes frightened snow geese inside the refuge. Some geese that flew outside the refuge to feed came under heavy fire, which caused flocks to separate and scatter (Prevett and MacInnes 1980). There are instances where hunting has caused birds to fly up and hit transmission lines, buildings, or windows as they try to escape the noise.

Hunting and other human-caused noise not only disrupt members of a population but also cause animals to avoid habitats. On staging grounds, snow geese were highly sensitive to hunters and noise (Berger 1977). Use of specific areas of the bay and daily flights by brant geese were disturbed by hunting. On days without hunting, brant geese left deep-water areas and flew to shallow water. Brant usually did not fly at other times, except when disturbed by aircraft, anglers, or boaters. Disturbance by hunters resulted in five to six times more flights than on nonhunting days. Flights were more frequent and occurred soon after hunting began (Kramer et al. 1979).

Animal behavior changes during hunting, evidenced by the effects of disturbance of Canada-goose roosting areas. Hunting disturbance resulted in an increased separation of family members, which may have contributed to the large number of geese shot during the hunting season (Kramer et al. 1979). This, in turn, disrupted the cohesiveness of family groups. As the season advanced, families (after hunting losses) and random groups used the same roosting areas less. Intact family groups used the same roosting areas at the same rate throughout the season. A disintegration of family structure seemed to be related to the extent of disturbance and hunting pressure (Bartelt 1987).

Loss of social structure subsequently has an impact on migration and future reproduction of the flock.

Wildlife Viewing

Wildlife viewing can have diverse impacts on wildlife. An example of how viewing can impact birds in migration involves the Platte River in central Nebraska. This is the staging area for some 500,000 sandhill cranes that move up from the southern part of the United States. Each spring cranes roost at night on small sand islands in the river; in the daytime they fly out to pick waste corn left in adjacent fields. They remain in this area for approximately six weeks. During this time they accumulate body fat, and as the weather begins to clear farther north, they begin migrating into the northern United States and Canada, where they breed (Norling et al. 1992).

Each year the North Platte area attracts visitors who come to view the cranes. When people approach feeding or roosting cranes, the birds become disturbed, and young birds seeming to be the most susceptible. Viewers moving through the area may startle them, causing birds to flush, expending unnecessary energy in disturbance flights. Cranes are also injured or killed when they fly into power lines. Thus people are affecting this important cohort of cranes.

Intrusion

Perhaps the major way that people have influenced wildlife populations is through encroachment into wildlife areas. This has increased dramatically over the past few decades, usually resulting in the elimination of many local populations, particularly the larger vertebrates (Eltringham 1990). Numerous studies document how water-related recreation activities cause waterfowl to avoid prime nesting areas or abandon their nests once eggs are laid. A study conducted in Wisconsin found that waters heavily used by recreationists were not used by breeding ducks, despite the fact that the habitat was suitable for nesting in all other ways (Jahn and Hunt 1964).

During the waterfowl breeding season, anglers contribute to a serious decline in breeding waterfowl. One study in Germany revealed a 90% decrease over 10 years. The investigator found that a single angler can prevent ducks from establishing territories or selecting nest sites when the area of open water is less than 1 ha. Disturbance was less of a problem on larger waters. Intensive angling reduced the number of waterfowl nests by 80%, and the remaining nests were found only in areas inaccessible to anglers. Breeding success was also much lower due to increase of predation on clutches by crows and black-billed magpies. Also, waves created by motorboats tipped over exposed nests (Reichholf 1976).

Walking near a nest, whether to fish or just look at the bird eggs or young, can attract predators to an area. Keith (1961) found that trails and tracks leading to nests and disturbance of nest cover caused predation on nests at levels intolerable to birds in Alberta wetlands.

Nesting geese suffered from elevated predation resulting from increased disturbance by human intruders. Predation levels on clutches of nestling Canada geese were observed from 1965 to 1969 on the McConnell River, Northwest Territories, Canada. Human disturbance seemed to contribute to the loss of partial clutches. Losses reached 55% of all lost eggs after repeated visits to individual nests. The proportion of completely destroyed nests did not vary among years, but the proportion of partial loss of eggs increased significantly following intrusion. The latter difference was due to changes in predator activity, particularly herring gulls, and interaction with humans, not just changes in human activity.

Greater proportions of small clutches were destroyed because the lost eggs represented a higher proportion of the initial clutch. In the absence of human disturbance, losses of eggs to predation would have been about 10% and would have varied little from year to year despite demonstrated changes in predator activity (MacInnes and Misra 1972; MacInnes et al. 1974). There are cases, however, in which a single visit to nest sites by people can cause nest abandonment (Glover 1956).

Results of disturbances, however, have not always been consistent. In New England, Coulter and Miller (1968) found variable responses to human disturbance. Hens occasionally deserted their clutches after the first disturbance, but only one of 30 trapped hens deserted her nest. Hen trapping was attempted at 291 nests and 223 (80%) hens were captured. Desertion of nests after a trap was set was 6% in ring-necked ducks, 9% in American black ducks and 26% in mallards. Ten of 18 desertions were in a group of especially intolerant mallards. Certain other mallards were more tolerant, and only 16% of 49 hens deserted nests at other islands. Abandonment of nests was not as common during the last week of incubation.

Some animals can be harmed by recreational activities during the time they are under parental care. Most waterfowl young form cohesive groups with their mothers and/or fathers. When these family units are broken up through separation or death of a parent, the whole family can suffer and the young may perish. On the Seney National Wildlife Refuge in northern Michigan, families of Canada geese are common in the late spring and early summer. Family ties of Canada geese are fragile during the first three to four weeks of life, and a brood can be easily divided. Vehicles on dikes surprised broods, causing them to panic and disperse in all directions. Some goslings got lost in the dense vegetation when parents headed for the pool, or parents swam off leaving goslings behind that could not follow (Sherwood 1965).

Human disturbance of summer-molting ducks may impair their selection of lakes for the flightless stage. Human disturbance to canvasbacks that were on lakes with little or no boating was negligible, while lakes used for recreation by boaters and anglers were no longer suitable for molting canvasbacks (Bergman 1973).

Disturbance of wildlife can alter the population structure and dynamics of a species. When many young are lost, a cohort may become very small or nonexistent. In some cases the species can no longer survive in an area. The decline in North American waterfowl can be attributed partially to a loss of isolated, disturbance-free nesting sites for birds.

In Missouri, Kaiser and Fritzell (1984) documented how different forms of recreation disturbed the behavior of green-backed herons. River boating resulted in a decrease of foraging on rivers. When boating was common, herons were more difficult to locate because they had moved into the backwaters. During times of reproduction, there was less food supply available for the herons' young, and they had to expend more energy to get food.

In north-central Minnesota, Fraser et al. (1985) examined the impact of human activities on breeding bald eagles. They found that eagles flushed as people began to approach an area. However, it was possible to locate control sites around eagle nests, where people could view the birds without causing excessive problems. They found that human activities had little impact on the reproductive success of the eagles when they implemented the proper management strategy at their particular study site.

During the construction of the trans-Alaskan oil pipeline, people intruded into bear habitat. Bears apparently became adjusted to the human presence and did not bother the pipeline builders. However, their refuse attracted the bears. Most of the bear-person encounters resulted when people attempted to feed the bears or when they happened to be near disposed garbage (Follman and Hechtel 1990). There have been many studies of how such dumps alter the habitat and population dynamics of bears. When bears eat at garbage dumps, the demographic consequences are complex. Bears that frequent dumps tend to be larger, have elevated reproductive rates, and live longer than bears that eat natural food. The elevated nutritional state also accelerates maturation and weaning.

In northeastern Minnesota, when natural foods were scarce, black bears with access to garbage dumps matured earlier (4.4 years) than bears with natural diets (6.3 years) (Rogers 1987). However, at dumps, juveniles are particularly vulnerable to being killed by other bears, because they are concentrated in small areas (Stringham 1989). When bears become accustomed to garbage dumps and then the dumps are closed, bears move into other areas, which causes human-wildlife conflicts. Two to four years after dump closure, bear populations tend to again be regulated by their natural food sources.

Stress

When people intrude into wildlife habitat, stress on wildlife populations is one result. Snowmobile activity is a particular problem as people move into wintering areas where animals may already be stressed. Airboats in Texas disturb and stress animals year-round. Mabie et al. (1989) found that the direct approach by airboats caused detectable differences in behavior patterns of four families of whooping cranes. The birds did, however, return to their normal activity after the airboats left.

It is difficult to evaluate completely the full impact of stress. Geist (1971) expressed concern about the effects of aircraft harassment on big game. He asserted that increased wildlife viewing in the Arctic could have serious adverse impacts because the yearly cycles of animals were interrupted. Pregnant animals suffered higher stress from wildlife viewers, causing some to abort. Geist used ruminants such as caribou and sheep to show how running increased their need for food. He pointed out that animals can be stressed to the point that they require more energy than they can take in, so they must use body reserves. Continuous stress from human recreation could eventually cause illness or death of an animal.

In waterfowl, energetic costs often occur when birds are flushed from wetlands (see Chapter 14). When birds are flushed prior to or during migration, the energy cost could be great enough that they might not make it to their destination, or they may be more susceptible to diseases. If they are flushed during their spring migration, loss of body weight could lead to reproduction problems.

Reduced caloric intake can profoundly increase the time it takes for waterfowl to replenish fat reserves. Reducing daily caloric intake by 19% (390 kcal per day) more than doubles the estimated time required for fat replenishment, but these estimates represent a simplified case and actual requirements may be different. Concentrated numbers of waterfowl create two problems that could adversely affect their energy budgets. First, food supplies become depleted more rapidly. Decreased food availability necessitates increased foraging time or longer foraging flights for species that forage in fields. Secondly, hunting pressure tends to increase in areas where waterfowl concentrate. Harassment by hunters could increase movements and reduce time for foraging, thereby increasing energy use (Fredrickson and Drobney 1979).

While all impacts on animals cannot be documented, it is clear that loss of body reserves has negative effects on the individuals concerned. When combined with other factors such as stressful winters, the animals could die or fail to reproduce. In such cases, populations would decline. When disturbance occurs over a large region for many years, the population may be unable to continue to reproduce and survive in the area.

Management Options for Coexistence

There are many ways in which management can control human disturbance of animals (see Chapters 3 and 20). Closing off areas that are particularly important for breeding animals can reduce some impacts. Calving areas in national parks are often closed, and areas of high waterfowl reproduction could be marked so that people will not enter during courtship and nesting periods. In wildlife viewing areas, blinds can be set up to provide an unobtrusive site for watching the animals (see Chapter 15). This has been done during the spring along the Platte River to minimize impact on the sandhill crane population. In general, disturbances should be strongly regulated, especially during times of natural stress such as winter or breeding seasons.

National parks and wildlife management areas can effectively manage people so that they remain out of the area. Boating access and use areas can be restricted. Fencing and other control measures can be used to maintain a recreation situation and to protect wildlife. Hikers or snowmobilers can be encouraged to stay on marked trails, leaving the trailless areas for wildlife.

When people go into territory occupied by wild animals, they should be aware of possible problems. National parks and national forests attempt to educate people about grizzly bears, but they are never completely successful because a number of people are attacked and harmed by grizzly bears every year—often when people are carrying food. Sometimes, people even attempt to approach too close to these dangerous animals to take photographs.

If people are to continue recreational activities where wildlife exists, they must understand their own impacts. They must take care to stay on trails, avoid feeding wild animals, avoid approaching too close to photograph, and consider the proper time of year to go into a particular area. Management agencies can successfully assist people who want to view wildlife by erecting viewing stands, which limit habitat encroachment and thereby reduce the impact of human recreational activities.

Knowledge Gaps

The impact of recreation on populations can result in alteration of the population, the habitat, or the community interactions in such a way that the number of individuals changes. Recreation may influence one or all stages of the life history of a population. They may affect one or both sexes. Each population is different, so we need to look at the whole life-history pattern to define the problem.

Population biologists often examine mortality rates using life tables (see Anderson 1991). This enables the biologist to follow a cohort through time. Life tables, although somewhat arbitrary, do indicate time frames in which the greatest numbers of deaths occur. If constructed for a population affected by recreational activity, a life table can indicate sensitive periods during which people are having an impact on the population.

The age structure of a population, or the number of individuals in each age category, is very important in determining its status. A population consisting primarily of old individuals will obviously die off soon. It is also more prone to impact by human intrusion. In contrast, populations that consist primarily of young individuals are likely to increase rapidly in size, forcing managers to introduce methods to contain the numbers. Data are needed on sex- and age-specific rates of growth, reproduction, migration, and survivorship, as well as on population density, sex ratios, age structure, social organization, and availability of the nutrient quality of natural fruits.

Currently, managers close big-game calving areas, yet little data exist to show how intrusion affects the raising and feeding of young in most species. Both birth and death rates can be expressed for the whole population or for different age classes. Data should be collected to assess whether wildlife viewers affect mortality. If recreationists prevent adults from getting food or being near their young, fewer young may survive to reproduce.

It is sometimes difficult to separate impacts on populations from impacts on habitat. We do not know how adaptable most populations are to people. Likewise, we do not know how adaptable most populations are to habitat alteration. When people build campgrounds, trails, and boat-ramps, they change the habitat and sometimes the species of wildlife. What is the limit of tolerance of different species?

Population interactions are important in maintaining wild populations. We know that predators depend on prey. We can see that disturbance of a duck nest by anglers can cause predation, but often we do not know how recreation affects predator-prey interactions. Comparative studies evaluating different degrees of disturbance are necessary to determine these effects.

Removal of some animals may change competition, and cause a population to decline. We need to examine the sex ratios and age structures of populations in an area when some species are removed. By following these populations for several years, we can draw inferences about the effects of recreation.

Stress can also be a factor in wildlife survival. While a single stress-causing incident is not bad, continual harassment of animals causes them to expend energy beyond what they can take in during the winter, so, some animals can die or fail to reproduce. Stress has been shown to be an important contributor to declining populations in some animals but such population-related work is rare. Again, we need to determine the number of young produced and young that survive in areas with different levels of recreation-induced stress.

To examine populations that might be impacted by recreationists, we need to use a life-table approach. Long-term studies comparing the longevity, sex ratios, age structure, etc. of disturbed and undisturbed populations would be ideal. This is a complex problem, but one method to obtain this information is to develop a series of small experiments based on different parts of a life table. The results of these projects could be put together in a model that would look at impacts on the population as a whole. If we were to summarize our knowledge about one population in a life table, we could see changes in sex ratios, age structure, and causes of mortality. Perhaps the current approach of conducting research in isolation should be changed. Perhaps groups of investigators should work together to answer different questions about recreation-induced impacts on a population. Then all aspects could be brought together to examine effects on the whole population.

Literature Cited

Anderson, S.H. 1991. *Managing our Wildlife Resources.* 2nd ed., Englewood Cliffs, New Jersey: Prentice Hall.

Bartelt, G.A. 1987. Effects of disturbance and hunting on the behavior of Canada goose family groups in east central Wisconsin. *Journal of Wildlife Management* 51:517–522.

Berger, T.R. 1977. The Berger report: northern frontier, northern homeland. *Living Wilderness* 41:4–33.

Bergman, R.D. 1973. Use of southern boreal lakes by post-breeding canvasbacks and redheads. *Journal of Wildlife Management* 37:160–170.

Coulter, M.W. and W.R. Miller. 1968. Nesting biology of black ducks and mallards in northern New England. Bulletin 68-2. Montpelier, Vermont: Vermont Fish and Game Department.

Eltringham, S.K. 1990. Wildlife carrying capacities in relation to human settlement. *Koedoe* 33:87–97.

Follman, E.H. and J.L. Hechtel. 1990. Bears and pipeline construction in Alaska. *Arctic* 43:103–109.

Fraser, J.D., L.D. Frenzel, and J.E. Mathisen. 1985. The impact of human activities on breeding bald eagles in north-central Minnesota. *Journal of Wildlife Management* 49:385–392.

Fredrickson, L.H. and R.D. Drobney. 1979. Habitat utilization by post-breeding waterfowl. In *Waterfowl and Wetlands—An Integrated Review*, ed., T.A. Bookhout, 119–131. LaCrosse, Wisconsin: North Central Section, The Wildlife Society, LaCrosse Printing Co.

Gasaway, W.C., R.O. Stephenson, J.L. Davis, P.E.K. Shepherd, and O.E. Burris. 1983. Interrelationships of wolves, prey and man in interior of Alaska. Wildlife Monograph 84. Washington D.C.: The Wildlife Society.

Geist, V. 1971. Is big game harassment harmful? *Oilweek* 22:12–13.

Glover, F.A. 1956. Nesting and production of blue-winged teal in northwest Iowa. *Journal of Wildlife Management* 20:28–46.

Jahn, L.R. and R.A. Hunt. 1964. Duck and coot ecology and management in Wisconsin. Technical Bulletin No. 33. Madison, Wisconsin: Wisconsin Conservation Department.

Kaiser, M.S. and E.K. Fritzell. 1984. Effects of river recreationists on green-backed heron behavior. *Journal of Wildlife Management* 48:561–567.

Keith, L.B. 1961. A study of waterfowl ecology on small impoundments in southeastern Alberta. Wildlife Monograph 6. Washington D.C.: The Wildlife Society.

Knight, R.L. and D.N. Cole. 1991. Effects of recreational activity on wildlife in wildlands. *Transactions of the North American Wildlife and Natural Resources Conference* 56:239–247.

Kramer, G.W., L.R. Rauen, and S.W. Harris. 1979. Populations, hunting mortality and habitat use of black brant at San Quintin Bay, Baja California, Mexico. In *Proceedings of the Symposium on Management and Biology of Pacific Flyway Geese*, eds., R.L. Jarvis and J.C. Bartonek, 242–254. Washington, D.C.: The Wildlife Society.

Liddle, M.J. and H.R.A. Scorgie. 1980. The effects of recreation on freshwater plants and animals: a review. *Biological Conservation* 17:183–206.

Mabie, D.W., L.A. Johnson, B.C. Thompson, J.C. Barron, and R.B. Taylor. 1989. Response of wintering whooping cranes to airboat and hunting activities on the Texas coast. *Wildlife Society Bulletin* 17:249–253.

McGowan, J.D. 1975. Effect of autumn and spring hunting on ptarmigan population trends. *Journal of Wildlife Management* 39:491–495.

MacInnes, C.D. and R.K. Misra. 1972. Predation on Canada goose nests at McConnell River, Northwest Territories. *Journal of Wildlife Management* 36:414–422.

MacInnes, C.D., R.A. Davis, R.N. Jones, B.C. Lieff, and A.J. Pakulak. 1974. Reproductive efficiency of McConnell River small Canada geese. *Journal of Wildlife Management* 38:686–707.

Nixon, C.M., M.M. McClain, and R.W. Donohoe. 1975. Effect of hunting and mast crop on a squirrel population. *Journal of Wildlife Management* 39:1–25.

Norling, B.S., S.H. Anderson, and W.A. Hubert. 1992. Roost site used by sandhill cranes staging along the Platte River, Nebraska. *Great Basin Naturalist* 52:253–261.

Prevett, J.P. and C.D. MacInnes. 1980. Family and other social groups in snow geese. Wildlife Monograph 71. Washington, D.C.: The Wildlife Society.

Reichholf, J. 1976. The influence of recreational activities on waterfowl. In *Proceedings of the International Conference on the Conservation of Wetlands and Waterfowl*, ed., M. Smart, 364–369. Slimbridger, England: International Waterfowl Research Bureau.

Rogers, L.L. 1987. Effects of food supply and kinship on social behavior, movement and population growth of black bears in northeastern Minnesota. Wildlife Monograph 97. Washington, D.C.: The Wildlife Society.

Sherwood, G.A. 1965. Canada geese of the Seney National Wildlife Refuge. Completion report for wildlife management studies 1 and 2. Seney National Wildlife Refuge, Seney, Michigan. Minneapolis, Minnesota: U.S. Fish and Wildlife Service.

Stringham, S.F. 1989. Demographic consequence of bears eating garbage at dumps, an overview. In *Bear-people conflicts, symposium on management strategies*, 35–42. Yellowknife, Northwest Territories, Canada: Northwest Territories Department of Renewable Resources.

U.S. Fish and Wildlife Service. 1993. *1991 National Survey of Fishing, Hunting, and Wildlife-Associated Recreation*. Washington, D.C.: U.S. Fish and Wildlife Service.

Wiseley, A.N. 1974. Disturbance of snow geese and other large waterfowl species by gas compressor sound simulation. In *Studies on Snow Geese and Waterfowl in the Northwest Territories, Yukon Territory, and Alaska 1973*, eds., W.W.H. Gunn, W.J. Richardson, R.E. Schweinburg, and T.D. Wright. Arctic Gas Biological Report, series 27. Yellowknife, NT, Canada.

CHAPTER 10

Recreational Disturbance and Wildlife Communities

Kevin J. Gutzwiller

Outdoor recreation has been recognized as an important factor that can reduce biosphere sustainability (Lubchenco et al. 1991, Fig. 4; 1993, Box 1). Indeed, recreational activities, including many that seem innocuous, can alter
vertebrate behavior, reproduction, distributions, and habitats (see review by
Boyle and Samson 1985). Recreational disturbance is quickly becoming a
dominant structuring force in many wildlife communities, and projections
indicate that the frequency and geographic extent of such disturbance will
continue to increase in natural landscapes in the years ahead (see Chapter 1;
Purdy et al. 1987; Holecek 1993).

Our understanding of how recreational activities influence communities is
just developing. Until recently, investigators have focused primarily on how
recreationists influence wildlife at the levels of individuals, family groups, and
populations. Only a few studies to date have addressed issues important for
understanding the effects of recreationists on community structure and dynamics. Accordingly, many community-level questions remain unanswered.
Until the influences of recreationists on wildlife, at the community level, are
better understood, efforts to achieve and sustain the coexistence of recreationists and intact wildlife communities will not be successful. My objectives here
are to: (1) describe what is presently known about the influences of recreationists on wildlife communities; (2) suggest management strategies by
which recreationists and wildlife communities can coexist; (3) identify major
voids in our knowledge about recreational impacts on wildlife communities;
and (4) recommend research approaches that will help fill these gaps.

Recreational Influences

Recreationists can directly alter competitive, facilitative, and predator-prey relations, three types of interaction that have the potential to affect community

169

structure and dynamics. Species richness, abundance, and composition in communities can be altered by displacement and through the indirect effects of recreationists on habitat structure.

Competitive and Facilitative Relations

In the Pacific Northwest region of the United States, a number of bird species feed during winter months on salmon that have spawned, died, and washed up on gravel bars (Knight et al. 1991). Salmon carcasses weigh too much for these species to be able to fly away with them, so the scavengers must feed on the ground, where they are susceptible to disturbance from anglers and other recreationists (Skagen et al. 1991). Knight et al. (1991) studied responses of an avian scavenging guild (comprised of bald eagles, common ravens, and American crows) to the presence of anglers on gravel bars. They found that, although most eagles and ravens typically foraged during early- and mid-morning hours, the presence of anglers caused an unusually high percentage to feed during late afternoon hours. In contrast, crows foraged in approximately equal proportions throughout the day when anglers were present, although crows usually fed early in the morning when anglers were absent. Disturbance from anglers differentially altered the normal feeding times for these scavengers, enabling crows to exploit carcasses during certain hours with less competition from eagles and ravens.

In a study of a similar salmon-scavenging guild (bald eagles, American crows, and glaucous-winged gulls), Skagen et al. (1991) detected a dominance hierarchy associated with body size. Eagles were dominant over both gulls and crows, and gulls were dominant over crows. At undisturbed salmon carcasses, eagles ate more than 50% of the salmon consumed by all of the guild members, whereas at carcasses that were disturbed, eagles accounted for only 5.7% of all salmon eaten. Human disturbance caused eagles to flush sooner than the other species, and eagles rarely returned to a carcass following disturbance. Gulls and crows, in contrast, permitted closer approaches by people before they flushed, and both gulls and crows returned to carcasses within an average of 7 and 16 minutes, respectively. Thus, in areas with frequent recreational disturbance, competitive relations among these three species can be altered because eagles will typically abandon salmon carcasses when disturbed by people. Skagen et al. (1991) predicted that in such situations, gull and crow densities would increase. Higher gull and crow densities would, in turn, disrupt normal guild and community structure and could threaten eagle persistence when food supplies were low or inaccessible. In the absence of disturbance by recreationists, all three species can coexist under typical winter conditions.

In the scavenging guild just discussed, eagles facilitate foraging by gulls and crows in a significant way. Only when eagles have torn the skin of salmon carcasses can these other species obtain appreciable amounts of food (Skagen et al. 1991). In other guilds or mixed-species groups involving birds, mammals, or both, certain species increase the foraging efficiency of others through such actions as flushing prey, opening carcasses, and locating food sources (see Skagen et al. 1991). Recreational disturbances will curtail the ecological services of facilitator species to the extent that such species are sensitive to recreationists. Bald eagles tend to avoid areas used by recreationists, and if large areas (hence many salmon carcasses) were not accessible to eagles, less food would be available for gulls and crows. The uncoupling of dependencies among species can alter the structure and dynamics of guilds and their associated communities.

Predator-Prey Relations

Predation, by influencing the presence or abundance of prey, can affect trophic structure and therefore the paths of energy and nutrient flow in wildlife communities. Intense predation on one species may increase competition among predators for the remaining prey species used in common. All such effects have the potential to influence community structure and dynamics. Recreational effects on predation rates have been documented for some species, and several examples will illustrate the basic impact patterns and processes.

Anderson and Keith (1980) found that when recreationists approached ground-nesting brown pelicans or walked among their nests, predation by western gulls and ravens increased substantially on pelican eggs and chicks left unattended by displaced parents. Similarly, recreationists and tour groups exacerbate predation by Heermann's gulls on young Heermann's gulls, elegant terns, and royal terns (Anderson and Keith 1980). Natural levels of predation, occurring in the absence of recreationists and other human intruders, typically have little effect on productivity in colonial birds (Anderson and Keith 1980).

Double-crested cormorants easily defend their eggs and chicks against herring gulls and great black-backed gulls, except when human intruders disturb them. Both cormorants and gulls left their nests at the approach of people, but the gulls returned sooner and ate undefended cormorant eggs and chicks (Kury and Gochfeld 1975). By defending their own nests and adjacent nests of conspecifics, cormorants can create a protective "neighborhood effect"; interspecific neighborhood effects have been documented as well (see brief review by Kury and Gochfeld 1975). Clearly, the disruption of such interactions by recreationists can increase predation and thereby influence community structure.

Young wildlife are quite susceptible to predation for some time after they have left their nest or den. Recreationists can directly increase the risk of predation in some species by increasing predator contacts, or by dispersing young, which causes stragglers to be left behind and then preyed on. For example, common eider ducklings encountered approximately five times as many predators within five minutes after anglers, hikers, windsurfers, and boaters disturbed them as they encountered five minutes before such disturbances. When disturbed, broods often ran, and ducklings that could not keep up had a higher chance of being eaten by herring gulls and great black-backed gulls (Keller 1991).

Juveniles that get displaced from familiar surroundings (e.g., home ranges) by recreationists may also be more susceptible to predation. Clarke et al. (1993) reported that young-of-the-year chipmunks released 100 meters from their main burrow (outside of their home range), and then chased by an investigator, used more than twice as much time to find a refuge as did those released and pursued within 10 meters of their main burrow (inside their home range). The proximity of their main burrow was not a factor because chipmunks rarely used main burrows as refuges. Virtually all juveniles released outside of their home range used trees for refuges, whereas most of those released within their home range used holes, which are more secure than trees against chipmunk predators (raccoons, ermine, red foxes, domestic cats, and dogs). Outside of their home ranges, juvenile chipmunks would be less capable of escaping predation than if they were attacked within their home range (Clark et al. 1993). Thus, young displaced from familiar habitat by recreationists may evade predators less successfully.

The activities of field researchers are comparable to those of nature viewers and wildlife photographers in that each group approaches wildlife closely, repeatedly, and sometimes for extended periods. An examination of investigator effects on predation should therefore be fruitful in assessing the potential for predator-related impacts caused by recreationists whose actions are similar to those of field researchers. In 22 of 29 studies of nesting success, investigator disturbance increased intra- and inter-specific predation on eggs and young (see Götmark 1992). Of 17 studies that examined researcher impacts on predation rates, 10 described higher losses of nests or young to predators as a result of investigator disturbance. Götmark (1992) found that the primary predators were larids and corvids, and that, contrary to conventional wisdom, there was little documented evidence of investigator-induced mammalian predation on nests.

Red-fox predation on piping plover nests that were monitored from 3 to 15 meters away was higher than that for nests studied from distances of less than 3 meters (MacIvor et al. 1990). The authors suggested that fresh human scent near nests may have repelled red foxes. Human scent may be a particularly strong deterrent for mammalian predators that are persecuted by people; alternatively, the paucity of evidence for mammalian predation at nests following investigator disturbance may be due to rapid scent dissipation and researchers' precautions, such as avoiding the creation of trails (Götmark 1992). Götmark and Åhlund (1984) determined that, for simulated common eider nests, covering the eggs with down significantly reduced researcher-induced predation by avian predators.

Nestling Cooper's hawks visited by investigators every two or three days had significantly lower postfledging survival than did those visited only one to three times during the breeding season (Snyder and Snyder 1974). Humans killed several of the hawks from frequently disturbed nests. From this, the authors suggested that individuals familiar with people may be more susceptible to persecution (Götmark 1992). Investigator disturbance apparently does not increase avian nest predation by snakes (Götmark 1992). Overall, these results logically indicate that recreationists who visit nests to watch or photograph them may increase losses of eggs and chicks to avian predators because the frequency and proximity of their visits are similiar to those of researcher visits. Given this, wildlife viewers and photographers are likely to induce comparable predation increases.

Species Richness, Abundance, and Composition

Species displacement caused by recreationists can alter species richness, abundance, and composition in wildlife communities. Displaced animals are forced out of familiar habitat and must then survive and reproduce in areas where they are not familiar with the locations of food, shelter, and other vital resources. They may also face conflicts with established conspecifics in these new areas. Displacement can alter community structure through immediate effects (via initial displacement of species from disturbed habitats) and long-term consequences (via lower persistence of individuals or species in unfamiliar or poor habitats). Species that are sensitive to the presence of people may be displaced permanently; accordingly, Hammitt and Cole (1987:87) ranked displacement as being more detrimental to wildlife than harassment or recreation-induced habitat changes. Depending on the species that are lost, or the interspecific interactions that are uncoupled by displacement, the presence or abundance of other species may also be affected. For example, hunters can displace predators, which in turn may influence abundance of other members of the food chain (Hammitt and Cole 1987:88).

Less intrusive disturbances can cause displacement as well. Densities for 8 of 13 breeding bird species were negatively associated with the intensity of recreation activity by park visitors, primarily pedestrians and cyclists (van der Zande et al. 1984). Each of these species exhibited a different degree of susceptibility to visitor disturbance, and recreation activities hundreds of meters away affected the densities of common species. Visitor activity also influenced the presence and absence of five common species (van der Zande et al. 1984). From these impacts, it follows that visitor activity also affected species evenness, richness, and composition. In a study of the effects of shoreline recreationists (boaters, cyclists, walkers, moped riders) on breeding bird abundance, van der Zande and Vos (1984) discovered that numbers for 11 of 12 species were lower in areas where recreation intensity was high than in areas with fewer visitors. The lower abundances were associated with recreation intensities that ranged from 8 to 37 people per hectare (maximum number of visitors present simultaneously). Finally, recreationists have displaced a variety of mammals as well (e.g., Dorrance et al. 1975; Tuttle 1979; Gerrodette and Gilmartin 1990).

The effects of recreationists on habitats are known to alter richness, abundances, and composition in wildlife communities (see Chapter 11). For instance, Blakesley and Reese (1988) found that habitat structure in campgrounds differed from that in noncampground areas. Tree, shrub, and sapling densities; cover of dead, woody vegetation; and natural litter depth were lower at campgrounds, while grass cover was greater at campground sites. These structural changes, probably resulting from trampling, soil compaction, and fuelwood gathering by campers, determined in part the bird species that could persist. Five of the seven species associated with campgrounds nested in trees, whereas six of the seven species associated with noncampground sites nested on the ground or in the understory. Campers simplified the horizontal and vertical structure of the ground and understory vegetation, which reduced abundances of ground and understory species at campground sites. Avian abundances and community composition were thus influenced indirectly by the effects of recreationists on habitat structure. Recreationists typically reduce environmental structure and complexity, and although some species may increase numerically under these conditions, typically species diversity and richness decline (Hammitt and Cole 1987:88).

Management Options for Coexistence

Based on available information, managers can begin to ensure the coexistence of recreationists and intact wildlife communities by minimizing recreation-induced changes to normal interactions among species; by reducing wildlife displacement; and by maintaining and restoring the floristic and structural heterogeneity of wildlife habitat.

Minimizing Changes in Species Interactions

As I described above, recreationists can disrupt such normal interactions as competition, facilitation, and predation. If these processes cease, natural, intact wildlife communities will not persist. To minimize recreation-induced changes in these relations, managers should attempt to reduce disturbances to species that are clearly linked or interdependent. This can be accomplished by establishing specific times, places, and modes of travel for public access that minimize problems for vulnerable species.

Minimizing recreational disturbance to salmon-scavenging guilds during the winter, for example, would improve the chances that normal competitive and facilitative interactions would continue and that all guild members would persist. If recreationists minimized the proximity, frequency, and duration of their visits to colony-nesting birds, fewer abnormal predation events would occur (Kury and Gochfeld 1975; Anderson and Keith 1980). Knowledge of when and where breeding occurs can be used to reduce contact between recreationists and the young of susceptible species, and thereby reduce recreation-caused predation on juveniles. Managers could educate nature viewers and wildlife photographers on how to modify their actions to avoid inducing abnormal levels of predation on eggs and chicks (see also Hammitt and Cole 1987:269–276). Natural richness, abundances, composition, and functions of communities are more likely to be sustained when normal levels of competition, facilitation, and predation occur than when recreationists disrupt these processes.

Reducing Wildlife Displacement

To reduce recreation-related displacement, managers should control the proximity, frequency, duration, and seasonal timing of disturbances, especially for species at high risk. Such management efforts may involve establishing buffer zones, reducing activities in places that are vital for wildlife reproduction and survival, or restricting access during crucial periods of a species' annual cycle (see Chapter 20). In some situations, managers may reduce displacement more effectively by concentrating recreational activities in busy, heavily altered sites, while not permitting expansion into previously undisturbed or only mildly impacted areas (see van der Zande et al. 1980, 1984).

Maintaining Wildlife Habitat

Species richness, abundance, and composition in a community depend in part on the presence and heterogeneity of specific habitat characteristics. When recreational activities alter habitat conditions, these community attributes can change. Communities with normal structure and function are more likely to persist if habitat modification by recreationists is minimized. Blakesley and Reese (1988), who detected lower abundances of ground- and understory-nesting birds at campground sites, recommended that managers maintain existing shrubs and saplings in campgrounds and that campground developers retain a variety of shrub species throughout new campsites. Damaged campsites could be rehabilitated by closing and revegetating them (Hammitt and Cole 1987:207).

Vegetation is not the only aspect of habitat that recreationists degrade. Because of a packing effect, snowmobiles reduce the insulation properties of snow for small mammals, and off-road vehicles can collapse burrows of desert mammals and reptiles (Webb and Wilshire 1983; Hammitt and Cole 1987:84–85). In areas where such impacts are numerous, managers should control the intensity and spatial scale of recreation.

Ideally, a balance should be struck between recreational activity and habitat protection so that recreationists can enjoy their activities, and wildlife habitat will not be destroyed. Visitor management is an essential strategy for protecting wildlife habitats and therefore wildlife communities. Currently employed techniques include zoning, party-size limits, information programs, and controls on amount of use, dispersal of use, length of stay, and seasonal use (Hammitt and Cole 1987:243–283).

Knowledge Gaps

Significant voids exist in our understanding of how recreationists influence wildlife community structure and processes. We must conduct research specifically designed to fill these knowledge gaps and create a more solid foundation from which to work.

Community Structure

In both terrestrial and aquatic systems, different types of stress (including commercial fishing, deforestation, pollutant discharges, habitat alteration, and introductions of exotics) often produce rather similar results for communities. Typical responses to such stresses include: increases in proportions of r-strategists; decreases in the size of organisms; decreases in lifespans of organisms; shortened food chains due to less energy flow at higher trophic levels, greater vulnerability of predators to stress, or both; declines in species diversity; increases in dominance by fewer species; increases in negative interactions such as parasitism; and decreases in beneficial interactions like mutualism (Odum 1985; Rapport et al. 1985). Immediate study is warranted to determine whether such changes in community structure are caused by stresses from recreationists.

We are beginning to understand how a few specific types of recreation can affect species' interactions and influence simple variables like richness, abundances, and composition, but I am unaware of any study that has explicitly addressed these other possible changes. Field experiments with temporal and spatial controls, randomized designs, covariates, and adequate replication will be necessary to assess cause-and-effect relations (Gutzwiller 1991). Because of the variability inherent in natural systems, particularly between years, studies lasting 3 to 10 years will probably be necessary to determine clearly whether recreational disturbance has impacts similar to those of other stresses. Serial management experiments, an adaptive approach with which managers regularly improve the effectiveness of management schemes and researchers continually update their understanding about causal relations, will, in many situations, be the most reliable and defensible strategy to minimize recreational impacts and to learn about their causes (Gutzwiller 1993).

Community Persistence

Recreational activities may affect the persistence of communities, yet there are no studies that have assessed recreational impacts on community stability, even though such information is crucial for managers who attempt to maintain intact wildlife communities. Conclusions about community stability often depend on the spatial scale, temporal scale, and level of taxonomic resolution (e.g., species, guild) examined (Rahel 1990). Interpretations about community stability also depend frequently on the degree of numerical resolution—absolute abundances, abundance rankings, or presence and absence, for example (Rahel 1990). Investigators studying the effects of recreationists on community persistence should consider these caveats and, whenever feasible, examine effects at various temporal, spatial, taxonomic, and numerical scales (see Wiens 1989; Rahel 1990).

Cumulative and Synergistic Impacts

Cumulative effects arise when two or more factors whose impacts, taken together, are more influential than any single component effect (e.g., Montopoli and Anderson 1991). If repeated in space or time, a single type of disturbance may also generate cumulative effects. Synergistic impacts occur when one or more factors exacerbate the effects of at least one other factor (see Myers 1987). Given the numerous types of recreational disturbances that can occur in wildlands, many wildlife communities may experience cumulative and synergistic effects. For example, activities in a national forest, such as hunting, fishing, snowmobiling, skiing, hiking, nature viewing, wildlife photography, horseback riding, mountain biking, camping, and off-road vehicles, may all be permitted in the same or adjacent areas.

At present, nothing is known about whether cumulative or synergistic effects at the community level result from recreational disturbances. Experiments or observational studies should be used to determine whether such impacts actually arise when two or more types of recreation occur at the same time, the same place, or both. Wildlife photography, hiking, and mountain biking, for instance, may collectively have cumulative or synergistic effects that are far more influential than any one of these disturbances alone. Standard experimental designs and statistical methods (e.g., analysis of variance, regression) are well-suited to answer these questions. Knowledge of such effects would be valuable in managing recreational activities so that recreationists and wildlife communities could coexist.

Recolonization, Recruitment, and Community Restoration

Recreational activities that disrupt relations among species, displace wildlife, or degrade habitats, may prevent or slow colonization of acceptable habitat. Annual recruitment of new individuals into populations may be affected similarly. Impacts on colonization and recruitment can influence community structure and stability through changes in species' abundances and their presence and absence. No one has determined whether recreational disturbance over several years actually affects colonization and recruitment. If it does, it would be valuable to know how much time without recreationists would be needed before colonization, recruitment, and community variables would return to normal. Because the presence or activities of individual animals often attract conspecifics to acceptable habitats (Smith and Peacock 1990), and because consistent recruitment can help stabilize community dynamics (cf. De-Sante 1990), it may be possible to restore wildlife communities by controlling recreational disturbances that thwart colonization and recruitment. Perhaps reductions in recreation intensity or slight changes in the seasonal timing of activities, rather than a complete ban of activities, would be sufficient.

No work has been done to determine how resilient wildlife communities are to recreational disturbances. A well-designed management experiment, in which managers temporarily banned recreational activities (or one of particular interest) on some sites (control sites) and permitted the usual level of activities on other sites (treated sites), would be very informative. With data for marked individuals of various species from control and treated sites during each of several years, one could determine whether recreation influenced colonization, recruitment, and community structure and dynamics. If recreation had an impact, a temporary ban of activities on all sites could be imposed. The manager could then determine how much time must elapse, without recreation, before community data from previously treated sites resembled data for the control sites.

Literature Cited

Anderson, D.W. and J.O. Keith. 1980. The human influence on seabird nesting success: conservation implications. *Biological Conservation* 18:65–80.

Blakesley, J.A. and K.P. Reese. 1988. Avian use of campground and noncampground sites in riparian zones. *Journal of Wildlife Management* 52:399–402.

Boyle, S.A. and F.B. Samson. 1985. Effects of nonconsumptive recreation on wildlife: a review. *Wildlife Society Bulletin* 13:110–116.

Clarke, M.F., K. Burke da Silva, H. Lair, R. Pocklington, D.L. Kramer, and R.L. McLaughlin. 1993. Site familiarity affects escape behaviour of the eastern chipmunk, *Tamias striatus*. *Oikos* 66:533–537.

DeSante, D.F. 1990. The role of recruitment in the dynamics of a Sierran sub-alpine bird community. *American Naturalist* 136:429–445.

Dorrance, M.J., P.J. Savage, and D.E. Huff. 1975. Effects of snowmobiles on white-tailed deer. *Journal of Wildlife Management* 39:563–569.

Gerrodette, T. and W.G. Gilmartin. 1990. Demographic consequences of changed pupping and hauling sites of the Hawaiian monk seal. *Conservation Biology* 4:423–430.

Götmark, F. 1992. The effects of investigator disturbance on nesting birds. In *Current Ornithology*, vol. 9, ed., D.M. Power, 63–104. New York: Plenum Press.

Götmark, F. and M. Åhlund. 1984. Do field observers attract nest predators and influence nesting success of common eiders? *Journal of Wildlife Management* 48:381–387.

Gutzwiller, K.J. 1991. Assessing recreational impacts on wildlife: the value and design of experiments. *Transactions of the North American Wildlife and Natural Resources Conference* 56:248–255.

Gutzwiller, K.J. 1993. Serial management experiments: an adaptive approach to reduce recreational impacts on wildlife. *Transactions of the North American Wildlife and Natural Resources Conference* 58:528–536.

Hammitt, W.E. and D.N. Cole. 1987. *Wildland Recreation: Ecology and Management*. New York: John Wiley and Sons.

Holecek, D.F. 1993. Trends in world-wide tourism. In *New Challenges in Recreational and Tourism Planning*, eds., H.N. van Lier and P.D. Taylor, 17–34. Amsterdam, The Netherlands: Elsevier Science Publishers B.V.

Keller, V.E. 1991. Effects of human disturbance on eider ducklings *Somateria mollissima* in an estuarine habitat in Scotland. *Biological Conservation* 58:213–228.

Knight, R.L., D.P. Anderson, and N.V. Marr. 1991. Responses of an avian scavenging guild to anglers. *Biological Conservation* 56:195–205.

Kury, C.R. and M. Gochfeld. 1975. Human interference and gull predation in cormorant colonies. *Biological Conservation* 8:23–34.

Lubchenco, J., A.M. Olson, L.B. Brubaker, S.R. Carpenter, M.M. Holland, S.P. Hubbell, S.A. Levin, J.A. MacMahon, P.A. Matson, J.M. Melillo, H.A. Mooney, C.H. Peterson, H.R. Pulliam, L.A. Real, P.J. Regal, and P.G. Risser. 1991. The sustainable biosphere initiative: an ecological research agenda. *Ecology* 72:371–412.

Lubchenco, J., P.G. Risser, A.C. Janetos, J.R. Gosz, B.D. Gold, and M.M. Holland. 1993. Priorities for an environmental science agenda in the Clinton-Gore Administration: recommendations for transition planning. *Bulletin of the Ecological Society of America* 74:4–8.

MacIvor, L.H., S.M. Melvin, and C.R. Griffin. 1990. Effects of researcher activity on piping plover nest predation. *Journal of Wildlife Management* 54:443–447.

Montopoli, G.J. and D.A. Anderson. 1991. A logistic model for the cumulative effects of human intervention on bald eagle habitat. *Journal of Wildlife Management* 55:290–293.

Myers, N. 1987. The extinction spasm impending: synergisms at work. *Conservation Biology* 1:14–21.

Odum, E.P. 1985. Trends expected in stressed ecosystems. *BioScience* 35:419–422.

Purdy, K.G., G.R. Goff, D.J. Decker, G.A. Pomerantz, and N.A. Connelly. 1987. *A Guide to Managing Human Activity on National Wildlife Refuges.* Fort Collins, Colorado: USDI Fish and Wildlife Service, Office of Information Transfer.

Rahel, F.J. 1990. The hierarchical nature of community persistence: a problem of scale. *American Naturalist* 136:328–344.

Rapport, D.J., H.A. Regier, and T.C. Hutchinson. 1985. Ecosystem behavior under stress. *American Naturalist* 125:617–640.

Skagen, S.K., R.L. Knight, and G.H. Orians. 1991. Human disturbance of an avian scavenging guild. *Ecological Applications* 1:215–225.

Smith, A.T. and M.M. Peacock. 1990. Conspecific attraction and the determination of metapopulation colonization rates. *Conservation Biology* 4:320–323.

Snyder, H.A. and N.F.R. Snyder. 1974. Increased mortality of Cooper's hawks accustomed to man. *Condor* 76:215–216.

Tuttle, M.D. 1979. Status, causes of decline, and management of endangered gray bats. *Journal of Wildlife Management* 43:1–17.

van der Zande, A.N. and P. Vos. 1984. Impact of a semi-experimental increase in recreation intensity on the densities of birds in groves and hedges on a lake shore in The Netherlands. *Biological Conservation* 30:237–259.

van der Zande, A.N., W.J. ter Keurs, and W.J. van der Weyden. 1980. The impact of roads on the densities of four bird species in an open field habitat— evidence of a long distance effect. *Biological Conservation* 18:299–321.

van der Zande, A.N., J.C. Berkhuizen, H.C. van Latesteijn, W.J. ter Keurs, and A.J. Poppelaars. 1984. Impact of outdoor recreation on the density of a number of breeding bird species in woods adjacent to urban residential areas. *Biological Conservation* 30:1–39.

Webb, R.H. and H.G. Wilshire, eds. 1983. *Environmental Effects of Off-Road Vehicles: Impacts and Management in Arid Regions*. New York: Springer-Verlag.

Wiens, J.A. 1989. Spatial scaling in ecology. *Functional Ecology* 3:385–397.

CHAPTER 11

Indirect Effects of Recreation on Wildlife

David N. Cole and Peter B. Landres

Most of this book focuses on direct impacts to wildlife that result from contact with people. The purpose of our chapter is to provide a broad overview of the indirect influences that recreation has on wildlife. Recreational activities can change the habitat of an animal. This, in turn, affects the behavior, survival, reproduction, and distribution of individuals. Although more difficult to isolate and study, these indirect impacts may be as serious and long-lasting as direct impacts for many species.

Recreational Influences

Virtually all types of recreation alter some characteristics of soil, vegetation, or aquatic systems. By directly impacting these components, people affect an animal's food supply and availability as well as shelter, or living space. In turn, impacts on food and living space influence behavior, survival, reproduction, and/or distribution. These relationships (Fig. 11.1) provide the structure with which our chapter describes research on (1) recreational impacts to soil, vegetation, and aquatic systems, (2) the effects of these habitat changes on animals, and (3) the effects of recreation on wildlife in situations where habitat change appears to be the primary mechanism of impact.

Recreational Impact on Animal Habitats

Though a wide variety of outdoor recreational pursuits impact soil, vegetation, and aquatic systems, we will not distinguish between the effects of different forms of recreation except in the few cases where impacts are caused only by a particular sport or related business. The significance and magnitude of any effect are related to the extensiveness, intensity, and timing of the activity. The vulnerability and rarity of the habitat, and its importance to wildlife, should also be considered.

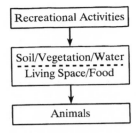

Figure 11.1   Simple conceptual model
of the indirect effects of recreation on
animals.

Numerous studies have documented the effects of recreation on vegetation
and soils. Most of these studies report results of human trampling caused by
hiking, camping, fishing, and nature study. Significantly fewer studies report
the consequences of horse and bicycle riding or that of off-road vehicles and
snowmobiles. Useful reviews of this literature include Liddle (1975), Manning
(1979), Cole (1987), Hammitt and Cole (1987), and Kuss et al. (1990).

Impacts on soil include loss of surface organic horizons (Burden and Randerson 1972), compaction of mineral soil (Iverson et al. 1981), reductions in
macro and total porosity (Monti and Mackintosh 1979), reductions in infil-
tration rates (James et al. 1979), and increases in soil erosion (Wilshire et al.
1978). Localized changes in soil chemistry have also been reported, but the
precise changes noted have been inconsistent (Chappell et al. 1971; Cole 1982;
Stohlgren and Parsons 1986). Other impacts include both reductions and increases in soil moisture (Settergren and Cole 1970; Blom 1976) and increases
in the diurnal and, perhaps, seasonal range of soil temperature (Liddle and
Moore 1974).

These changes in soil characteristics adversely affect the germination, establishment, growth, and reproduction of plants. Compaction reduces the
heterogeneity of soil surfaces and, therefore, the density of favorable germination sites (Harper et al. 1965). Compaction increases the mechanical resistance
of the soil to root penetration and can reduce the emergence of seedlings. Reduced macroporosity can result in oxygen shortages and less water being available to plants. These physical changes, along with reductions in organic matter
and changes in soil microbiota, can seriously disrupt ecosystem processes
(e.g., decomposition). They can impede soil-plant-animal interactions (e.g.,
nutrient cycling), causing decreased primary productivity.

The most obvious direct impacts on vegetation come from the crushing,
bruising, shearing, and uprooting of vegetation that often accompanies recreational use. Various changes in individual plant characteristics occur, including reductions in plant height, stem length, leaf area, flower and seed pro-

duction, and carbohydrate reserves (Liddle 1975; Hartley 1976). Plants are often killed outright. Those that survive typically are not as vigorous and reproduce less successfully. Consequently, recreation areas characteristically have vegetation that is less abundant (reduced density and cover), of a reduced stature, and with a different species composition from undisturbed areas (Cole 1982; Luckenbach and Bury 1983; Cole 1993).

Species composition and vegetation structure change because species and growth forms differ in their ability to tolerate recreational disturbance (Dale and Weaver 1974; Cole 1982). They vary in their ability to resist being damaged, in their ability to recover from damage, and in their ability to flourish in the soil and microclimatic conditions that occur on disturbed sites. Some of the characteristics that individually, or in combination, make plants tolerant of recreational disturbance include: (1) being either very small or very large; (2) growing flat along the ground or in dense tufts; (3) having leaves that are tough and/or flexible; (4) having growth points at or below the ground surface; (5) having a rapid growth rate; (6) producing numerous seeds; and (7) being an annual (Frenkel 1970; Kuss 1986; Cole 1993). Composition also changes due to propagules of exotic species carried into the area by recreationists. Such species are often well-adapted to periodic disturbance.

Vegetation cover can also be reduced and species composition altered by intensive grazing of meadows where camping trips are supported by horses and other pack animals (Cole 1987). This can reduce the availability of forage that is important to indigenous herbivores.

Generally, species richness and diversity declined where recreational impact was pronounced (Young 1978; Cole 1993); however, diversity may be greatest at low to moderate levels of disturbance (Liddle 1975; Slatter 1978). The complexity of vertical structure of vegetation may respond in a similar manner. The vulnerability of vegetation of moderate height means that heavily used recreation areas often lose intermediate vegetation layers, even if it is possible to maintain an overstory and some resistant groundcover. Tree saplings and pole-sized trees are also thinned out, both purposely by management and by recreationists (Bratton et al. 1982). Low levels of disturbance, however, may result in increases in the complexity of both vertical structure and the spatial pattern of vegetation, for example, by creating canopy openings and areas of nutrient enrichment (Dale and Weaver 1974).

The implications of recreational impacts on vegetation, then, vary with intensity of disturbance. With few exceptions, the abundance and total biomass of vegetation will decline as disturbance increases. However, habitat diversity can increase at low to moderate levels of disturbance due to increases in vertical stratification, spatial heterogeneity, and even species richness.

Aboveground dead vegetation, from standing dead trees to fallen logs and

brush piles, tends to decline in recreation areas. It can be pulverized under-
foot, collected as firewood, or cleared out either for aesthetic reasons or to reduce fire hazard. Loss of downed woody material can adversely affect water
and nutrient conservation on the site, as well as biological activity (Maser et al.
1988).

Snow-based recreation can also affect soils and vegetation. The most pronounced impacts are those associated with ski-resort development, which can
involve substantial tree felling and ground-surface leveling, in addition to facility construction. Such developments can fragment and reduce the availability of critical habitat. Of the more extensive activities, the impacts of snowmobiling appear most pronounced. Snowmobiling damages shrubs and
saplings (Neumann and Merriam 1972), reduces vegetative standing crop,
and changes species composition (Keddy et al. 1979).

Water is impacted by both water-based recreational activities, such as motorboating and canoeing, and by land-based activities, such as fishing, hiking,
and off-road vehicle travel. Trampling and other recreational impacts to the
shoreline can alter flow regimes and eliminate the protective cover afforded by
overhanging banks. It can result in increased sedimentation and turbidity
caused by runoff across denuded surfaces. Loss of vegetation can reduce organic matter input to water bodies. Streambeds and lake bottoms can be disturbed by vehicular travel and also by wading.

Motorboating and shore-based activities can alter the chemical qualities of
water. Outboard motors discharge oil and gasoline. These certainly contribute
to chemical changes, though the effects on aquatic organisms are, as yet,
poorly understood (Jackivicz and Kuzminski 1973). Shore-based activities
can contribute nutrient influxes, such as the exceptionally high level of phosphorus recorded at a semi-wilderness lake in Canada that had experienced a
20-fold increase in recreational use (Dickman and Dorais 1977).

Aquatic plants can also be altered by recreational use. It is sometimes a result of direct mechanical disturbance by boats, but more often due to nutrient
influxes. For example, in King's Canyon National Park, California, Taylor and
Erman (1979, 1980) found that heavily used lakes had less nitrate, more iron,
and more aquatic plants than lightly used lakes. They postulate that many
years of heavy camping contributed trace elements, such as iron, that had restricted plant growth in the past. Heightened plant growth would have increased nitrogen uptake, decreasing nitrate levels. They also found more
abundant insects, aquatic worms, and clams on the bottoms of the heavily
used lakes. Chemical changes, along with biological and physical changes
(such as increased turbidity), can have ripple effects throughout the entire
aquatic community. This may include increases or decreases in productivity
and nutrient cycling.

Effects of Habitat Change on Animals

Recreation impacts discussed in the preceding section influence important characteristics of animal habitats (Table 11.1), in turn affecting the quality and quantity of food and living space of animals (Fig. 11.1). The type of recreation activity, its location and areal extent, the severity or magnitude of impact, and its timing (interval, frequency, and predictability) all shape the "habitat template" (Southwood 1977) and an animal's subsequent response (Knight and Cole 1991). The following discussion focuses on selected examples that illustrate how habitat characteristics sensitive to recreational activities affect the behavior, distribution, survivorship, and reproductive ability of individual animals. Over a longer time period, impacts also occur to the population, community, and ecosystem; these are discussed at the end of this section.

FOOD

Habitat changes that alter the type, distribution, and food amount available will impact animals, whether in water or on land. For example, in soil, any activity that removes or reduces the overlying organic layers, or organic material

Table 11.1
Primary Recreational Impacts
on Animal Habitat

Soil characteristics
Loss of surface organic horizons
Reduced soil porosity
Altered soil chemistry
Altered soil moisture and temperature
Altered soil microbiota

Vegetation characteristics
Reduced plant density/cover
Altered species composition
Altered vertical structure
Altered spatial pattern
Altered individual plant characteristics

Aquatic system characteristics
Altered bank/shoreline characteristics
Altered bed/bottom characteristics
Altered flow regimes
Increased sedimentation/turbidity
Altered organic matter content
Altered water chemistry

within the soil, reduces a primary food source for many species. Seastedt (1984) found that about 95% of the 300,000 microarthropods living in one square meter of soil in northern coniferous and deciduous forests were mites and collembolans (organisms feeding exclusively on decaying organic matter). Recreation disturbances that remove this organic matter remove the food source for these primary decomposers in temperate climate ecosystems (Seastedt 1984).

Different plant species present different types of food (including leaves, flowers, fruit, and seeds) at different times of the year. They also exhibit different chemical and morphological defenses and attractants. Each of these features significantly impacts food availability. For example, Holmes and Robinson (1984) found that all 10 of the foliage-gleaning birds in a New Hampshire forest foraged preferentially on yellow birch trees. They avoided beech and sugar maple. Differences in tree preferences were partly attributable to differences in insect abundance among the three trees. Yellow birch had a higher density of all arthropods and lepidopteran larvae. Recreation that changes plant species composition will very likely change availability of food for animals.

Disturbances created by recreation also favor the germination, establishment, and growth of exotic annual plant species (Mack 1986), substantially altering food availability within a habitat. In Manitoba, Canada, for example, Wilson and Belcher (1989) found that exotic grasslands with a high proportion of exotic Eurasian grasses had a significantly different resident bird community than native grasslands. Differences in resident birds were attributable to differences in food supply and habitat structure.

For many streams, allochthonous organic input from riparian vegetation is a primary source of food and nutrients for the entire aquatic ecosystem (Gregory et al. 1991). For example, Fisher and Likens (1973) showed that 98% of the organic matter in Bear Brook, New Hampshire, came from the surrounding forest. The quantity of riparian vegetation entering a stream strongly affects the invertebrate community. For example, biomass of shredders feeding on vascular plant tissues typically reaches a maximum at the time of greatest litter availability (Cummins et al. 1989). Consequently, recreation that alters natural litter input to streams may substantially change the animal community.

In addition, recreational activities that reduce riparian vegetation will increase soil erosion; resulting sedimentation can cover and kill the aufwuchs community of periphyton, bacteria, and fungi on the surface of rocks (Cordone and Kelly 1961; Murphy et al. 1981). These are an important food source for arthropods, many amphibians, and fish. Sedimentation also increases the turbidity of water; this caused up to a 50% reduction in bluegill feeding rates

(Gardner 1981). Both the concentration of suspended sediments and the duration they are in the water will have a large impact on aquatic ecosystems (Newcombe and MacDonald 1991).

LIVING SPACE

Changes in habitat that alter living space, whether for breeding, feeding, or any other use, will have a large impact on wildlife communities. In soil, for example, many organisms live in the pore spaces between mineral particles, and require pores of sufficient size to survive and reproduce. Soil compaction reduces the size of pore spaces, altering the soil fauna. Wallwork (1970) found that collembolan species were restricted to certain soil types based on the size of available pores; smaller species were found in compacted soil in comparatively smaller pore spaces than the larger species. In addition, pore size can become so minimal that water completely fills the pore space, reducing oxygen levels sufficiently to inhibit aerobic metabolism. This causes a shift in composition toward anaerobic species (Paul and Clark 1989). Recreation that compacts soil is, therefore, likely to cause dramatic shifts in the distribution and species composition of the soil fauna.

Organic matter lying on the soil's surface provides a variety of microhabitats for vertebrate and invertebrate animals. In general, the greater the variety of habitats, the greater the variety of soil organisms. For example, Anderson (1978) found a strong, positive correlation between microhabitat diversity related to organic matter and mite species diversity. Recreational activities that remove organic layers from the soil will remove these microhabitats and their associated animal communities.

The physiognomy, density, and spatial pattern of plants strongly affect the living space of terrestrial animals. Numerous studies have shown the influence of vegetation physiognomy on birds (Holmes et al. 1986), small mammals (Harney and Dueser 1987), and insects (Lawton 1983). For example, plants with greater structural diversity offer insects a greater variety of microclimates, oviposition sites, predator refugia, overwintering sites, and food sources than plants of simpler structure (Strong et al. 1984). Well-known relationships also exist between the vertical distribution of vegetation, or "foliage height diversity," and the distribution of birds within a habitat. Stauffer and Best (1980) studied microhabitat selection of 41 birds nesting in riparian vegetation in Iowa. Based on regression of bird counts to microhabitat variables, Stauffer and Best predicted that removing woody vegetation would eliminate 32 of the bird species, while reducing woody vegetation to narrow strips along stream margins would eliminate six species and decrease the abundance of 16 others.

Plant density and other attributes of a habitat closely related to plant density, such as canopy cover, basal area of trees, shrub density, and fruit or cone density, can affect animal distribution. Lundquist and Mariani (1991) found that chestnut-backed chickadees and Vaux's swifts were "strongly (positively) correlated with density of total live trees" in the Cascade Range of southern Washington. Lundquist and Mariani attributed these correlations to a greater number of either insects or cones for the chickadees, and a greater abundance of nesting sites and insects for the swifts. For some species, plant density provides protection from predators; Dwernychuk and Boag (1972) found an inverse correlation between egg loss in artificial nests in relation to the quantity of plant cover.

Spatial pattern of plants is an important component of the living space of animals. Increasing the heterogeneity or patchiness of vegetation traditionally was considered "good" for animal habitat because it increased the proportion of "edge." Although increasing heterogeneity benefits some species, notably game species, several studies have recently shown the importance of contiguous, undisturbed habitat for many native species. For example, Robinson (1988) found that increasing habitat heterogeneity caused local reproductive failure of rare, endemic bird species. The causes were heightened competition for food between edge and endemic species, elevated brood parasitism, and higher nest predation. Recreational pursuits impacting the spatial pattern of plants may strongly affect animal assemblages.

Riparian vegetation greatly determines the living space of aquatic organisms. Riparian vegetation and overhanging banks provide protective cover for many fish and salamanders (Hawkins et al. 1983). By shading the water, streamside vegetation controls water temperature and solar radiation; this, in turn, directly governs primary and secondary productivity. For example, heavily shaded stream reaches have lower densities and relative abundances of herbivores than comparable stream reaches with open riparian canopies (Hawkins and Sedell 1981). In addition, woody debris that falls into streams provides important habitat for invertebrates and fish as well as retention of organic material (Harmon et al. 1986; Gregory et al. 1991). Speaker et al. (1988), for example, found that woody debris dams were the most efficient structures for retaining leaves, and that stream reaches in Oregon with debris dams retained four times more organic matter than reaches without debris dams.

Water flow regime (e.g., velocity, turbulence, and temporal pattern), strongly influences the behavior and metabolism of all major groups of aquatic organisms. Statnzer et al. (1988) found that mean water velocity and complex hydraulic characteristics were major factors in the distribution of microorganisms, macroinvertebrates, and fish within a stream reach. In addition, the size, surface area, texture, heterogeneity, and stability of substrate particles affect the living space of aquatic organisms (Minshall 1984) by providing, for example, attachment and oviposition sites. Malmqvist and Otto (1987) experimentally altered substrate stability, causing changes in macroinvertebrate composition and densities. Recreation activities that alter flow regime or substrate size will therefore have a large impact on the aquatic organisms within a stream.

Long-Term Effects of Habitat Change

When food or living space are altered, short-term effects on the behavior, survival, reproduction, and distribution of individual animals will likely cause long-term reactions throughout an animal community. These "trophic cascades" generally occur because habitat alterations either allow a variety of parasites, pathogens, and competitors into an area; or, they eliminate native species, often with long-term effects on the remaining species. For example, recreational activities that create clearings in forest provide favorable habitat for species such as the brown-headed cowbird that requires open foraging areas. Brown-headed cowbirds are nest parasites, laying their eggs in the nests of other birds. Several studies have shown decreases in the abundance of native birds caused by brown-headed cowbirds (e.g., Brittingham and Temple 1983).

Aquatic ecosystems may show time lags regarding the influence of recreational activities. For example, reducing streamside vegetation could cause increased erosion after substantial rainfall. Pollutants produced by recreational activities (e.g., gasoline and oil leaked by off-road vehicles) or sewage effluent may take considerable time to flow into groundwater or be flushed from the soil surface to streams or lakes. Aquatic systems also demonstrate that impacts to one species may ripple widely throughout the aquatic community (Power et al. 1988), even extending to terrestrial species. Spencer et al. (1991) showed how inadvertent introduction of opossum shrimp to Flathead Lake, Montana, caused the decline of zooplankton populations. This contributed to the collapse of kokanee salmon, an important food source for bald eagles and many other terrestrial vertebrates. Fall concentrations of feeding eagles dropped precipitously a few years after shrimp introduction.

If recreation affects the density or distribution of species that functionally dominate a community or ecosystem, resulting impacts will be especially severe. For example, Brown and Heske (1990) showed that experimental removal of kangaroo rats caused Chihuahuan desert shrub habitat to be converted into arid grassland. Similarly, beaver have been shown to dramatically alter the hydrology, biogeochemistry, and productivity of forest streams (Naiman et al. 1986). Consequently, activities that adversely impact species like the kangaroo rat and beaver can be expected to dramatically alter the present vegetation and animal community.

Indirect Impacts on Animals

In contrast to the sizeable literature of direct effects on wildlife, very few studies have documented impacts resulting from habitat changes induced by recreational activities. Accordingly, our review is based on limited empirical data.

Several studies have documented declines in numbers of a wide variety of soil invertebrates in trampled places (Chappell et al. 1971; Duffey 1975). One study found decreases in soil mites on footpaths, while springtails were most abundant in moderately trampled soils (Newton and Thomas 1979). Declines were caused, to some extent, by direct impact to soil invertebrates, but alteration of living spaces and food sources was important as well (Bayfield 1979).

Several studies have reported indirect effects of off-road vehicles (ORVs) on animal populations. In the Algodones Dunes, California, control plots had 1.8 times the number of lizard species, 3.5 times the number of individual lizards, and 5.9 times the lizard biomass of ORV-impacted areas. Controls also had 1.5 times the number of mammal species, 5.1 times the number of individual mammals, and 2.2 times the mammal biomass of ORV-impacted areas (Luckenbach and Bury 1983). In addition, control plots had 40 times the shrub volume of ORV-impacted areas, and cover and volume of perennial vegetation were both positively correlated with the number of individuals and number of species of lizards and mammals. This suggests that at least some of the decline in animal populations was indirectly the result of vegetation damage, although the precise causal mechanisms remain unknown.

Several studies compared animal populations in campgrounds to those in adjacent undisturbed areas to ascertain the effects of campground development on animals. Blakesley and Reese (1988) found seven bird species to be positively associated with campgrounds and another seven species to be associated with noncampgrounds. Changes in both food sources and living space provided likely explanations for differential species responses. Those species associated with the campgrounds nested in trees. Most of the noncampground associates, in contrast, nested on the ground, in shrubs, or in small trees, where disturbance was more severe. Ground foragers attracted to human food sources (such as the American robin) were associated with campgrounds, while ground foragers wary of humans (such as the fox sparrow) were associated with noncampgrounds. Mixed hardwood campgrounds in Wisconsin had a greater density of birds than adjacent noncampground forests (Guth 1978). They had slightly greater species richness, but less equitability. Moreover, a greater proportion of campground species were widespread species, in contrast to the number of rare forest species that were found only away from campgrounds.

Small mammal populations were significantly greater in campgrounds at

Canyonlands and Arches National Parks, Utah, than in adjacent controls (Clevenger and Workman 1977). The authors attribute this to increased food supplies due to camping use. In campgrounds at Yosemite National Park, California, "the animal populations studied display the full range of direct vs. secondary (indirect) and positive vs. negative visitor impacts" (Foin et al. 1977). Deer mouse populations responded positively to human use of campgrounds, while montane vole populations were unrelated to the distribution of human use. Some bird species, such as the junco, were negatively impacted both directly and indirectly. They nested and foraged on the ground and were influenced directly by human disturbance and indirectly through vegetation removal.

In sum, the indirect effects of recreation on animal populations are likely to be substantial, but there is little rigorous documentation of these impacts. Recreational activities clearly have substantial and generally adverse influences on terrestrial vegetation and soil, and on aquatic systems. Since these provide living space, shelter, and food for wildlife, animals are affected by these changes. For invertebrates, amphibians, reptiles, small birds, small mammals, and many fish, these indirect effects are likely to be more substantial than direct impacts of recreationists.

Management Options for Coexistence

Indirect impacts differ from direct impacts in two ways: (1) Indirect impacts are inevitable, occurring wherever and whenever recreational use occurs; and (2) they generally occur over long periods of time, with effects that are long-lasting and that may take place only after a time lag. Consequently, the timing of activities has less influence on indirect than on direct impacts.

These differences suggest that the appropriateness of various management strategies will vary given the nature of the activity. In particular, strategies that restrict the amount, type, and spatial distribution of use, as well as those that enhance site durability, seem well-suited for managing indirect impacts. Strategies that emphasize visitor education and temporal restrictions, while worthwhile in some situations, are less effective on indirect impacts than on direct ones.

Restrictions on Amount of Use
The severity of most recreational impacts on animal habitat is influenced by the amount of use that occurs. Since impact levels generally increase as use levels increase, indirect influences on wildlife could be limited by controlling the amount of recreation allowed. Numerous studies show, however, that the

relationship between amount of use and amount of impact is highly curvilinear; as use levels increase, additional use has less and less effect on amount
of impact (refer to Cole 1987 and Kuss et al. 1990 for reviews). This suggests
that limiting recreation is effective in reducing indirect impacts only when
usage can be virtually eliminated.

Restrictions on Type of Use

The nature and severity of recreational impacts are influenced by the type of
recreational activities involved. Limiting or prohibiting particularly destructive types of use is an approach with considerable promise. This approach is
most commonly used to manage motorized recreational activities, which are
generally much more disruptive than nonmotorized activities. Motorized use
is often prohibited in an area of concern (a campground or a nesting area); or,
it is restricted to particular trails or locations while nonmotorized use is
allowed anywhere. This latter strategy (zoning), where certain uses are allowed
only in certain places, is a common means of avoiding extensive impact caused
by particularly destructive pursuits. Other examples include creating nature
preserves that allow nonconsumptive uses while prohibiting consumptive
activities.

Restrictions on the Spatial Distribution of Use

Because impact is almost synonymous with use, impacts can be reduced by
limiting the spatial extent of use. This confinement strategy is one of the most
commonly employed techniques in recreation management. Visitors are required to stay on trails, keep out of meadows, and camp only in designated
campsites. Motor vehicles are required to stay on trails and in designated "sac-
rifice areas." Recreational use can also be concentrated more subtly by developing small high-use areas that provide for visitor needs, such as water and
picnic facilities (Usher et al. 1974). Other examples include providing a ramp
for launching boats or a raised boardwalk for viewing birds. The zoning
strategy described as a restriction on type of activity also involves restricting
use spatially.

The effectiveness of this approach can be amplified by confining use to sites
that are particularly durable and able to withstand repeated disturbance, and
by keeping use away from habitat that is rare or critical to animals. To illustrate, trails and campgrounds might be situated away from critical strips of riparian vegetation, while periodic opportunities for visitors to access the wa-
tercourses they find so attractive are maintained.

The success of attempts to employ spatial control as a management technique can be greatly increased through thoughtful site design. A good design
will meet the needs and aspirations of recreationists, while minimizing both

the extent and severity of impact. Activity zones can be interspersed with buffer zones and truly undisturbed zones (McEwen and Tocher 1976) to provide for the needs of a variety of animal species. Barriers, signs, and attention to the attractiveness and location of trails and facilities can all be used to minimize the proportion of an area that is frequently disturbed (Hammitt and Cole 1987).

Enhancing Site Durability

Maintenance and hardening of recreation sites can also reduce impact to animal habitats. Sites may need to be hardened, for example, to avoid increasing sediment inputs to streams. At off-road vehicle areas in California, debris basins have been built to trap sediment eroded from the area (Smith 1980). Active intervention may be necessary to maintain a diverse vertical structure and appropriate spatial pattern for vegetation in frequently disturbed places such as campgrounds (Blakesley and Reese 1988). In some cases, active rehabilitation must be engaged to restore the habitat requirements of animal species. This "hands-on" approach has costs as well as benefits, which must be carefully considered.

Knowledge Gaps

The preceding review should make it clear that animals are impacted by recreational disturbance of habitat. Two important questions remain, however: how significant are these impacts to wildlife populations and communities, and which habitat disturbances are most damaging to wildlife? Answers to these questions will require research designs radically different than the short-term correlational analyses that characterized previous research on indirect impacts.

Clearly, individual animals and localized populations can be impacted indirectly by recreation; but are these impacts significant when considered from the perspective of a larger spatial scale? Significant impacts might be those that are both extensive and severe, those that affect rare or important habitat, and those that affect rare, threatened, or keystone species. For example, impacts to small aquatic ecosystems might be more significant than those to large terrestrial ecosystems because more of the ecosystem is altered. Off-road vehicle impacts on dunes occupied by a rare herpetofauna might be more significant than camping impacts on forests occupied by a relatively common avifauna. Other examples of significant impacts include those that are long-lasting and those that impact populations and communities rather than just the behavior of individuals.

Research on these questions should begin by developing criteria for evaluating the significance of different impacts. It might proceed by identifying species, communities, and habitats that are particularly important or rare. Specific studies of the impacts of recreation on these species and habitats could then be conducted, including an assessment of the proportion of animals and habitat being affected. These studies should be long-term and attempt to identify changes at the population and community levels. Long-term changes at these higher levels of the biological hierarchy are particularly important because they incorporate a broad range of ecosystem interactions and processes. For such studies, research designs must be capable of providing the large spatial and long temporal perspectives that are needed.

Understanding the importance of various habitat disturbances will depend on a better understanding of (1) cause-and-effect relationships and (2)the importance of individual factors that influence the severity of impact. To unravel cause-and-effect, studies need to adopt experimental designs (Gutzwiller 1991), in contrast to the correlational analyses commonly used because of their relative ease. For specific disturbances, it is necessary to conduct experiments that separate effects on food sources from effects on living space.

An in-depth discussion of the factors that influence the severity of impact is beyond the scope of this review. A general treatment of the subject is presented in Chapter 5. Characteristics of the recreational disturbance (frequency, type of activity, timing, location, etc.), vulnerability of the habitat, and vulnerability of the animals all determine the impact of any recreational activity. Factorial research designs are capable of contributing to our knowledge about the importance of individual factors. For example, recreation areas with similar environments but different use levels can be compared to assess the relationship between amount of use and amount of impact. Alternatively, similar amounts of recreational use can be studied in different habitats to assess variation in the vulnerability of those habitats. Since it is often virtually impossible to control all these variables in the real world, the alternative is simulation of recreational disturbance under controlled experimental situations. Cole and Bayfield (1993) describe a simple experimental procedure for assessing the effects of both amount of trampling and vegetation vulnerability on vegetation impact. This approach could be extended to the secondary effects of vegetation change on animals.

The indirect impacts of recreation on wildlife are clearly substantial but even more poorly understood than the direct impacts. Reasons for this lack of understanding include the difficulty of unraveling cause-and-effect and the lack of interest in those animals most affected in indirect ways. Nevertheless, these less conspicuous species are important, and indirect impacts are extensive. So, we need to better understand recreational impacts resulting from habitat change and to find improved means of minimizing these impacts.

Literature Cited

Anderson, J.M. 1978. Inter- and intra-habitat relationships between woodland cryptostigmata species diversity and the diversity of soil and litter microhabitats. *Oecologia* 32:341–348.

Bayfield, N. 1979. Some effects of trampling on *Molophilus ater* (Meigen) (Diptera, Tipulidae). *Biological Conservation* 11:219–232.

Blakesley, J.A. and K.P. Reese. 1988. Avian use of campground and noncampground sites in riparian zones. *Journal of Wildlife Management* 52:399–402.

Blom, C.W.P.M. 1976. Effects of trampling and soil compaction on the occurrence of some *Plantago* species in coastal sand dunes. I. Soil compaction, soil moisture and seedling emergence. *Oecologia Plantarum* 11:225–241.

Bratton, S.P., L.L. Stromberg, and M.E. Harmon. 1982. Firewood-gathering impacts in backcountry campsites in Great Smoky Mountains National Park. *Environmental Management* 6:63–71.

Brittingham, M.C. and S.A. Temple. 1983. Have cowbirds caused forest songbirds to decline? *BioScience* 33:31–35.

Brown, J.H. and E.J. Heske. 1990. Control of a desert-grassland transition by a keystone rodent guild. *Science* 250:1705–1707.

Burden, R.F. and P.F. Randerson. 1972. Quantitative studies of the effects of human trampling on vegetation as an aid to the management of semi-natural areas. *Journal of Applied Ecology* 9:439–457.

Chappell, H.G., J.F. Ainsworth, R.A.D. Cameron, and M. Redfern. 1971. The effect of trampling on a chalk grassland ecosystem. *Journal of Applied Ecology* 8:869–882.

Clevenger, G.A. and G.W. Workman. 1977. The effects of campgrounds on small mammals in Canyonlands and Arches National Parks, Utah. *Transactions of the North American Wildlife and Natural Resources Conference* 42:473–484.

Cole, D.N. 1982. Wilderness campsite impacts: effect of amount of use. USDA Forest Service Research Paper INT–288, Ogden, Utah. Intermountain Research Station.

Cole, D.N. 1987. Research on soil and vegetation in wilderness: a state-of-knowledge review. In *Proceedings—National Wilderness Research Conference: Issues, State-of-Knowledge, Future Directions*, comp., R.C. Lucas, 135–177. USDA Forest Service General Technical Report INT–220, Ogden, Utah. Intermountain Research Station.

Cole, D.N. 1993. Trampling effects on mountain vegetation in Washington, Colorado, New Hampshire, and North Carolina. USDA Forest Service Research Paper INT-464, Ogden, Utah. Intermountain Research Station.

Cole, D.N. and N.G. Bayfield. 1993. Recreational trampling of vegetation: standard experimental procedures. *Biological Conservation* 63:209–215.

Cordone, A.J. and D.E. Kelly. 1961. The influence of inorganic sediment on the aquatic life of streams. *California Fish and Game* 47:189–228.

Cummins, K.W., M.A. Wilzbach, D.M. Gates, J.B. Perry, and W.B. Taliaferro. 1989. Shredders and riparian vegetation. *BioScience* 39:24–30.

Dale, D. and T. Weaver. 1974. Trampling effects on vegetation of the trail corridors of north Rocky Mountain forests. *Journal of Applied Ecology* 11:767–772.

Dickman, M. and M. Dorais. 1977. The impact of human trampling on phosphorus loading to a small lake in Gatineau Park, Quebec, Canada. *Journal of Environmental Management* 39:335–344.

Duffey, E. 1975. The effects of human trampling on the fauna of grassland litter. *Biological Conservation* 7:155–174.

Dwernychuk, L.W. and D.A. Boag. 1972. How vegetative cover protects duck nests from egg-eating birds. *Journal of Wildlife Management* 36:955–958.

Fisher, S.G. and G.E. Likens. 1973. Energy flow in Bear Brook, New Hampshire: an integrative approach to stream ecosystem metabolism. *Ecological Monographs* 43:421–439.

Foin, T.C., E.O. Garton, C.W. Bowen, J.M. Everingham, R.O. Schultz, and B. Holton, Jr. 1977. Quantitative studies of visitor impacts on environments of Yosemite National Park, California, and their implications for park management policy. *Journal of Environmental Management* 5:1–22.

Frenkel, R.E. 1970. Ruderal vegetation along some California roadsides. *University of California Publications in Geography* 20:1–163.

Gardner, M.B. 1981. Effects of turbidity on feeding rates and selectivity of bluegills. *Transactions of the American Fisheries Society* 110:446–450.

Gregory, S.V., F.J. Swanson, W.A. McKee, K.W. Cummins. 1991. An ecosystem perspective of riparian zones. *BioScience* 41:540–551.

Guth, R.W. 1978. Forest and campground bird communities of Peninsula State Park, Wisconsin. *Passenger Pigeon* 40:489–493.

Gutzwiller, K.J. 1991. Assessing recreational impacts on wildlife: the value and design of experiments. *Transactions of the North American Wildlife and Natural Resources Conference* 56:248–255.

Hammitt, W.E. and D.N. Cole. 1987. *Wildland Recreation: Ecology and Management.* New York: John Wiley and Sons.

Harmon, M.E., J.F. Franklin, F.J. Swanson, P. Sollins, S.V. Gregory, J.D. Lattin, N.H. Anderson, S.P. Cline, N.G. Aumen, J.R. Sedell, G.W. Lienkaemper, K. Cromack, and K.W. Cummins. 1986. Ecology of coarse woody debris in temperate ecosystems. *Advances in Ecological Research* 15:133–302.

Harney, B.A. and R.D. Dueser. 1987. Vertical stratification of activity of two *Peromyscus* species: an experimental analysis. *Ecology* 68:1084–1091.

Harper, J.L., J.T. Williams, and G.R. Sagar. 1965. The behavior of seeds in soil.

I. The heterogeneity of soil surfaces and its role in determining the establishment of plants from seed. *Journal of Ecology* 53:273–286.

Hartley, E.A. 1976. Man's effects on the stability of alpine and subalpine vegetation in Glacier National Park, Montana. Ph.D. dissertation. Durham, North Carolina: Duke University.

Hawkins, C.P. and J.R. Sedell. 1981. Longitudinal and seasonal changes in functional organization of macroinvertebrate communities in four Oregon streams. *Ecology* 62:387–397.

Hawkins, C.P., M.L. Murphy, N.H. Anderson, and M.A. Wilzbach. 1983. Density of fish and salamanders in relation to riparian canopy and physical habitat in streams of the northwestern United States. *Canadian Journal of Fisheries and Aquatic Science* 40:1173–1185.

Holmes, R.T. and S.K. Robinson. 1984. Tree species preferences of foraging insectivorous birds in a northern hardwoods forest. *Oecologia* 48:31–35.

Holmes, R.T., T.W. Sherry, and F.W. Sturges. 1986. Bird community dynamics in a temperate forest: long-term trends at Hubbard Brook. *Ecological Monographs* 56:201–220.

Iverson, R.M., B.S. Hinckley, R.M. Webb, and B. Hallet. 1981. Physical effects of vehicular disturbances on arid landscapes. *Science* 212:915–917.

Jackivicz, T.P. and L.N. Kuzminski. 1973. A review of outboard motor effects on the aquatic environment. *Journal of the Water Pollution Control Federation* 45:1759–1770.

James, T.D.W., D.W. Smith, E.E. Mackintosh, M.K. Hoffman, and P. Monti. 1979. Effects of camping recreation on soil, jack pine, and understory vegetation in a northwestern Ontario park. *Forest Science* 25:333–349.

Keddy, P.A., A.J. Spavold, and C.J. Keddy. 1979. Snowmobile impact on old field and marsh vegetation in Nova Scotia, Canada: an experimental study. *Environmental Management* 3:409–415.

Knight, R.L. and D.N. Cole. 1991. Effects of recreational activity on wildlife in wildlands. *Transactions of the North American Wildlife and Natural Resources Conference* 56:238–247.

Kuss, F.R. 1986. A review of major factors influencing plant responses to recreation impact. *Environmental Management* 10:637–650.

Kuss, F.R., A.R. Graefe, and J.J. Vaske. 1990. Visitor impact management: a review of research. Washington D.C.: National Parks and Conservation Association.

Lawton, J.H. 1983. Plant architecture and the diversity of phytophagous insects. *Annual Review of Entomology* 28:23–29.

Liddle, M.J. 1975. A selective review of the ecological effects of human trampling on natural ecosystems. *Biological Conservation* 7:17–36.

Liddle, M.J. and K.G. Moore. 1974. The microclimate of sand dune tracks: the relative contribution of vegetation removal and soil compression. *Journal of Applied Ecology* 11:1057–1068.

Luckenbach, R.A. and R.B. Bury. 1983. Effects of off-road vehicles on the biota of the Algodones Dunes, Imperial County, California. *Journal of Applied Ecology* 20:265–286.

Lundquist, R.W. and J.M. Mariani. 1991. Nesting habitat and abundance of snag-dependent birds in the southern Washington Cascade Range. In *Wildlife and Vegetation of Unmanaged Douglas-Fir Forests*, tech. coord., L.F. Ruggiero, K.B. Aubrey, A.B. Carey, and M.H. Huff, 221–240. USDA Forest Service General Technical Report GTR-PNW-285. Portland, Oregon.

Mack, R.N. 1986. Alien plant invasion into the Intermountain West: a case history. In *Ecology of Biological Invasions of North America and Hawaii*, eds., H.A. Mooney and J.A. Drake, 191–213. New York: Springer-Verlag.

Malmqvist, B. and C. Otto. 1987. The influence of substrate stability on the composition of stream benthos: an experimental study. *Oikos* 48:33–38.

Manning, R.E. 1979. Impacts of recreation on riparian soils and vegetation. *Water Resources Bulletin* 15:30–43.

Maser, C., R.F. Tarrant, J.M. Trappe, and J.F. Franklin, eds. 1988. From the forest to the sea: a story of fallen trees. USDA Forest Service General Technical Report PNW-229. Portland, Oregon: Pacific Northwest Research Station.

McEwen, D. and S.R. Tocher. 1976. Zone management: key to controlling recreational impact in developed campsites. *Journal of Forestry* 74:90–93.

Minshall, G.W. 1984. Aquatic insect-substrate relationships. In *The Ecology of Aquatic Insects*, eds., V.H. Resh and D.M. Rosenburg, 358–400. New York: Praeger Publishers.

Monti, P. and E.E. Mackintosh. 1979. Effects of camping on surface soil properties in the boreal forest region of northwestern Ontario, Canada. *Soil Science Society of America Journal* 43:1024–1029.

Murphy, M.L., C.P. Hawkins, and N.H. Anderson. 1981. Effects of canopy modification and accumulated sediment on stream communities. *Transactions of the American Fisheries Society* 110:469–478.

Naiman, R.J., J.M. Melillo, and J.M. Hobbie. 1986. Ecosystem alteration of boreal forest streams by beaver (*Castor canadensis*). *Ecology* 67:1254–1269.

Neumann, P.W. and H.G. Merriam. 1972. Ecological effects of snowmobiles. *Canadian Field-Naturalist* 86:207–212.

Newcombe, C.P. and D.D. MacDonald. 1991. Effects of suspended sediments on aquatic ecosystems. *North American Journal of Fisheries Management* 11:72–82.

Newton, J. and M.P. Thomas. 1979. The effects of trampling on the soil Acari and Collembola of a heathland. *International Journal of Environmental Studies* 13:219–223.

Paul, E.A. and F.E. Clark. 1989. *Soil Microbiology and Biochemistry.* San Diego, California: Academic Press.

Power, M.E., R.J. Stout, C.E. Cushing, P.P. Harper, F.R. Hauer, W.J. Matthews, P.B. Moyle, B. Statzner, and I.R. Wais de Badgen. 1988. Biotic and abiotic controls in river and stream communities. *Journal of the North American Benthological Society* 7:456–479.

Robinson, S.K. 1988. Reappraisal of the costs and benefits of habitat heterogeneity for nongame wildlife. *Transactions of the North American Wildlife and Natural Resources Conference* 53:145–155.

Seastedt, T.R. 1984. The role of microarthropods in decomposition and mineralization processes. *Annual Review of Entomology* 29:25–46.

Settergren, C.D. and D.M. Cole. 1970. Recreation effects on soil and vegetation in the Missouri Ozarks. *Journal of Forestry* 68:231–233.

Slatter, R.J. 1978. Ecological effects of trampling on sand dune vegetation. *Journal of Biological Education* 12:89–96.

Smith, T.C. 1980. ORV's and the California Department of Parks and Recreation resource management efforts: a summary. In *Off-Road Vehicle Use: A Management Challenge,* eds., R.N.L. Andrews and P.F. Nowak, 169–172. Washington D.C.: USDA Office of Environmental Quality.

Southwood, T.R.E. 1977. Habitat, the template for ecological strategies. *Journal of Animal Ecology* 46:337–365.

Speaker, R.W., K.J. Luchessa, J.F. Franklin, and S.V. Gregory. 1988. The use of plastic strips to measure leaf retention by riparian vegetation in a coastal Oregon stream. *American Midland Naturalist* 120:22–31.

Spencer, C.N., B.R. McClelland, and J.A. Stanford. 1991. Shrimp stocking, salmon collapse, and eagle displacement. *BioScience* 41:14–21.

Statnzer, B., J.A. Gore, and V.H. Resh. 1988. Hydraulic stream ecology: observed patterns and potential applications. *Journal of the North American Benthological Society* 7:307–360.

Stauffer, D.F. and L.B. Best. 1980. Habitat selection by birds of riparian communities: evaluating effects of habitat alterations. *Journal of Wildlife Management* 44:1–15.

Stohlgren, T.J. and D.J. Parsons. 1986. Vegetation and soil recovery in wilderness campsites closed to visitor use. *Environmental Management* 10:375–380.

Strong, D.R., J.H. Lawton, and T.R.E. Southwood. 1984. *Insects on Plants: Community Patterns and Mechanisms.* Cambridge, Massachusetts: Harvard University Press.

Taylor, T.P. and D.C. Erman. 1979. The response of benthic plants to past levels of human use in high mountain lakes in Kings Canyon National Park, California, USA. *Journal of Environmental Management* 9:271–278.

Taylor, T.P. and D.C. Erman. 1980. The littoral bottom fauna of high elevation lakes in Kings Canyon National Park. *California Fish & Game* 66:112–119.

Usher, M.B., M. Pitt, and G. DeBoer. 1974. Recreational pressures in the summer months on a nature reserve on the Yorkshire Coast, England. *Environmental Conservation* 1:43–49.

Wallwork, J.A. 1970. *Ecology of Soil Animals*. London, England: McGraw-Hill.

Wilshire, H.G., J.K. Nakata, S. Shipley, and K. Prestegaard. 1978. Impacts of vehicles on natural terrain at seven sites in the San Francisco Bay Area. *Environmental Geology* 2:295–319.

Wilson, S.D. and J.W. Belcher. 1989. Plant and bird communities of native prairie and introduced Eurasian vegetation in Manitoba, Canada. *Conservation Biology* 3:39–44.

Young, R.A. 1978. Camping intensity effects on vegetative ground cover in Illinois campgrounds. *Journal of Soil and Water Conservation* 33:36–39.

CHAPTER 12

Nature Tourism: Impacts and Management

Leslie HaySmith and John D. Hunt

Rural communities and wildlands throughout the world are hosts to a growing entourage of camera-toting nature tourists. In 1991, the World Travel and Tourism Council estimated that at the end of the 1980s, gross spending for tourism worldwide was $2.5 trillion. It generated 112 million jobs—one out of every 15 jobs in the world is in the tourism industry. In the United States in 1991, tourism was the nation's third-largest retail industry, second only to health services in employment.

Growth and economic impact of tourism have increased in many nations of the world. In 1989, Costa Rica hosted 376,000 visitors while in 1992, visitation had increased to 610,000. In Kenya, wildlife tourism accounts for almost $350 million in tourist receipts annually (Whelan 1991). In numerous locations worldwide, governments are embracing tourism development, welcoming foreign capital and the promises of more to come. Tourism can bring precious foreign exchange to developing countries with the accompanying "multiplier effect" wherein money spent on commodities or services multiplies into other regions and communities (Sherman and Dixon 1991). However, because nature tourism is a new phenomenon in many places, impacts are poorly documented and regulations regarding tourist behavior in wildlands are frequently nonexistent or unenforced. Tourism industry leaders and natural resource managers will face significant challenges in promoting sustainable development of tourism to help support protected areas and rural communities and in managing for reduced impacts on fauna and flora. Referring to conservation in the twenty-first century, Ugalde (1989) stated, ". . . nature tourism will flourish . . . and will impose challenging situations and opportunities for the national park services of the future."

In this discussion, we are defining nature tourism as domestic or foreign travel activities that are associated with viewing or enjoying natural ecosystems and wildlife for educational or recreational purposes. However, this definition is somewhat arbitrary and may not conform with other uses or meanings of the terms. For example, others may choose to define most any recreational or travel-related activity taking place in natural settings such as hunting, fishing, river running, or backpacking as nature tourism.

In recent years, a sort of taxonomy of tourism has begun to emerge. The terms used tend to classify tourism by market segments, activities, or place-settings such as nature tourism, ecotourism, ethnic tourism, cultural tourism, gaming tourism, heritage tourism, rural tourism, urban tourism, and more. Still others use terms that tend to describe the manner in which tourism is promoted, developed, or managed, such as responsible tourism or sustainable tourism. Needless to say, there appears to be little agreement on what these terms mean. This is not surprising in light of the fact that researchers and mar-keters have not been able to agree on a universal definition for tourism over the past four decades (Hunt and Layne 1990).

In recent years the term ecotourism has come into vogue to describe travel activities associated with nature and the outdoors. It has been described as a type of tourism where tourists and the industry are more sensitive to the environment, more socially responsible, and actively attempt to accentuate the positive and mitigate the negative impacts that may accrue from tourist use and development (Western 1993). Unfortunately, as the interest in travel to experience nature and the environment has grown, the definition of eco-tourism has become broader. We have not attempted to distinguish nature tourism from ecotourism. They are interrelated and may be used interchange-ably. On the other hand, some people draw distinctions between them. Re-gardless of the terms selected to describe any aspect of tourism, the material in our chapter addresses travel that is dependent on nature-oriented activities such as wildlife-watching and the use of nature trails. We recognize that any kind of social activity may impact the environment and its inhabitants. Like-wise, we adhere to the philosophy that such activities, be they tourism or any other activity, should be environmentally sensitive and socially responsible.

The distinction between tourism and recreation is highly flexible; many claim there is no difference. We feel that drawing a distinction between recre-ation and nature tourism is also very arbitrary. However, for the purposes of this book we will make the distinction with the understanding that it must be tentatively defined by the researcher. We are defining nature tourism as leisure or educational activities that normally involve *groups of people* participating in nature-oriented activities who are usually part of an *organized tour* (e.g., a tour group with a leader, guide, and itinerary). In contrast, recreationists may travel as individuals or small groups and are not part of an organized tour.

Another distinction between recreationists and nature tourists is that, with the exception of hunting and fishing, recreationists may not interact with wildlife intentionally, but rather accidentally. The pursuit of recreation is often outdoor activities that are not intended to view wildlife. On the other hand, nature tourism focuses on wildlife (or other aspects of nature) as the center of activities, with the intention of interacting with species or ecosystems (Edington and Edington 1986). Unfortunately, here again the distinctions become confused as most outdoor recreation activities take place in natural settings that are, more often than not, essential to the character of the activities.

Still another distinction between tourism and recreation may surround the type of activities in which people participate when they recreate or travel. That is, as people travel physically and culturally farther from the environment in which they live, their activities appear to change from generally active, extractive, or consumptive to more passive, appreciative, and nonconsumptive activities (Hunt 1968; McCool 1978). Thus, nature tourists may engage more frequently in activities that involve learning about or appreciating aspects of the natural environment. Recreation may require more personal physical contact with the environment. Here again, there are exceptions. For example, many people, who are clearly tourists, travel half-way around the world to culturally exotic and remote wild places to hunt, fish, trek, or engage in other activities that are physically strenuous or consumptive.

Finally, nature tourism may require various infrastructure, facilities, and services that are unique and are less important to many recreation activities. That is, the necessity to accommodate groups of people who are away from home for extended periods of time may require more hotels, food services, transport, and related infrastructure than are necessary for recreationists. The impact of this development may be greater than that of the development required to accommodate many recreation uses. However, the degree of impact or intrusion on the environment and wildlife may be variable, and the type of impact may be quite different from that expected. For example, a single recreationist hunting, camping, or trekking in an area may have decidedly more impact than a group of tourists housed and fed in an urban hotel and subsequently floated over the area in hot-air balloons. Relatively large numbers of nature tourists restricted to a bus, boat, or trail may result in minimal impacts. In fact, accommodations and transportation may be located in such a fashion that they provide for a large volume of tourists and result in less impact than that caused by indiscriminate or dispersed recreational use.

Obviously, the lines between nature tourism and recreation are blurred. In the final analysis, we are not able to draw clear and defensible borders to separate the two. However, we have attempted to adhere to the above described distinctions throughout the chapter, including the research summaries. Nonetheless, some authors do not clarify the exact type of visitor use in references cited, and despite our attempts to exclude nonapplicable cases, there may be some overlap.

Recreational Influences

Impacts of nature tourism on wildlife is a topic of growing interest to researchers and managers as increasing numbers of nature tourists visit wild areas. It has been a topic of much speculation but little quantitative research. However, the general type of impacts from tourism and recreation on wildlife are similar, with specific differences due to group size, visitor activities, and behavior, among other factors.

Impacts on wildlife from nature tourism are varied, and are often difficult to observe and interpret. Reactions of animals to visitors are complex. Initially, some species or individuals of a species retreat from visual or auditory stimuli caused by humans but become habituated over time. Other species or individuals that are more sensitive may alter their behavior and activities to completely avoid contact with visitors, with potentially long-term effects. Other animals cannot escape the disturbance and may be negatively affected, directly injured (e.g., manatees and boat traffic in Florida; see Chapter 18) or killed.

In some cases, nature tourism may be blatantly invasive toward wildlife, for example, where hundreds of observers congregate to view one rare animal. In other cases, negative interactions result from artificial feeding to draw animals for tourist observations, which can result in habituation and other problems.

Another type of biological impact from nature tourism is disturbance to relationships between species. Alterations in relationships may cause increased vulnerability to competitors and/or predators (see Chapter 10) or disturbance of relationships between members of the same species. The following sections will further investigate these biological and ecological aspects of impacts from nature tourism on wildlife.

Direct Impacts from Nature Tourism on Wildlife

Direct effects on individuals can occur through harassment. Harassment is a form of disturbance that can cause physiological effects, behavioral modifications, or death. Physiological responses have not been adequately studied, are complex, and are highly variable. For example, researchers have documented an increase in heart rate in different species when approached by visitors, which can subsequently initiate other physiological effects caused by stress, including death (see Chapter 7). Other effects from nature tourism can be divided into impacts on species, populations, or communities. They may take the form of impacts on behavior or on population parameters such as reproduction and survival.

Species-focused impacts often manifest themselves in changes in behavior. Nature tourist activity can cause displacement of individuals and redistribution of animal home ranges, or wildlife movement. During two annual seasons in Manu National Park in Peru, Groom (1990) documented flight behavior of several species of birds from river corridors of tourist travel within the park. Species of wildlife respond very differently to boat traffic. Of 59 nonpasserine bird species along the river, 19 species were considered sensitive because they fled from approaching boats, 18 species were considered moderately sensitive because they fled 35% to 65% of the time, and 22 species were insensitive because they rarely fled from boats. Bird species of the families Laridae, Cracidae, Psittacidae, and Alcenidae were most sensitive, and Caprimulgids, migrant shorebirds, and raptors were least sensitive.

Most mammal species in the Manu study appeared to be disturbed by boat traffic because they fled into the forest at the sound of a boat motor. Primates (eight species) were the only group that appeared to be affected very little by passing boats. Among reptiles, black caiman, spectacled caiman, and side-necked turtles usually fled only if boats reduced speed. Other disturbance to wildlife in Manu has occurred through independent guides, who have reportedly dug up turtle nests, and chased giant otters, swimming jaguars, and tapir in an effort to give their clients better viewing opportunities (Groom et al. 1991).

Some species appear to be undisturbed by tourist boat activity. In Royal Chitwan National Park in Nepal, researchers observed behavioral responses of the ruddy shelduck to groups of canoeists on the Rapiti River (Hulbert 1990). During the four-month study, the average length of time a bird was disturbed by tourists was 11 minutes, which represented only 2.6% of total daily activities. The study was initiated because managers felt that tour groups disturbed the feeding behavior of several wildlife species. However, ruddy shelduck were chosen for the study because they are the most common species on the river; therefore they may not be the most sensitive.

A study in Glacier National Park documented behavioral responses of mountain goats, a species popular among tourists. Researchers evaluated tourist impacts at two important locations, a mineral lick and a goat underpass (Pedevillano and Wright 1987). Goats at mineral licks did not appear to be disturbed by groups of visitors, despite the fact that the site received over 150,000 visitors during the two-year study. However, mountain goats attempting to cross goat underpasses appeared to be adversely affected by visitor presence on the highway. Goat use of underpasses was negatively correlated to the number of vehicles on the highway.

A behavioral study that evaluated a popular target species in Africa was conducted in Amboseli National Park by Western, Henry, and other researchers. Henry (1980) evaluated visitor viewing patterns, use, and capacity of Amboseli. Results indicated that cheetah behavior was affected by increasing levels of tourist use and harassment. Level of vehicle use accounted for 25% of the variance of cheetah activity patterns. Several daily routine and hunting activities occurred only when an average of less than one vehicle was present. When subjected to harassment, cheetah tried to avoid vehicles or waited until they left before engaging in hunting or other routine activities.

As previously mentioned, tourism may cause impacts between species; for example, hyenas may use minibuses as camouflage to steal prey from cheetah (Edington and Edington 1986), and lions may use minibuses as blinds to stalk prey (Henry pers. comm.).

Another type of impact on wildlife from nature tourism are impacts on behavioral mechanisms that are related to reproduction and survival. A study in Canada investigated tourism impacts on harp seal behavior during two whelping seasons (1986 and 1987) in the Gulf of St. Lawrence. Kovacs and Innes (1990) documented significant effects on all aspects of mother-pup behavior. Female attendance to pups declined when tourists were present, and those mothers remaining with pups spent significantly less time nursing and more time on alert. When tourists were present, pups were more active, rested less, and changed locations more frequently.

The impacts of tourism on juvenile survival can also occur due to disruption of the critical bonding period between mother and young. In East Africa, several observations have been made by visitors and park staff of tourist vehicles separating wildebeest and zebra mothers from their young, which can result in rejection and death of the young (Edington and Edington 1986). Another effect on reproductive survival from tourist activities occurs among sea turtle hatchlings. Lights from roads, facilities, or tourist flashlights can disorient the hatchlings, and cause them to migrate towards the lights and away from the ocean, often resulting in mass mortality (Edington and Edington 1986).

Other impacts on reproductive behavior have been observed with Thomson's gazelles. Females will readily leave the breeding territory when disturbed by tourist vehicles, which results in separation of sexes for longer periods of time (Edington and Edington 1986). In the Galapagos, strict territorial systems of male land iguanas were also observed to be broken down after tourists began feeding them (de Groot 1983).

Reproductive impacts on birds can result in decreased success or failure of nests, thereby affecting population growth. Groom (1990) documented a 10% loss in nesting effort among shore-nesting birds in Manu National Park, Peru, from nature-tourist disturbance. Black skimmers, large-billed terns and yellow-billed terns, pied lapwings, sand-colored nighthawks, and collared plovers nest in mixed-species groups on roughly 60% of the river beaches. The birds suffered significant egg or hatchling mortality from tourist presence, in addition to trampling of nests and handling of eggs or nestlings. Black skimmers and the two species of terns abandoned nests if tourists were present for a few hours. Once the most sensitive species abandoned nests, remaining nests were more vulnerable to predation. Enhanced vulnerability is due to the sand-colored nighthawk's dependence on the defensive behavior of the skimmer and terns, and the interdependence of the other five species on each other's defensive abilities. Therefore, the complexity of tourism disturbance may be more amplified with wildlife assemblages or communities due to inter-specific interactions.

Other studies investigating tourism impacts reported negative effects on seabird reproduction. At Punta Tombo in Patagonia, king shags and magel-lanic penguins experienced significant increases in egg loss to predatory gulls after visitors entered breeding colonies (Kury and Gochfield 1975). Similar observations were made on brown pelican colonies in Mexico by Anderson and Keith (1980), where the pelicans experienced greatly reduced breeding success following visitor disturbance.

A study at Ding Darling National Wildlife Refuge in Florida evaluated be-havioral impacts on the wading-bird community (Klein 1989). Results indi-cated that displacement of wading birds from foraging areas was correlated to changing visitation levels along a dike in the reserve, although bird species demonstrated differential responses. Migrants were more sensitive than resi-dent species, and wading birds were the least sensitive. Herons, egrets, brown pelicans, and anhingas were most likely to habituate to humans. Shorebirds showed intermediate sensitivity. Mottled ducks and several of the Ardeids showed "split-responses," with some individuals becoming habituated and others remaining sensitive to disturbance.

One of the few studies investigating tourist impacts on population parame-ters of reptiles was conducted by Cott (1969). In Murchison Falls National Park, Uganda, tourists in boats caused female crocodiles who guard nests on the river bank to retreat into the water. Their retreats were followed by attacks on the nestlings and eggs by olive baboons and monitor lizards (Edington and Edington 1986).

A study reporting impacts from tourism development on primate densities occurred in Cabo Blanco Nature Reserve, Costa Rica. This study reported a correlation between primate density declines and an increase in nature tourist density over a two-year period (Lippold 1990). Censuses before and after the reserve was opened to unregulated nature tourism show a 40% decline in howler monkeys and a 27% decline in white-throated capuchins.

Researchers at Yosemite National Park evaluated impacts from visitors on the bird and mammal community (Foin et al. 1977). They demonstrated a positive correlation between voles and field mice density and distance to trails. However, the latter species also showed a negative correlation between density and visitor trails in another location. The bird community also demonstrated variable responses. Generally, some species increased (e.g., Brewer's blackbird, mountain chickadee, and brown-headed cowbird) while other species declined (e.g., Oregon junco). The authors concluded that an understanding of individual species biology was necessary to interpret the findings, as well as data on long-term trends of visitor use and species responses. This study demonstrates the complications that arise from investigating and interpreting impacts of nature tourism on community parameters.

In summary, nature tourism can have impacts on various levels of biological organization including individuals, species, populations, and communities. The types of impacts may affect behavior, reproduction, and survival. As demonstrated, there is a high variability of wildlife responses to nature tourists, and differences occur between individuals and between species. Differences may be due to physiological, biological, or ecological factors in addition to differences in human behavior and activities. While some of these can be quantitatively defined, others have proven difficult to measure and predict.

Indirect Impacts on Wildlife Habitat

Nature tourism can impact an entire ecosystem in a wildland area. Impacts occur to aquatic and terrestrial systems including nutrient processes, inorganic compounds, soils, and vegetation. Impacts to soils include soil compaction and density, moisture, organic matter and litter cover, soil fauna and microflora, and nutrient availability. Direct and indirect impacts also occur to vegetation (see Chapter 11). These may include seed germination and seedling establishment, and plant growth processes (e.g., photosynthesis and carbohydrate metabolism).

Indirect effects can occur due to contamination of air, soil, water, and vegetation from pollutants. In addition to contamination, inadequate disposal of organic wastes from sewage in tourist areas can cause dense algal growth, which can affect aquatic ecosystems. In Oahu, Hawaii, sewage effluent caused blooms of algae, which in turn caused large-scale damage to the coral reef (Edington and Edington 1986). Habitat modifications may include deforestation of sites for buildings, facilities, and roads. Indirect effects may also occur from development of trail networks and picnic areas, which not only remove habitat, but increase habitat edge. The edge effect creates a different micro- and macro-biotic regime along habitat edges and opens these areas for colonization by exotic or colonizing species of plants and animals.

Solid-waste disposal not only contaminates an area, but can modify habitat and attract scavenging animals. Polar bears and grizzly bears have been noted to scavenge in refuse dumps, which alters their normal foraging behavior. Flocks of scavenging seagulls can cause danger to light aircraft. The refuse dumps usually attract these animals from their normal ranges, perhaps causing permanent displacement and disruption to normal habitat use. Refuse dumps create ideal habitats for disease vectors (e.g., rats, flies, mosquitos), potentially causing harm to humans in addition to wildlife. In the Great Barrier Reef, Australia, habitat modification due to refuse dumps has caused an increase in silver gull populations, which has resulted in increased predation by silver gulls on eggs and chicks of crested terns (Edington and Edington 1986).

Another type of indirect effect on wildlife from tourism activities occurs worldwide and is extremely complex due to social, economic, and ecological factors. In Nepal, Yonzon and Hunter (1991) reported a chain reaction effect on red pandas by tourism-driven markets. Langtang National Park provides habitat to red panda in addition to fodder and fuelwood for the local communities. Several farmers produce milk for cheese, which is consumed by tourists. The demand for cheese by tourists is high, which has created a substantial market for milk. In addition, the cheese production factory uses fuelwood taken from the park to process milk into cheese. Driven to supply the market demand for milk, local farmers overgraze fields and forest with chauri (yak-like cattle). Chauri cause habitat destruction from excessive grazing, which inhibits tree regeneration and trampling, which causes soil compaction. Over a two-year study, most of the red panda deaths by known causes (57%) were due to the presence of these large herds of chauri, in addition to dogs and herders. Only about 40 red pandas are left in four isolated populations in the park. Several other species of wildlife in the park have been displaced by livestock including serew, musk deer, barking deer, black bear, tahr, and pigs. This example illustrates that habitat destruction and associated declines in wildlife can be caused by complex socioeconomic problems originating from nature-tourism development.

Management Options for Coexistence

The following section briefly presents management options for coexistence of nature tourism and wildlife, which include alternatives for guides and industry, and wildlife and protected areas.

Tour Guides and Interpretation

Strategies to manage tourists are not different than those designed to manage other recreationists (see Chapter 20). However, tourists may visit areas in larger groups, be less familiar with the areas they visit, and concentrate their use in fewer locations. The fact that many nature tourism groups may be accompanied by a guide, transported in mass conveyances, and/or confined to specific locations provides the opportunity to enhance coexistence with wildlife through education and interpretation. At least information can be targeted and distributed efficiently. Wallace (1993) listed some lessons learned about visitor management in Galapagos National Park, one of the world's premier nature-tourism destinations.

The ability to inform and educate tourists through tour guides or interpretive media depends on both the medium and the message. In the case of tour guides, training and ability to communicate are critical. Government or industry guide regulations, training, and certification programs can help to enhance visitor experiences and reduce impacts on the environment. Enforcement of regulations and periodic recertification are important.

Signing and other traditional interpretive media may also help nature tourists avoid or reduce their potential negative impacts. Audio and visual media may be used on buses and other forms of group transportation to educate visitors.

Industry Operators and Suppliers

Managing the tourism industry to enhance coexistence with wildlife may entail government or self-imposed regulation, education, and coordination. Guidelines or restrictions on pollution, solid waste, campfires, firearms, vehicles, and other activities and equipment may be imposed on tour operators and their guests. In other cases, emphasis is being placed on the design of facilities to reduce environmental impacts and enhance visitors' satisfaction and awareness of the environment (Andersen 1993). The type, amount, and location of uses can be controlled or managed through lotteries, licensing, permits, and regulations. The "greening" of the industry through recycling, low-impact activities, and other environmentally and socially responsible actions can reduce certain costs associated with nature tourism. In fact, the growing concern for the environment among many tourists has increased the demand for products, services, and accommodations that are environmentally sensitive.

Some organizations are developing and promoting environmental "codes of ethics" and other guidelines for tourists, tour companies, and other suppliers in the tourism industry (Blangy and Wood 1993).

Areas of Ecological Sensitivity

Some ecosystem types are considered more sensitive to human impacts than others, often requiring long time periods to recover. For example, alpine areas are considered highly sensitive and show significant impacts after minor use. Other areas may be ecologically sensitive because of their overall importance to organisms in the food chain. Mangroves, for example, support numerous species of invertebrates and vertebrates and are an important nursery ground. Through adequate biological surveys, areas that are ecologically sensitive can be identified, protected, and managed to reduce use by tourists. In such cases, nature tourist activity can be concentrated in areas that are less ecologically sensitive and can sustain higher visitation levels.

Critical Foraging and Breeding Grounds

Similar to sensitive ecological zones are areas important to one or several species because they provide critical foraging or breeding habitat. These areas often have aggregations of wildlife species and attract large numbers of visitors. Impacts from tourists at these sites can result in decreased reproduction and recruitment, and increased mortality. For instance, in the previously discussed example in Manu National Park, Groom (1990) documented high nest failure of birds on beaches exposed to tourism. A simple management action, which prohibited tourists from camping on sensitive beaches and provided clear markings of nest colony locations, resulted in a 10% reduction in nest failures in 1986, followed by a 3% reduction the second year and a 1% reduction a year later.

In a similar case in Monteverde, Costa Rica, Powell and colleagues have investigated the reproductive behavior of the resplendent quetzal. Through radio-telemetry studies, they have documented home ranges, habitat use, and nesting areas of quetzals and relationships of these to nature tourist activities (Powell pers. comm.). By providing this information, they convinced the management committee at Monteverde to temporarily close trails where quetzals were building nests, because quetzals are less tolerant to tourists during this part of the breeding season. After females began incubating eggs (when their tolerance to visitor presence was higher), trails could be reopened. Such management actions are relatively simple to administer, but require critical information on tourism impacts on wildlife species before they can be implemented.

Activities of tourists in wildlife foraging areas should be carefully regulated, and some activities should be completely prohibited. The previously discussed research by Klein (1989) of foraging waterbirds at Ding Darling Reserve in Florida demonstrated that foraging can be negatively impacted in some species. One of the difficulties in wildlife research is demonstrating that a perceived change in wildlife behavior or populations is a result of nature tourism activities. Klein's study is particularly important because she used experimental manipulations and demonstrated causal relationships.

Munn (1992) described the unique opportunities for wildlife-based nature tourism in Manu Biosphere Reserve in Peru. Several species of macaws, in

flocks of hundreds, are observed daily at mineral licks in the reserve. The resource is apparently very important to macaw biology, as the same individuals may visit the clay mineral licks several times a day. While Munn did not
discuss potential impacts on the macaw or parrot populations, the case represents an important foraging/resource area to large aggregations of macaws,
and if not managed properly could result in negative impacts on the bird
populations.

Sensitive Species

The issue of species sensitivity is of concern to numerous researchers and biologists, but little quantitative information exists that evaluates species sensitivity for different nature tourist activities. Numerous issues confound our
ability to assess impacts such as habituation and differences in ecological,
physiological, nutritional, and reproductive status among species, which may
significantly affect wildlife responses to tourists.

Researchers and managers need to monitor and observe wildlife species for
behavior or population changes. Strategies should be standardized so that
wildlife can be monitored over long periods of time by different researchers.
Systematic and standardized monitoring and observation strategies would
also allow researchers to make comparisons between sensitive species in different geographical regions. The importance of conducting biological surveys
cannot be underestimated. Without monitoring, we have no barometer of
change to warn managers of the need for management action.

Zones and Standards

Appropriate planning and design of protected areas could significantly reduce
negative impacts. Plans should incorporate zoning for different types and degrees of use. Planning should involve land-classification systems such as ROS
(Recreation Opportunity Spectrum), a system used by the United States
Forest Service in National Forests (Driver and Brown 1978). Zones may include areas of strict preservation without tourist use (e.g., critical feeding or
breeding grounds), limited visitor use (e.g., seasonal, times of the day, or restriction on visitor numbers), and unlimited visitor use subject to regulations.
The Man and the Biosphere program initiated by UNESCO also promotes
zoning in the biosphere reserve network.

Several other progressive techniques are being used in the United States and
abroad to develop management strategies for nature tourism in wildlands. For
example, Geographical Information Systems (GIS) are being used to identify
ecologically sensitive areas and to plan tourism development. The Limits of
Acceptable Change (LAC) system (Stankey et al. 1985) is being used to monitor and set standards for acceptable levels of impact. LAC follows the carrying

capacity concept but sets no magical number on volume of visitors allowed. Instead, LAC sets quantifiable standards of impact levels beyond which management actions will be implemented. A fourth method is the Visitor Impact Management (VIM) process, which was developed by the U.S. National Park Service (Graefe et al. 1990). VIM calls for monitoring, and selection of indicators and standards of change (see Chapter 3). These techniques, however, are not mutually exclusive. They can be complementary in managing wildland systems to evaluate and implement strategies to decrease impacts from nature tourism.

Viewing Points and Tourist Mobility

Another solution is to develop fixed viewing points where tourists are confined to specific observation areas in carefully designed facilities (see Chapter 15). Some of these sites are supplemented with food or water to attract wildlife. This, of course, can result in problems, but if managed properly can avoid the more dispersed impacts of nonpoint viewing (Edington and Edington 1986).

The desire by tourists to view mobile wildlife has been solved in some areas by providing mobility to tourists (which must be subject to regulations). In Royal Chitwan National Park, Nepal, visitors view rhinoceros from the backs of elephants. In the east African examples previously discussed, tourists use minibuses, which unfortunately can present numerous problems to wildlife and vegetation. Strict regulations on minibus access appears to be the best management solution. Regulations on access for whale watching have also been implemented by the National Marine Fisheries Service (Edington and Edington 1986). However, regulatory measures require consistent monitoring and enforcement, which is infeasible in many areas due to economic constraints.

Knowledge Gaps

In general, information on the impact of tourism on the environment and people is limited. The literature does not yield a concise list of research needs in nature tourism. The knowledge gaps expand as questions become more focused.

Research needs and knowledge gaps can be placed into four categories: (1) impacts of tourists and tourism management on wildlife; (2) impacts on wildlife habitat; (3) tourist level of knowledge about wildlife; and (4) human values toward wildlife, perceived costs and benefits, and socioeconomic impacts of nature tourism.

In the first category, knowledge gaps exist regarding physiological effects, wildlife behavior, adaptability, intra- and inter-specific interactions, and effects on population parameters in relation to human activity, both in the short and long run. Other gaps in knowledge exist for how tourist infrastructure and superstructure, including support services and transportation corridors, affect wildlife populations and movements. There is a need for more information on both target species (e.g., those that are focal for tourists or the wildlife management entity) and nontarget species (e.g., small mammals, birds, many reptiles, and amphibians).

Knowledge gaps exist about the impact of nature tourism on wildlife habitat, including edge effects from trails, roads, and infrastructure development. In addition, we need to understand the impacts of facilities or management actions that were intended to be corrective measures toward previous environmental problems and nonsite impacts (e.g., downstream effects, water pollution). The question of differences in visitor impacts between high and low seasons, and how these factors relate to biological seasons, is also poorly understood.

There is a clear lack of knowledge about tourist and tourist operator understanding of the ecology of wildlife and wildlands, animal behavior, and dangers associated with some animals. This includes questions about tourist levels of knowledge in relation to different tourist types or segments, tourist experience, and motivations.

Finally, there is a need for a better understanding of the willingness of tourists to accept various management strategies and regulations. Likewise, greater insight is needed into how to quantify costs and benefits of tourism. Questions and topical areas include: Who benefits from nature or wildlife-based tourism? Where does the money go? How much is lost through leakage? What is the percentage that goes to international, national, and local economies? How can it be assured that the local economy and protected area receive economic benefits? What jobs are created and how can they supplement other economic activities? What is the tourist's willingness to pay? Issues of the social impacts and local attitudes toward tourism, and long-term benefits of tourism to both the community and the tourist, appear to be particularly important areas of inquiry in reducing local and regional impacts from tourism on wildlands and wildlife.

Summary

The trend of natural resource transformation and degradation has a new element known as "nature tourism" that may be viewed as a threat to wildlife and therefore, as a management challenge. The consequence of deforestation worldwide creates a fragmented landscape with differently sized "islands" of protected areas. The "islands" are under increasing pressure to produce economic benefits in the face of debt crisis, including demand by visitors for wild places and wildlife. The pressure to develop these habitat fragments may be slightly relieved by nature tourism, which can offer considerable economic benefit for protecting these areas. However, several case studies demonstrate that nature tourism may represent another threat to wildlife, wildlands, and local economies, if not managed for ecological and social factors. If properly planned and managed, however, it may serve as an important relief pressure valve from economic and other extractive pressures placed on wildlife and protected areas.

Literature Cited

Andersen, D.L. 1993. A window to the natural world: the design of ecotourism facilities. In *Ecotourism: A Guide for Planners and Managers,* eds., K. Lindberg and D.E. Hawkins, 116–133. North Bennington, Vermont: The Ecotourism Society.

Anderson, D.W. and J.O. Keith. 1980. The human influence on seabird nesting success: conservation implications. *Biological Conservation* 18:65–80.

Blangy, S. and M.E. Wood. 1993. Developing and implementing ecotourism guidelines for wildlands and neighboring communities. In *Ecotourism: A Guide for Planners and Managers,* eds., K. Lindberg and D.E. Hawkins, 32–54. North Bennington, Vermont: The Ecotourism Society.

Cott, H.B. 1969. Tourists and crocodiles in Uganda. *Oryx* 10:153–160.

de Groot, R.S. 1983. Tourism and conservation in the Galapagos. *Biological Conservation* 26:291–300.

Driver, B.L. and P.J. Brown. 1978. The opportunity spectrum concept and behavioral information in outdoor recreation supply inventories: a rationale. In *Integrated Inventories and Renewable Natural Resources: Proceedings of the Workshop,* eds., Lund, H.G. et al., 24–31. General Technical Report RM-55. Fort Collins, Colorado: U.S. Department of Agriculture, Forest Service, Rocky Mountain Forest and Range Experiment Station.

Edington, J.M. and M.A. Edington. 1986. *Ecology, Recreation and Tourism.* Cambridge, England: Cambridge University Press.

Foin, T.C., E.O. Garton, C.W. Bowen, J.M. Everingham, R.O. Schultz, and B. Holton, Jr. 1977. Quantitative studies of visitor impacts on environments of Yosemite National Park, California, and their implications for park management policy. *Journal of Environmental Management* 5:1–22.

Graefe, A.R., F.R. Kuss, and J.J. Vaske. 1990. *Visitor Impact Management.* Vols. I and II. Washington, D.C.: National Parks and Conservation Association.

Groom, M.J. 1990. Management of ecotourism in Manu National Park, Peru:

controlling negative effects on beach-nesting birds and other riverine animals. *Proceedings from the Second International Symposium: Ecotourism and
Resource Conservation,* ed., J. Kusler, 532–540. November 27–December 2,
1990. Miami Beach, Florida: Association of Wetland Managers.

Groom, M., R.D. Podolsky, and C.A. Munn. 1991. Tourism as a sustained use
of wildlife: a case study of Madre de Dios, Southeastern Peru. In *Neotropical
Wildlife Use and Conservation,* eds., J.G. Robinson and K.H. Redford,
393–412. Chicago: The University of Chicago Press.

Henry, W. 1980. Relationships between visitor use and capacity for Kenya's
Amboseli National Park. Ph.D. dissertation. Fort Collins, Colorado: Colorado State University.

Hulbert, I.A.R. 1990. The response of ruddy shelduck, *Tadorna ferruginea,* to
tourist activity in the Royal Chitwan National Park of Nepal. *Biological
Conservation* 52:113–123.

Hunt, J.D. 1968. Tourist vacations—planning and patterns. Utah Agricultural
Experiment Station Bulletin 474. Logan, Utah: Utah State University.

Hunt, J.D. and D. Layne. 1990. Evolution of travel and tourism terminology
and definitions. *Journal of Travel Research* 29:7–11.

Klein, M. 1989. Effects of high levels of human visitation on foraging waterbirds at J.N. "Ding" Darling National Wildlife Refuge, Sanibel, Florida.
Masters thesis. Gainesville, Florida: University of Florida.

Kovacs, K.M. and S. Innes. 1990. The impacts of tourism on harp seals (*Phoca
groenlandica*) in the Gulf of St. Lawrence, Canada. *Applied Animal Behavior
Science* 26:15–26.

Kury, C.R. and M. Gochfield. 1975. Human interference and gull predation in
cormorant colonies. *Biological Conservation* 8:23–24.

Lippold, L. 1990. Primate population decline at Cabo Blanco Absolute Nature
Reserve, Costa Rica. *Brenesia* 34:145–152.

McCool, S.F. 1978. Recreation activity packages at water-based resources.
Leisure Sciences 1:163–173.

Munn, C. 1992. Macaw biology and ecotourism or "when a bird in the bush is
worth two in the hand." In *New World Parrots in Crisis: Solutions from Conservation Biology,* eds., S.R. Beissinger and N.F.R. Snyder, 47–72. Washington, D.C.: Smithsonian Institution Press.

Pedevillano, C. and R.G. Wright. 1987. The influence of visitors on mountain
goat activities in Glacier National Park, Montana. *Biological Conservation*
39:1–11.

Sherman, P.B. and J.A. Dixon. 1991. The economics of nature tourism: determining if it pays. In *Nature Tourism: Managing for the Environment,* ed., T.
Whelan, 39–131. Washington, D.C.: Island Press.

Stankey, G.H., D.N. Cole, R.C. Lucas, M.E. Petersen, and S.S. Frissell. 1985.

The limits of acceptable change (LAC) system for wilderness planning. General Technical Report INT-176. Ogden, Utah: U.S. Department of Agriculture, Forest Service, Intermountain Forest and Range Experiment Station.

Ugalde, A. 1989. An optimal parks system. In *Conservation for the 21st Century*, eds., D. Western, and M. Pearl, 145–149. Oxford, England: Oxford University Press. 365 pp.

Wallace, G.N. 1993. Visitor management: lessons from Galapagos National Park. In *Ecotourism: A Guide for Planners and Managers*, eds., K. Lindberg and D.E. Hawkins, 55–81. North Bennington, Vermont: The Ecotourism Society.

Western, D. 1993. Defining ecotourism. In *Ecotourism: A Guide for Planners and Managers*, eds, K. Lindberg and D.E. Hawkins, 7–11. North Bennington, Vermont: The Ecotourism Society.

Whelan, T., ed. 1991. *Nature Tourism: Managing for the Environment*. Washington, D.C.: Island Press.

Yonzon, P.B. and M.L. Hunter, Jr. 1991. Cheese, tourists, and red pandas in the Nepal Himalayas. *Conservation Biology* 5:196–202.

PART III

Case Studies

CHAPTER 13

Recreation and Bald Eagles in the Pacific Northwest

Robert G. Anthony, Robert J. Steidl, and Kevin McGarigal

Bald eagle populations have been adversely impacted by human activities both directly and indirectly over the last century. The effects of direct impacts such as shooting, poisoning, and electrocutions are often obvious and comparatively easy to identify (Prouty et al. 1977). In contrast, the results of indirect impacts, such as recreational activities, may be more difficult to identify and quantify (Fraser et al. 1985; McGarigal et al. 1991). Recreational activities usually only disrupt an eagle's environment and behavior temporarily, and they are usually short-term disturbances (McGarigal et al. 1991). In the long term, however, repeated short-term disturbances may affect individual fitness through effects on survival and reproductive success. Hence, recreational activities can potentially have both short-term behavioral and long-term ecological impacts on bald eagles.

Populations of breeding (Isaacs et al. 1983; McAllister et al. 1986) and wintering (Stalmaster 1987:82–84; Keister et al. 1987) bald eagles are substantial in the Pacific Northwest, and their prominence has led to considerable focus on proper management of areas they inhabit. Because of the diversity of topography and the abundance of public lands, the Pacific Northwest also provides numerous opportunities for recreational activities including skiing, fishing, boating, hiking, hunting, white-water rafting, and swimming. Many of these recreational activities are focused on or around major water bodies where bald eagles nest and forage; therefore the potential for human-eagle interactions is high during all parts of the year in this region. The purpose of this chapter is to review previous studies and provide a framework for managing recreational activities in and around bald eagle habitats.

Recreational Influences

Short-Term Effects on Behavior

Although recreational activities have short-term effects on eagles in a variety of environments (e.g., Stalmaster and Newman 1978; Knight and Knight 1984; Fraser et al. 1985; McGarigal et al. 1991), specific recreational activities cannot be classified simply as disturbing or nondisturbing to eagles. The occurrence of disturbance depends not only on the presence of a nearby eagle, but on the context within which the human-eagle encounter takes place, the eagle's physical and behavioral state, the nature of the human activity, and the time and location of the encounter. For example, the eagle's current activity, hunger level, age, reproductive status, previous experiences, and perch height may influence its response to human activities (see Chapter 5). An eagle's response may vary with time of day, weather, tide cycle, and season, and will depend on the number of humans and recreational vehicles and whether the disturbance is moving or stationary, loud or quiet, and fast or slow. Lastly, the location of the human-eagle encounter and whether it is close to a nest or foraging area may influence the likelihood of disturbance.

This multitude of factors affecting the response of bald eagles to humans makes it difficult to generalize about the impacts of recreational activities on eagles. However, there are some distinctive forms of recreational disturbance and patterns in eagle response behavior that are consistent in their effects and are important to note. McGarigal et al. (1991) described two general forms of human-eagle interactions that apply to foraging areas and nest sites: active and passive displacement of eagles from their habitats.

ACTIVE DISPLACEMENT

In active displacement, the eagle-use area consists of a relatively narrow river corridor where eagles nest or forage and where boaters come into close contact with eagles. In this situation, humans actively approach or pass by eagles, and if humans come too close, eagles generally react by flushing. Active displacement is typical of wintering populations on rivers where eagles are quite mobile and rely on salmon carrion for food (Stalmaster and Newman 1978; Knight and Knight 1984), as well as where eagles nest along narrow river corridors. In this environment, the research approach has been to study eagle responses to approaching boats or pedestrians and quantify their reactions in terms of flush response rate and flush distance. Flush response is defined as the percentage of all human-eagle encounters within a specified distance that result in an eagle flushing. Flush distance is the distance between an eagle and human at which an eagle flushes.

Flush response rates vary, in part, because of the different human activities involved, differences in encounter distances, and the context within which human-eagle encounters occur. For example, McGarigal et al (1991) reported that only 5% of all breeding eagles flushed when boats approached within 500 m on the Columbia River estuary. In contrast, Skagen (1980) found that 43% of wintering eagles on the Skagit River in Washington flushed when boats, pedestrians, or land-based vehicles approached within 500 m; Stalmaster et al. (unpublished report) found similar flush rates of 26% to 51% for eagles perched in trees and 86% to 88% for eagles perched on the ground along the Skagit River. McGarigal et al. (1991) found that only 6% of all breeding and nonbreeding eagles on the Columbia River estuary flushed from passing boats.

This variability in flush response rates may reflect differences between wintering and breeding birds. Territorial, resident eagles may be more secure or have more at stake in their surroundings than nonresident, wintering eagles. Further, these populations may have been exposed to different frequencies and intensities of disturbance and consequently develop different tolerances to humans (see Chapter 6; Fraser 1981; Buehler et al. 1991). For example, wintering eagles were more tolerant of boating activities on the Skagit River where human activity was relatively high than they were on the Nooksack River, where human activity was relatively low (Knight and Knight 1984). Another explanation for the variability in flush response rate is that response rates for each population were based on different numbers of encounters at close distances. If most human-eagle encounters occurred at far distances, the overall response rate would be lower compared to a population where most encounters occurred at closer distances. A much higher proportion of the human-eagle encounters occurred beyond 200 m on the Columbia River compared to the Skagit River; thus the response rates by eagles on the Columbia River were comparatively low. Clearly, response rates are a meaningful indicator of how disturbing recreational activities are to eagles; however, they are only meaningful if they are evaluated relative to the frequency distribution of encounter distances.

Reported flush distances for bald eagles vary from 25 to 990 m (Vian 1971; Steenhof 1976; Nye and Suring 1978; Stalmaster and Newman 1978; Skagen 1980; Knight and Knight 1984; Fraser et al. 1985; McGarigal et al. 1991) and represent the range in responses by individual eagles. The range in mean population flush distances is less variable, and flush distances obtained under like conditions have been surprisingly similar. Mean flush distance was 197 m for breeding eagles responding to boating activities on the Columbia River estuary (McGarigal et al. 1991); 196 m for wintering adult eagles in response to pedestrians on the Nooksack River (Stalmaster and Newman 1978); 168 m and 150 m for wintering birds perched in trees when they responded to boating disturbances on the Skagit and Nooksack rivers, respectively (Knight and Knight 1984); 137 m for eagles responding to boating disturbances in North Carolina (Smith 1988); and 215 m for eagles of all ages and seasons responding to boats along Chesapeake Bay (Buehler et al. 1991). The overall similarity in these distances suggests that there may be a general tolerance threshold for foraging eagles. When using flush distances for management decisions, one should evaluate frequency distributions and manage human activities at distances at which only a small proportion, rather than a population mean, of eagles are expected to flush.

Several factors seem to consistently explain flush response rate and flush distance, while other factors have yielded conflicting results—possibly because we have little understanding of how learning influences an eagle's tolerance or sensitivity to human activities. For example, Stalmaster and Newman (1978) noted that adult eagles flushed at greater distances from humans than immatures, suggesting that learning had taken place. However, others have not detected these age-class differences in flush response rate or flush distance (Russell 1980; Knight and Knight 1984; Buehler et al. 1991; McGarigal et al. 1991). Incubating eagles flushed at greater distances when disturbed repeatedly (Fraser et al. 1985), whereas the flush distance of winter migrants did not change when disturbed repeatedly (Stalmaster and Newman 1978). These conflicting observations may result from habituation of eagles to humans in some circumstances, differences in research methodologies, or variations among the populations studied.

For other factors, eagle responses have been more consistent. Eagle perch height has a particularly strong and consistent influence on response rates and distances. Eagles that are perched on or close to the ground flush in response to approaching boats and pedestrians more often and at greater distances than do eagles perched higher in trees (Jonen 1973; Stalmaster 1976; Skagen 1980; McGarigal et al. 1991; Knight et al. 1991; Steidl 1994). This implies that human disturbance is most serious for eagles that depend on large fish or mammal carcasses as their major food source. Knight and Knight (1984) and McGarigal et al. (1991) recorded nearly identical flush distances even though the populations, environments, and disturbance contexts were different.

Other factors relating to the disturbance context also influence flush response. Eagles flushed more often than expected when boats approached slowly or were loud than when boats approached rapidly or were quiet (McGarigal et al. 1991). Slow-moving boats disrupted eagle feeding activity more than fast-moving boats (Stalmaster et al. unpublished report). Fraser (1981) observed a similar response to pedestrian disturbances of incubating and brooding eagles. Grubb and King (1991) found that eagles were flushed more often from perches than from their nests, and that pedestrians (hikers, anglers, hunters) made up the most disturbing group of 13 categories of human activity. McGarigal et al. (1991) noted that eagles were largely unaffected by fast-moving, land-based vehicles, but became increasingly agitated as vehicles slowed to a stop. Lastly, time of day also seems to influence flush response; eagles flushed more often in response to human activities before 1000 hours; therefore human activities during early morning were potentially more disturbing to foraging eagles (McGarigal et al. 1991).

PASSIVE DISPLACEMENT

In contrast to active displacement, passive displacement occurs when the eagle-use area, particularly a foraging area, consists of a large body of water such as a lake, estuary, or coastal area, and human-eagle encounters are not restricted to a narrow corridor. In this context, eagles hunt from shoreline trees, pilings, or tidal mud flats, and boats often remain stationary. Human activities influence the eagle's environment and cause eagles to change their foraging locations and behaviors. For many eagle populations, this form of disturbance is more prevalent and potentially more disturbing than active displacement, yet this form of human disturbance has not been widely investigated. We are aware of only two experimental investigations of passive displacement (McGarigal et al. 1991; Steidl 1994), although a number of studies have assessed passive displacement indirectly by comparing eagle activity in areas receiving varying levels of human activity. McGarigal et al. (1991) found that breeding adult eagles avoided an area with an average radius of 400 m (range 200–900 m per pair) from an experimental stationary boat in high-use foraging areas. In most cases, eagles spent less time and made fewer foraging attempts in the high-use foraging area during the experimental disturbance; this may have also resulted in reduced foraging efficiency. The results of McGarigal et al.'s (1991) experimental disturbances confirmed that boating activities have the potential to significantly change eagle spatial use patterns and demonstrated that these temporary changes were the direct result of the human activity. Steidl (1994) experimentally found that the presence of campers within 100 m of a nest tree reduced the amount of food delivered to young and the amount of time adults spent feeding young.

Observations on the foraging behavior of nonbreeding eagles are similar to those on breeding populations. Skagen (1980) noted that winter feeding activity of eagles was significantly reduced for periods of up to 30 minutes following human activity. Similarly, Stalmaster and Newman (1978) noted that humans temporarily displaced eagles from foraging areas and restricted the population to a smaller area; birds avoided the same feeding area for long periods following the disturbance. Walter and Garrett (1981) noted that wintering eagles temporarily avoided humans such as waterfowl hunters.

Similarly, Smith (1988) noted that breeding and migrating eagles shifted spatial use patterns in response to weekend influxes of people. These studies, encompassing summer and winter populations throughout much of the bald eagle's range, combined with the few experimental findings, confirm that human activities can elicit dramatic changes in spatial use patterns of bald eagles.

Long-Term Effects on Fitness

Most research on the impacts of human activities on bald eagles has focused on short-term effects. There are, however, potential long-term effects that must be considered if we are to maintain healthy populations of eagles. We believe that recreation can have deleterious effects on eagle populations through reductions in survival, especially during winter, or in reduced reproductive success, if the effects due to recreation are cumulative (Montopoli and Anderson 1991).

Bald eagle populations are influenced most by changes in rates of juvenile (Sherrod et al. 1976) and adult survival (Grier 1980); therefore any reduction in survival could cause a long-term reduction in the population. Juvenile eagles have higher energy demands, are less efficient foragers, and must spend considerably more time and energy acquiring food than adults (Craig et al. 1988). Consequently, juvenile eagles are more likely to be adversely impacted by human disturbances. Because eagles of all ages may be food-stressed in winter, Stalmaster and Gessaman (1984) predicted that high levels of human disturbance during this time could increase energy demands and result in increased mortality rates.

There are a number of complex and interacting factors (pesticides, food, weather, proximity of other pairs) that influence reproductive success of bald eagles; therefore use of productivity as a measure of human disturbance is likely to be inappropriate. Studies that have attempted to measure human impacts on productivity have reported mixed results, probably because the type and intensity of human activities are often measured subjectively and vary considerably among studies and populations. Several studies have attributed lower productivity and territorial abandonment to various types of human disturbances (e.g., Thelander 1973; Weeks 1974; Grubb 1981), although other studies have failed to show any effects of human activities (e.g., Mathisen 1968; Grier 1969; Fraser 1981). Therefore, measures other than productivity are likely to be more effective in assessing the influence of human activities on eagles.

Besides direct reductions in survival and productivity, long-term effects on fitness may be manifested in more subtle ways. If human activity displaces eagles from preferred habitats, then decreased foraging efficiencies in less suitable habitats may result in reduced reproductive success or survival. Some research upholds this possibility. Bald eagles nest farther than expected from centers of human activity (Fraser 1981; Andrew and Mosher 1982), change nest trees within territories to avoid human activities (Anthony and Isaacs 1989), and avoid areas of high human activity (Stalmaster and Newman 1978; Buehler et al. 1991). Bortolloti (1989) found that nestlings raised in an area of higher prey abundance had higher growth rates than those raised in an area of lower prey abundance. Functionally, a reduction in available prey could occur if eagles are not able to forage in productive areas because of human activity. In some species, nestling growth rates have been positively correlated with postfledgling survival and recruitment (Coulson and Porter 1985). Hence, any human activity that displaces eagles to an area of lower habitat quality could result in decreased survival or reproduction and have long-term impacts.

Management Options for Coexistence

As we have outlined above, the responses of bald eagles to human recreational activities are variable and depend on an eagle's physiological and behavioral state, the type of human disturbance, and the context within which eagles and humans interact. This creates an enormous challenge for the manager who must accommodate increasing numbers of recreationists and provide the necessary resources and security for bald eagle populations. The resources that are essential to bald eagles, and thus must be managed for, include nest sites, communal night roosts, foraging areas, and perch sites. All of these resources are fundamental to the survival and successful reproduction of the species and are essential for the recovery and delisting of bald eagles under the Pacific States Bald Eagle Recovery Plan (USFWS 1986). The methods a manager can employ include spatial and temporal restrictions on human activities and habitat management designed to conceal human activities from eagles (see Chapter 20).

Spatial Restrictions

Management of human activities to protect bald eagles with spatial restrictions has been practiced for about 15 years (see Mathisen et al. 1977), and this concept has been adopted or modified in more recent publications (USFWS 1981; Stalmaster et al. 1985; Stalmaster 1987). These spatial restrictions, called buffer zones, were a good first attempt at dealing with the effects of all kinds of human activities on bald eagles. Typically, they established a primary zone of 100 m and a secondary zone of 200 m around active nests, within which there were restrictions on various kinds of human activities. However, many biologists and managers have recognized that the response of bald eagles to recreation is extremely variable due to the many factors that influence human-eagle interactions (see above). Therefore, a more flexible strategy for spatial restrictions is needed.

DUAL DISTURBANCE MODEL

Because of the variability in eagle response to human recreation, McGarigal et al. (1991) developed the Dual Disturbance Threshold Model to explain how eagles respond to human activities. The model is useful for evaluating disturbances in populations subject to any level of human disturbance, and it is particularly helpful when temporary disturbances of foraging areas are of concern, or when considering flush distance as a measure of disturbance reaction (Hediger 1968:40–41). Under any specific set of conditions, an eagle has at least two distinguishable disturbance thresholds (i.e., dual disturbance thresholds). For any human-eagle encounter, a relationship exists between the human-to-eagle distance and the eagle's behavior. As a human approaches an eagle, an eagle's behavior changes at two points (Fig. 13.1): (1) the *agitation distance*, which is defined as the distance at which an eagle experiences physiological or psychological changes (e.g., increased heart rate, diverted attention); and (2) the *flush distance*, which is defined as the distance at which an eagle flushes. Agitation distance is always greater than or equal to flush distance and is generally much greater. The model provides a conceptual basis for establishing spatial and temporal restrictions around nesting, roosting, and foraging areas of eagles.

McGarigal et al. (1991) also suggested that at distances between agitation and flush distances, the human-to-eagle distance and the time required to elicit flushing are related (Fig. 13.1). If a human approaches an eagle and remains at some distance between the two thresholds, the eagle's disturbance level increases over time until eventually the eagle flushes. In effect, the eagle's flush distance gradually increases if the disturbance continues. Naturally, the closer the human is to the eagle's flush distance, the less time is required to elicit a flush response. Although the mathematical shape of this relationship usually is unknown, the curvilinear pattern depicted in Figure 13.1 illustrates the concept.

The model also applies to interactions involving stationary human activities that elicit passive displacement behavior. The existence of an agitation distance is based on the assumption that an eagle will not subject itself to disturbing conditions for prolonged periods, and it is recognized as the avoidance of a stationary human activity. The flush distance in this case exemplifies that an eagle will not approach a human within this distance. Although

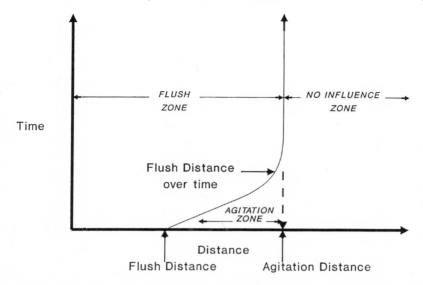

Figure 13.1   Relationship between the time required to elicit a flush response and the distance between human and bald eagle for a single human-eagle encounter (Dual Disturbance Threshold Model).

flush distance implies an actively approaching human, we use the term in the passive sense for consistency.

Each human-eagle encounter is unique and is influenced by a multitude of factors described as the disturbance context. Thus, the relationship for a single human-eagle encounter has limited management utility. However, if we determine the relationship for a number of encounters over a range of meaningful conditions (i.e., many human-eagle encounters), we can establish response curves pertaining to a specific breeding pair or population of interest. Because each encounter establishes a unique agitation and flush distance, we can produce frequency distributions for the disturbance thresholds (Fig. 13.2). These distributions can be used to determine disturbance thresholds for pairs or populations, which is useful information for managers. Further, this information can be used to develop site-specific territory management plans that are sensitive to the behavior and needs of the eagles occupying the territory. The dual disturbance model has the potential to fulfill these needs because it is more applicable to a variety of disturbances and different eagle populations. However, the model requires the collection of data on a variety of human disturbances and eagle populations in different environmental situations.

Application of the dual disturbance model would require the manager to have information from the published literature, reports, or data from the local area to establish the frequency distributions of flush and/or agitation distances

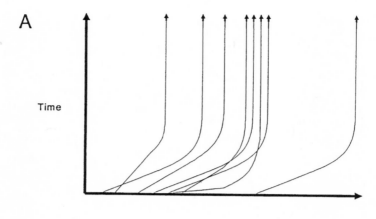

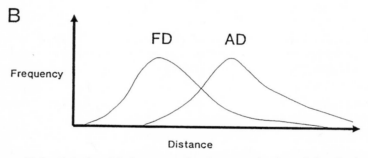

Figure 13.2   Relationship between the time required to elicit a flush response and the distance between human and bald eagle for several human-eagle encounters (A) and the frequency distributions of the disturbance thresholds (B). (FD is flush distance; AD is agitation distance.)

for a particular type of disturbance and specific population of eagles. These frequency distributions could be illustrated as in Figure 13.2B, and a decision made on what percentage (e.g., 50%, 75%, 95%) of the population to protect from human recreational activities. Because bald eagles are threatened and endangered, and we are striving to delist populations, we recommend the selection of the distance at which 95% of the population will not flush as a result of a given disturbance. This value is a percentile of the flush distances (population) and is not to be interpreted as a 95% confidence interval. This also would allow some extreme distances of flushing behavior in the data set to be excluded as "outliers." This distance would be used to establish spatial restrictions on human recreation during critical times of the year.

An important consideration in establishing spatial restrictions is the sample size upon which recommendations are made. Determination of an adequate sample size to estimate the distance at which 95% of the population will not flush can be computed by Stein's two-stage sample procedure (Steel and Torrie 1980:119). The procedure determines an adequate sample size of flush distances to estimate a mean with a confidence interval that is guaranteed to be no longer than a prescribed length.

Spatial restrictions depend also on the type of resource (i.e., nest sites, foraging areas, communal roosts, perch sites) that the manager is trying to protect from recreational activities. Specific considerations pertaining to these resources are addressed below.

NEST SITES AND ASSOCIATED FORAGING AREAS

Breeding bald eagles nest and forage in relatively small areas, and they tend to defend a portion of these areas against other members of the species, particularly nearest-neighbor breeding pairs (Garrett et al. 1993). Garrett et al. (1993) used the harmonic mean method (Dixon and Chapman 1980) to estimate home-range size and describe high-use areas for bald eagles on the Columbia River estuary. Active nests and high-use areas, including important foraging areas, were included usually within the 50% utilization distribution of a specific pair of eagles. The mean size of these areas during the breeding season was 1.3 km^2, and some were as large as 4.2 km^2. However, there was considerable variability in the size and location of high-use foraging areas in relation to active nests and size of the home range, so observations on each nesting area were required to develop management recommendations. The size of home ranges and high-use areas should be estimated and illustrated by one of the nonparametric methods for analyzing movement data (see Dixon and Chapman 1980; Anderson 1982; Whorton 1989), because breeding bald eagles do not use their home range uniformly (Garrett et al. 1993). Also, shoreline perches are important for foraging and resting, so they should be identified.

NIGHT COMMUNAL ROOSTS

Night communal roosts are prominent features in the ecology of bald eagles during the fall and winter months in the Pacific Northwest (Anthony et al. 1982; Keister et al. 1987; Isaacs and Anthony 1987; Stalmaster et al. 1985), and the same areas are used traditionally year after year. These roost areas have unique characteristics and should be managed to eliminate human disturbance within the roost and restrict human activities around the perimeter of the roost. Foraging areas used by bald eagles during the nonbreeding season are often some distance from communal roosts and necessitate different management strategies; therefore, management of foraging areas for wintering eagles will be discussed separately (see below). Because the boundaries of individual communal roosts are usually similar from year to year, fixed spatial restrictions are often acceptable and relatively simple to implement. However, the width of buffer zones around roosts depends on the amount of visual screening that is available from surrounding vegetation (Stalmaster 1987:168). Where human activity is screened from the eagle's view by vegetation, 100 m may be a sufficient buffer zone, but 250-300 m may be necessary in open country (Stalmaster 1987:169). In the Klamath Basin of southern Oregon, one of the largest communal roosts is used by 300 to 400 eagles on a given night, and this roost is protected by a 1,000 m buffer zone (USDI 1978). These buffer widths provide a range of values that may be applicable to other communal roosts.

Although buffer zones similar to those above are likely to be appropriate in most circumstances, there is a need for flexibility in determining buffer zones around individual roosts. Appropriate widths may depend on topography, surrounding vegetation, and responses of eagles to recreation. When such situations arise, we recommend the use of the dual disturbance model as described above and collection of site-specific data on eagle responses to human activities.

WINTER FORAGING AREAS

Food is a vital resource for survival of wintering bald eagles, yet there are very few protective measures for managing human activities around important foraging areas. This is partly due to the variety and ephemeral nature of food, particularly salmon runs and waterfowl concentrations, for wintering bald eagles in the Pacific Northwest. Therefore, the strategies for managing human activities must be flexible in space and time. Foraging habitat requires protection so eagles can feed without being disturbed by humans. This protection is critical because eagles that are foraging on the ground are most sensitive to humans (Stalmaster and Newman 1978; Knight and Knight 1984, 1986). Expansive, open areas are needed for eagles when they forage in the winter, and buffer zones around these areas should be at least 450 meters (Stalmaster 1987:171). We recommend the collection of data on eagle response to recreation and application of the dual disturbance model to establish spatial restrictions.

PERCH SITES

Bald eagles are generally a sit-and-wait predator, so hunting from a perch is one of their most common foraging methods. Bald eagles also capture most of their prey over water (Frenzel 1984; Watson et al. 1991), so protection and enhancement of perches close to water is important. Stalmaster (1987:170) recommended buffer zones of 75–100 m where vegetation screens eagles from human activities and zones of 250–300 m where no screening vegetation is present. As above, these strategies should be flexible, depending on the characteristics of the vegetation, type of human recreation, and response of eagles.

Temporal Restrictions

Temporal restrictions of human activity generally are used in conjunction with spatial restrictions to manage human activities near bald eagles. Restrictions on human activities need only be in effect during times when eagles are using essential resources. We recommend the use of temporal restrictions in combination with spatial restrictions to provide for undisturbed use of important resources by eagles during important times of the year and to lessen restrictions on human activities during other parts of the year. An exception to this strategy would be for breeding pairs that are resident on their territories the entire year, as described by Garrett et al. (1993) and Frenzel (1984). In these cases, spatial restrictions should be developed for the breeding and non-breeding seasons separately.

For the breeding season, Isaacs et al. (1983) recommend the use of buffer zones during the period of 1 February to 31 August to protect breeding pairs from human disturbance. Likewise, protection of wintering populations is most critical from 1 November to 31 March, when communal roosts and associated foraging areas are in use. However, these periods can be shortened if breeding attempts fail early in the breeding season or use of communal roosts or associated foraging areas occur during a more restricted part of the fall or winter. Temporal restrictions are most important in the early morning hours, because eagle foraging, during all times of the year, is usually most intense between dawn and 1000 hours for breeding (McGarigal et al. 1991; Watson et al. 1991) and wintering populations (Knight and Knight 1984; Stalmaster et al. unpubl. rep.).

Habitat Management

Several articles have been written on general habitat management for nest sites (Anthony et al. 1982; Anthony and Isaacs 1989; Stalmaster et al. 1985) and communal roosts (Anthony et al. 1982; Keister and Anthony 1983; Stalmaster et al. 1985) of bald eagles in the Pacific Northwest. Our purpose here is to address more specific strategies for habitat management to create visual screens for human activities. McGarigal et al. (1991) suggested that the variation in response of breeding pairs of bald eagles to stationary boats was related to differences in habitat characteristics (i.e., size of foraging areas, number of alternate perch trees) near their foraging areas. Stalmaster and Newman (1978) found that flush distances of eagles were less where vegetative cover concealed human activities. These articles indicate that high quality and quantity of vegetation can ameliorate the effect of recreation by (1) concealing human activities from line-of-sight contact with eagles, and (2) providing eagles with several alternate choices of foraging and resting perches when they are disturbed by humans.

Because foraging and resting perches are usually close to water, riparian habitat should be managed to provide visual screens to human activities from suitable perch trees. Older forests with large, open-branched trees and snags should by maintained for perching habitat within 100 m of water bodies used by bald eagles. Forests that have a high diversity of tree species and vertical structure are most preferred by bald eagles (Anthony and Isaacs 1989; Stalmaster et al. 1985). Conifer forests with a well developed understory are particularly effective in providing visual screens from human activities. Visual screens can be important in reducing the influence of human activities around nesting areas, communal roosts, foraging areas, and perch sites. Where there is a paucity of perches the construction and/or placement (i.e., pilings) of perches may enhance foraging opportunities for eagles.

Management Plans

Management plans have been used effectively to provide recommendations for management of habitat, eagle populations, and human activities of all kinds. Management plans for nest sites and associated foraging areas, communal roosts, winter foraging areas, and perch sites that address recreation should be developed. Where nest sites are numerous or wintering populations are large and widely distributed, plans should provide a regional perspective so there is a comprehensive plan to manage all nest sites, communal roosts, foraging areas, and perch sites that are important to the eagle population (Stalmaster 1987:165–167). Management plans require the cooperation of wildlife biologists, foresters, private landowners, and agency personnel to protect habitat and reduce the effects of recreation on eagle populations.

Information and Education

Humans often disturb wildlife unknowingly, so information and education programs may be just as effective as creating very restrictive spatial and temporal restrictions on human activities (see Chapter 3). Information and education should be an integral part of any management strategy concerning recreation around bald eagles and their habitat. Information and education can be provided in a variety of forms including newspaper articles, films, slide presentations, pamphlets, booklets, reports, or posters, which can alert humans to the potential effects of recreation on eagles. This aspect of eagle management has not been developed to the extent that it should be.

Knowledge Gaps

Studies that address specific situations or conflicts between humans and eagles will be most productive in increasing our knowledge of human-eagle interactions. Managers and researchers should identify jointly the various situations in which human-eagle interactions occur and design studies around these disturbance contexts. Experimental studies that simulate the relevant disturbance in the appropriate context will provide the most useful information for managers. Such studies will require replication and sufficient controls so that management decisions will be based on sound research.

Human-eagle interactions are complex and variable phenomena. Types of human disturbances and the responses of eagles vary in form as well as in effect. Indirect disturbances, although difficult to study and quantify, can have significant ecological consequences. Hence, we believe that most progress in understanding human-eagle interactions can be gained by experiments on eagle behavior that are designed to address specific management concerns under natural field conditions. For example, we know little about the effects of human activities on eagle survival and reproduction. We also need more information on the influence of food abundance and availability on the response of bald eagles to humans. Lastly, the importance of agitation behaviors on eagle fitness is poorly understood. Such information will be available only with radio transmitters that measure heart and respiration rates, which can be used in models of eagle energetics (see Stalmaster and Gessaman 1984). With the results from a collection of studies, we will be able to formulate better strategies to manage bald eagles and recreational activities.

Literature Cited

Anderson, J. 1982. The home range: a new nonparametric estimation technique. *Ecology* 63:103–112.

Andrew, J.M. and J.A. Mosher. 1982. Bald eagle nest site selection and nesting habitat in Maryland. *Journal of Wildlife Management* 46:382–390.

Anthony, R.G. and F.B. Isaacs. 1989. Characteristics of bald eagle nest sites in Oregon. *Journal of Wildlife Management* 53:148–159.

Anthony, R.G., R.L. Knight, G.T. Allen, B.R. McClelland, and J.I. Hodges. 1982. Habitat use by nesting and roosting bald eagles in the Pacific Northwest. *Transactions of the North American Wildlife and Natural Resources Conference* 47:332–342.

Bortolloti, G.R. 1989. Factors influencing the growth of bald eagles in North Central Saskatchewan. *Canadian Journal of Zoology* 67:606–611.

Buehler, D.A., T.J. Mersmann, J.D. Fraser, and J.K.D. Seegar. 1991. Non-breeding bald eagle communal and solitary roosting behavior and roost habitat on the northern Chesapeake Bay. *Journal of Wildlife Management* 55:273–281.

Coulson, J.C. and J.M. Porter. 1985. Reproductive success of the kittiwake *Rissa tridactyla:* the roles of clutch size, chick growth rates and parental quality. *Ibis* 127:450–466.

Craig, R.J., E.S. Mitchell, and J.E. Mitchell. 1988. Time and energy budgets of bald eagles wintering along the Connecticut River. *Journal of Field Ornithology* 59:22–32.

Dixon, K.R. and J.A. Chapman. 1980. Harmonic mean measure of animal activity areas. *Ecology* 61:1040–1044.

Fraser, J.D. 1981. The breeding biology and status of the bald eagle on the Chippewa National Forest. Ph.D. thesis. St. Paul, Minnesota: University of Minnesota.

Fraser, J.D., L.D. Frenzel, and J.E. Mathisen. 1985. The impact of human activities on breeding bald eagles in northcentral Minnesota. *Journal of Wildlife Management* 49:585–592.

Frenzel, R.W. 1984. Environmental contaminants and ecology of bald eagles in southcentral Oregon. Ph.D. thesis. Corvallis, Oregon: Oregon State University.

Garrett, M.G., J.W. Watson, and R.G. Anthony. 1993. Bald eagle home range and habitat use in the Columbia River estuary. *Journal of Wildlife Management* 57:19–27.

Grier, J.W. 1969. Bald eagle behavior and productivity responses to climbing nests. *Journal of Wildlife Management* 41:438–443.

Grier, J.W. 1980. Modeling approaches to bald eagle population dynamics. *Wildlife Society Bulletin* 8:316–322.

Grubb, T.G. 1981. An evaluation of bald eagle nesting in western Washington. In *Proceedings of the Washington Bald Eagle Symposium,* eds., R.L. Knight, G.T. Allen, M.V. Stalmaster, and C.W. Servheen, 87–104. Seattle, Washington: The Nature Conservancy.

Grubb, T.G. and R.M. King. 1991. Assessing human disturbance of breeding bald eagles with classification tree models. *Journal of Wildlife Management* 55:500–511.

Hediger, H. 1968. *The Psychology and Behavior of Animals in Zoos and Circuses.* New York: Dover Publishing, Inc. 166 pp.

Isaacs, F.B. and R.G. Anthony. 1987. Abundance, foraging, and roosting of bald eagles wintering in the Harney Basin, Oregon. *Northwest Science* 61:114–121.

Isaacs, F.B., R.G. Anthony, and R.J. Anderson. 1983. Distribution and productivity of nesting bald eagles in Oregon, 1978–1982. *Murrelet* 64:33–38.

Jonen, J.R. 1973. The winter ecology of the bald eagle in west-central Illinois. Masters thesis. Macomb, Illinois: Western Illinois University.

Keister, G.P., Jr. and R.G. Anthony. 1983. Characteristics of bald eagle communal roosts in the Klamath Basin, Oregon and California. *Journal of Wildlife Management* 47:1072–1079.

Keister, G.P., Jr., R.G. Anthony, and E.J. O'Neill. 1987. Use of communal roosts and foraging areas by bald eagles wintering in the Klamath Basin. *Journal of Wildlife Management* 51:415–420.

Knight, R.L. and S.K. Knight. 1984. Responses of wintering bald eagles to boating activity. *Journal of Wildlife Management* 48:999–1004.

Knight, R.L. and S.K. Skagen. 1987. Effects of recreational disturbance on birds of prey: a review. In *Proceedings of the Southwest Raptor Management Symposium and Workshop*, eds., R.L. Glinski et al. 355–359. Washington, D.C.: National Wildlife Federation. Sci. Tech. Ser. 11.

Knight, R.L., D.P. Anderson, and N.V. Marr. 1991. Responses of an avian scavenging guild to anglers. *Biological Conservation* 56:195–205.

Knight, S.K. and R.L. Knight. 1986. Vigilance patterns of bald eagles feeding in groups. *Auk* 103:263–272.

Mathisen, J.E. 1968. Effects of human disturbance on nesting bald eagles. *Journal of Wildlife Management* 32:1–6.

Mathisen, J.E., D.J. Sorenson, L.D. Frenzel, and T.C. Dunstan. 1977. A management strategy for bald eagles. *Transactions of the North American Wildlife and Natural Resources Conference* 42:86–92.

McAllister K.R., T.E. Owens, L. Leschner, and E. Cummins. 1986. Distribution and productivity of nesting bald eagles in Washington, 1981–1985. *Murrelet* 67:45–50.

McGarigal, K., R.G. Anthony, and F.B. Isaacs. 1991. Interactions of humans and bald eagles on the Columbia River estuary. *Wildlife Monograph* 115:1–47.

Montopoli, G.J. and D.A. Anderson. 1991. A logistic model for the cumulative effects of human intervention on bald eagle habitat. *Journal of Wildlife Management* 55:290–293.

Nye, P.E. and L.H. Suring. 1978. Observations concerning a wintering population of bald eagles on an area in southeastern New York. *New York Fish and Game Journal* 25:91–107.

Prouty, R.M., W.L. Reichel, L.N. Locke, A.A. Belisle, E. Cromartie, T.E. Kaiser, T.G. Lamont, B.M. Mulhern, and D.M. Swineford. 1977. Residues of organochlorine pesticides and polychlorinated biphenyls and autopsy data for bald eagles, 1973–74. *Pesticides Monitoring Journal* 11:134–147.

Russell, D. 1980. Occurrence and human disturbance sensitivity of wintering bald eagles on the Sauk and Suiattle rivers, Washington. In *Proceedings of the Washington Bald Eagle Symposium*, eds., R.L. Knight, G.T. Allen, M.V. Stalmaster, and C.W. Servheen, 165–174. Seattle, Washington: The Nature Conservancy.

Sherrod, S.K., C.M. White, and F.S.L. Williamson. 1976. Biology of the bald eagle on Amchitka Island, Alaska. *Living Bird* 15:143–182.

Skagen, S.K. 1980. Behavioral response of wintering bald eagles to human activity on the Skagit River, Washington. In *Proceedings of the Washington Bald Eagle Symposium,* eds., R.L. Knight, G.T. Allen, M.V. Stalmaster, and C.W. Servheen, 231–241. Seattle, Washington: The Nature Conservancy.

Smith, T.J. 1988. The effect of human activities on the distribution and abundance of the Jordan Lake–Falls Lake bald eagles. Masters thesis. Blacksburg, Virginia: Virginia Polytechnic Institute and State University.

Stalmaster, M.V. 1976. Winter ecology and effects of human activity on bald eagles in the Nooksack River Valley, Washington. Masters thesis. Bellingham, Washington: Western Washington State College.

Stalmaster, M.V. 1987. *The Bald Eagle.* New York: Universal Books. 227 pp.

Stalmaster, M.V. and J.A. Gessaman. 1984. Ecological energetics and foraging behavior of overwintering bald eagles. *Ecological Monographs* 54:407–428.

Stalmaster, M.V. and J.R. Newman. 1978. Behavioral responses of wintering bald eagles to human activity. *Journal of Wildlife Management* 42:506–513.

Stalmaster, M.V., J.L. Kaiser, and S.K. Skagen. Effects of recreational activity on wintering bald eagles. Unpublished report, 130 pp. + 78 illust.

Stalmaster, M.V., R.L. Knight, B.L. Holder, and R.J. Anderson. 1985. Bald eagles. In *Management of Wildlife and Fish Habitats in Forests of Western Oregon and Washington,* ed., E.R. Brown, 269–290. USDA Forest Service. Publ. No. R6-F&WL-192-1985, Portland, Oregon.

Steel, R.G.D. and J.H. Torrie. 1980. *Principles and Procedures of Statistics.* New York: McGraw-Hill Co.

Steenhof, K. 1976. The ecology of wintering bald eagles in southeastern South Dakota. Masters thesis. Columbia, Missouri: University of Missouri.

Steidl, R.J. 1994. Human impacts on the ecology of bald eagles in interior Alaska. Ph.D. thesis. Corvallis, Oregon: Oregon State University.

Thelander, C.G. 1973. Bald eagle reproduction in California. 1972–1973. California Department of Fish and Game, Wildlife Branch Administrative Report No. 73-5. 17 pp.

U.S. Department of Interior. 1978. Acquisition of the Bear Valley National Wildlife Refuge, Klamath County, Oregon. Environmental Impact Assessment. Portland, Oregon: U.S. Fish & Wildlife Service.

U.S. Fish & Wildlife Service. 1981. Bald eagle management guidelines, Oregon-Washington. Portland, Oregon: U.S. Fish & Wildlife Service pamphlet, 8 pp.

U.S. Fish & Wildlife Service. 1986. *Pacific Bald Eagle Recovery Plan.* Portland, Oregon: U.S. Fish & Wildlife Service, 160 p.

Vian, W.E. 1971. The wintering bald eagle (*Haliaeetus leucocephalus*) on the

Platte River in southcentral Nebraska. Masters thesis. Kearney, Nebraska: Kearney State College. 60 pp.

Walter, H. and K.L. Garrett. 1981. The effect of human activity on wintering bald eagles in the Big Bear Valley, California. Unpubl. report to the U.S. Forest Service. 89 pp.

Watson, J.W., M.G. Garrett, and R.G. Anthony. 1991. Foraging ecology of bald eagles in the Columbia River estuary. *Journal of Wildlife Management* 55:492–499.

Weeks, F.M. 1974. A survey of bald eagle nesting attempts in southern Ontario, 1969–1973. *Canadian Field-Naturalist* 88:415–419.

Whorton, B.J. 1989. Kernel methods for estimating the utilization distribution in home range studies. *Ecology* 70:164–168.

CHAPTER 14

Hunting and Waterfowl

Luc Bélanger and Jean Bédard

Over the last several decades, outdoor recreation on inland and coastal areas in North America has increased considerably. Depending on the motivation, recreational disturbances in these areas may originate from an intentional (hunting or birdwatching) or an unintentional action (fishing, boating, etc.) (Dahlgren and Korschgen 1992). These recreation-induced disturbances are not the only disturbances caused by humans; other activities (e.g., transportation) also affect wildlife. Consequently, a better understanding of the various impacts of outdoor activities on waterfowl distribution, behavior, and energetics is crucial for scientists and managers at both the population and community level (Drent et al. 1979; Frederick et al. 1987; Korschgen and Dahlgren 1992).

The effects of human disturbance on breeding and wintering waterfowl have been well studied in the past (Hume 1976; Batten 1977; Anderson and Keith 1980; Tuite et al. 1983; Bell and Austin 1985; Johnson et al. 1987; Henson and Grant 1991), yet the consequences of human activities to waterfowl on their staging areas has only recently received attention (Korschgen et al. 1985; Madsen 1985; Kahl 1991; Havera et al. 1992).

Waterfowl may be vulnerable to human activities in ways other than just habitat destruction, hunting, or various hazard mortalities. Disturbance can also (1) modify the distribution and use of various habitats by birds (Owens 1977; White-Robinson 1982; Madsen 1985); (2) affect their activity budget and reduce their foraging time and consequently their ability to store fat reserves necessary both for migration and breeding (Raveling 1979; Thomas 1983); and finally, (3) disrupt pair and family bonds and contribute to increased hunting pressure (Bartelt 1987). Therefore, human disturbance (one cause alone or many types acting synergistically) may reduce the overall carrying capacity of a given staging area for waterfowl and other waterbirds (e.g., Pfister et al. 1992). Disturbance is detrimental to birds if it reduces energy intake so much that it cannot be compensated by either (1) increasing rate of food intake during undisturbed periods; (2) avoiding disturbance by night-time feeding; and/or (3) using a temporary, adjacent undisturbed wetland habitat of similar ecological value.

With this in mind, we studied the impacts of human-induced disturbance on staging greater snow geese in Québec (Bélanger and Bédard 1989, 1990) from 1985 to 1987. Objectives of our research were (1) to identify the causes of disturbance to geese during their spring (no hunting season) and fall (hunting season) stopovers; (2) to investigate the effects of disturbance on goose activities and distribution, and finally; (3) to examine energetic consequences of the birds' responses to disturbance, particularly during the hunting season. Here we summarize the results we obtained during these studies.

Recreational Influences

Greater snow geese stop along the St. Lawrence estuary in Québec for five to seven weeks during spring and fall migrations between their Arctic breeding grounds and their wintering quarters on the Atlantic coast of the United States. Choice of a staging site by birds in this part of the river is influenced by the abundance of three-square bulrush (Giroux and Bédard 1988a), its location relative to neighboring sites, its area and accessibility to geese under high hunting pressure—particularly areas closed to hunting (Giroux and Bédard 1986, 1988), and tradition and site fidelity (Maisonneuve and Bédard 1992). It was also obvious to us even before we began our study that human disturbance had an influence on goose use of staging sites, particularly during the hunting season.

Our study was conducted at the Montmagny Bird Sanctuary, 70 kilometers east of Québec City along the south shore of the St. Lawrence River in Québec, Canada. Established in 1969 by the Canadian Wildlife Service, this no-hunting area is critical during fall when it is used by roughly 10% of the greater snow goose migrating through the estuary (Maisonneuve and Bédard 1992). The sanctuary is surrounded by agricultural and urban activities (see Bélanger and Bédard 1989 for a location map). There is a small airport 3 km east of the sanctuary, and aircraft (Cessna type) are used both for sightseeing tours of the area and for transporting residents to offshore islands. A marina is located on the eastern edge of the sanctuary and a ferry boat, linking Mont-magny and Isle-aux-Grues, use the eastern boundary of the sanctuary. During fall, hunting pressure is high as hunters used blinds or pits along firing lines in the lower marsh and in adjacent agricultural fields (Giroux and Bédard 1986).

We watched geese during daylight hours under all tide conditions in fall and only during low tides in spring. During spring high tides, geese moved to adjacent upland agricultural fields to feed. We had 287 hours of observation, averaging 5.4 hours/day in 1985 and 4.8 hours/day in 1986. We recorded only disturbances that caused a part or all of the observed flock to take flight. Natural flights of snow geese are easily distinguished from those caused by disturbance, because when disturbed, all birds take off simultaneously. Only the latter were considered. The cause of a disturbance was classified as human-related, natural, or unidentified. Human-related disturbance was then subdivided as hunting activities (gun shots, movements of hunters to or from their pits, and movements in the marsh to retrieve shot birds), nonhunting activities (passage of humans such as bird watchers or photographers in or near the sanctuary), agricultural or urban activities related to the presence or rapid passage of cars, trucks, or agricultural implements, and transport activities related to aircraft (e.g., planes, helicopters) overflights, passage of the ferry boat, and passage of small yachts or motorboats. For the purpose of our paper, we also distinguished between disturbances induced by recreational activities (hunting, nonhunting, marina) and others such as those related to transportation and surrounding agricultural, industrial, and residential activities. Because we could not distinguish whether aircraft-induced disturbance was recreation or nonrecreation in nature, these activities were assigned to the nonrecreation category. Finally, disturbances caused by predators were not included. Additional information on the study area, terminology, and sampling procedures can be found elsewhere (Bélanger and Bédard 1989; 1990).

Bird Responses to Cause and Rate of Disturbance

During the three years of our study, we recorded 652 disturbances in 471 hours of observation (Bélanger and Bédard 1989). Because we recorded only disturbances severe enough to cause geese to take flight, our data represent a minimum estimate of the effects of disturbance to staging geese. We found that mean disturbance rate was higher during fall (hunting) than during the spring (nonhunting) (Table 14.1). Human-related disturbances were more frequent than natural or unidentified disturbances in both spring (72% vs. 28%) and fall (81% vs. 19%). Transport activities were the most important cause of disturbance in both seasons (≥45%) and aircraft overflights ranked first among these (Fig. 14.1).

The impact of recreation activities differed greatly between staging seasons (6% and 41% of all cases of the disturbances in spring and fall, respectively). Hunting-related activities caused 27% of disturbances in fall, whereas nonhunting activities (mainly birdwatching) caused less than 5% in all seasons. Among hunting activities, gun shots accounted for 20% and hunter movements for 7% of all cases of disturbance.

The impact of human disturbance varies among waterbird species and depends on their wariness and their capability of habituating to certain disturbing events (Burger 1981; Klein 1993). Moreover, the disturbance behavior

Table 14.1

Synthesis of the Results of the Authors' Study[a] on the
Impact of Human-Induced Disturbance to Staging Greater
Snow Geese during the Hunting (Fall) and Nonhunting
(Spring) Seasons, St. Lawrence Estuary, Québec, 1985–1987

Characteristic of	Staging season	
the disturbance	Hunting	Nonhunting
Disturbance rate	1.46 ± 0.11	1.02 ± 0.09
% of disturbance that caused the entire flock to take flight	12.0	56.0
Mean time in flight per disturbance	56.0 ± 2.9	77.0 ± 10.4
Mean time to resume feeding after a disturbance (sec)	726.3 ± 139.1	122.4 ± 79.5
% of time disturbed flock used the same feeding site after a disturbance	83.3	74.7
% of disturbance after which geese stop feeding	40.4	44.6
% of disturbance that caused departure of birds	4.0	44.6

[a] See Bélanger and Bédard (1989, 1990).

of gregarious bird species on staging areas is largely determined by the individual behavior of the most nervous flock members, because the take-off of only a few birds may cause the entire flock to take flight. In support of this, we found that waterfowl were more likely to flush with increasing flock size in fall. We also observed that waterfowl were easily disturbed earlier during the hunting season than later. We believed that decreased hunting pressure at the end of the season and/or habituation of geese to gunfire (Owens 1977) may explain this pattern.

The amplitude of the response to disturbance by waterfowl varied depending upon the disturbance type and when it occurred. We observed that an entire goose flock took flight in 20% of disturbances, however this varied with season. During the hunting season, 49.8% of disturbances affected less

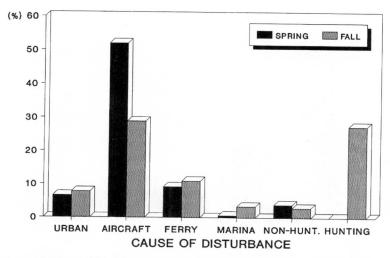

Figure 14.1   Relative importance of various causes of human-induced disturbance to staging greater snow geese, Montmagny bird sanctuary, Québec, 1985–1987.

than 20% of the flocks, and ≥70% of them affected ≤40% of the flock (Fig. 14.2). However, in spring, ≥50% disturbances caused the entire flock to take flight (Table 14.1). Disturbances affecting ≥60% of the flock were mainly caused by human activities related to transport, and those particularly in spring. Disturbances related to hunting and nonhunting outdoor recreation affected only a small proportion (≤40%) of the flocks.

Geese spent more time in flight/disturbance during the hunting than the nonhunting season, for a mean difference of about 20 seconds (Table 14.1). Following post-disturbance flight, geese continued to feed in 59.6% and 55.4% of cases in fall and spring, respectively. When they resumed feeding, they fed on the site used immediately before disturbance 83.3% and 74.7% of the time in fall and spring, respectively. When geese stopped feeding (about 40% of all cases), the mean time to resume feeding after disturbance was seven times higher in fall compared to spring. In both seasons, geese spent more time flying after transport-related disturbances than they did after any other type of disturbance. Time in flight was longer after aircraft passage during the spring than the fall (109.7 ± 19.4 and 65.6 ± 5.0 sec, respectively). Hunting disturbance caused geese to take flight for about 50 seconds in fall. Time to resume feeding was not influenced by the type of disturbance but was generally greater during the hunting than the nonhunting season (Table 14.1) for most disturbance types, mainly for aircraft-induced ones. Hunting and transport-related activities caused the greatest loss of feeding time in fall (≥15 min/disturbance) whereas nonhunting activities caused less than 550 sec.

The impacts of disturbance on waterfowl was also related to disturbance

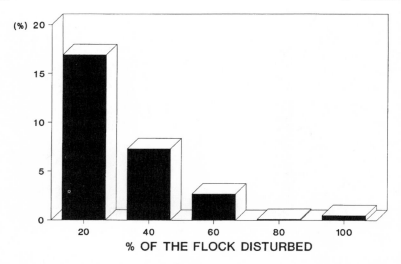

Figure 14.2   Extent of disturbance caused to greater snow geese (percent of the flock disturbed, time in flight, and time to resume feeding after the disturbance) by hunting-related activities within the Montmagny bird sanctuary, Québec, 1985–1987.

frequency. An extreme response of waterfowl to severe and frequent disturbance was to abandon a disturbed site (Thornburg 1973; Hume 1976; Tuite et al. 1983). Birds may learn to identify the danger associated with particular sites and try to avoid them at certain times (Owens 1977; Madsen 1985). Because of the high hunting pressure and the presence of firing lines around the sanctuary, only 4% of all cases of disturbance resulted in the departure of geese from the sanctuary during the hunting season (Table 14.1). In spring, however, geese were almost 10 times more likely to abandon the area.

We suspected geese might show a delayed response to high levels of disturbance during the hunting season, so we compared the rate of disturbance observed one day with the difference recorded in flock size the next day. During the hunting season, flock size significantly decreased with disturbance rates; at low rates of disturbance, no change was noted in flock size the next day but when disturbance rate reached or exceeded two per hour, the flock size decreased by about 50%, which roughly corresponds to a drop of approximately 4,000 birds. This decrease was not observed during the nonhunting season.

Effects of Disturbance on Bird Energetics

A comprehensive understanding of energetic requirements of waterfowl in relation to disturbance is essential for management and conservation, particularly in protected areas. If disturbance level is too high, a site could be unsuitable even if it constitutes a highly productive habitat. However, disturbance will be detrimental to birds if it decreases energy balance so drastically that it cannot be made up for by the compensatory mechanisms mentioned earlier. By using a simple empirical model (see Bélanger and Bédard 1990), we compared different disturbance regimes and responses by geese. In particular, we were interested in comparing the energetic cost of two major responses displayed by birds—*Response A:* fly away but promptly return to the foraging site and resume feeding, and *Response B:* fly away, leave the foraging site for a roost site, and interrupt feeding. We also modeled and compared two extreme consequences of Response B: (1) a net loss of foraging time as disturbed birds simply increase resting, preening time, or both (*passive reaction*); and (2) the loss of feeding time is partly integrated into the normal goose activity budget by reducing time normally allocated to resting (e.g., increase night feeding, increasing daytime ingestion rate [*compensatory reaction*], or a combination of both.

In our study area, the daylight foraging time of geese during the hunting season decreased depending on disturbance levels and bird responses (Bélanger and Bédard 1990). For instance, there would be a maximum reduction of 4% in feeding time under Response A, but under Response B, a similar reduction of 7.7% for a compensatory reaction and as much as 51% in the case of passive reaction by geese.

Daytime activity budgets of geese under different disturbance levels were then transformed into energetic values in terms of hourly metabolizable energy intake (HMEI) and hourly energy expenditure (HEE) (see Bélanger and Bédard 1990 for the exact procedures). We estimated that in Response A, each 0.5/hour increment in the disturbance rate reduced HMEI by about 1.2 kJ, for a 1.6% decrease. HEE increased with disturbance at an average rate of 2.7 kJ/hr per 0.5 disturbance/hr (Fig. 14.3). At the average disturbance rate recorded during the hunting season (1.46/hr), HEE rose by 5.3% as a result of the additional time spent in flight alone.

In Response B, we observed that variations in HMEI and HEE were more important. During passive reaction, a mean decrease of 15.0 kJ/hr (19.4%) was observed in HMEI for each 0.5 unit in disturbance rate, and there was very little variation in HEE. However, in case of a compensatory reaction by geese, a 2.3 kJ/hr decrease in HMEI (2.9%) was observed, and HEE increased by 2.5 kJ/hr (3.4%). We then calculated the energy balance (HMEI minus HEE) for each bird response and for different disturbance levels. In both responses A and B, an energy deficit of ≥ 7.5 kJ/hr (i.e., ¼ of the basal metabolic rate) was observed for rates ≥ 1/hr. Therefore, neither responses A nor Response B allowed geese to balance their daytime energy budget.

An increase in night feeding as a compensatory mechanism for loss of daytime feeding opportunities due to disturbance has been suggested (Thornburg 1973; Pedroli 1982; Tuite et al. 1983), but quantitative assessments have

RESPONSE A

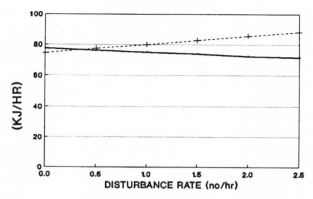

RESPONSE B
COMPENSATORY REACTION

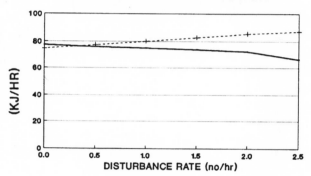

RESPONSE B
PASSIVE REACTION

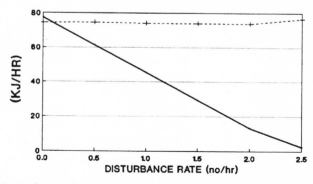

Figure 14.3 Estimated daytime, hourly, metabolizable energy intake (HMEI; solid line) and hourly energy expenditure (HEE; dashed line) of staging greater snow geese under different disturbance rates during the hunting season within the Montmagny bird sanctuary, Québec, 1985–1986.

not been performed. Nocturnal feeding in waterfowl is a common phenomenon often associated with heavy hunting pressure (Owen 1970; Owens 1977; Burton and Hudson 1978). At the average disturbance rate (1.46 disturbances/hr) in our study area during the hunting season, we found that a 4% increase in night feeding could compensate for energy losses caused solely by disturbance flights (Response A), compared to a 32% increase in nighttime feeding required to restore energy losses incurred in Response B. We should recall at this point that about 55% of the undisturbed nighttime budget of geese is already spent foraging (Gauthier et al. 1988). Therefore, at very high level of disturbance rates (>2/hr for instance), we estimated that an increase in night feeding as a behavioral compensatory mechanism could not counterbalance energy lost during the day.

An increasing rate of food intake during undisturbed daytime periods is another behavioral mechanism by which geese could compensate energy loss due to disturbance. We found no difference, however, in time allocated to different foraging activities of snow geese, particularly in time spent feeding among days with increasing disturbance levels (Fig. 14.4). Furthermore, feeding rate or feeding success (number ingestions/min) did not increase significantly with disturbance rates. Thus, geese could not compensate for a loss in feeding time by increasing their daily foraging behavior to maximize food intake during undisturbed periods.

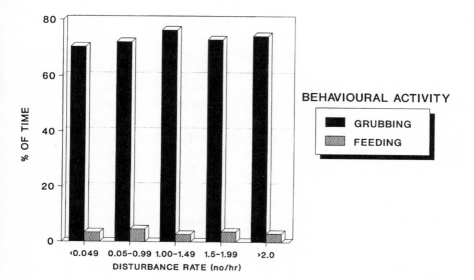

Figure 14.4   Activity budget of marsh-foraging greater snow geese under different disturbance levels within the Montmagny bird sanctuary, Québec, 1985–1986. The total number of observations is five.

Finally, the last possible compensatory mechanism is the use by geese of adjacent undisturbed feeding habitats (habitat shift). We reported earlier that during the hunting season flock size significantly decreased with disturbance rates within our study area. Therefore, when the mean daily disturbance level was too high, geese might balance their energy budget by foraging in another area. For this strategy to be effective, however, these so-called "emergency" foraging areas should (1) be near (<10 km for instance); (2) have good feeding conditions; and (3) be free of disturbance. Our sampling design did not allow us to determine (1) where disturbed geese went when they left the sanctuary, and (2) what were their behavioral activities and energy budget at these sites.

Management Options for Coexistence

About 80% of the entire population of greater snow geese stops in Québec during their fall migration (Maisonneuve and Bédard 1992). Bulrush marshes, their preferred staging habitats at this time, cover <4,000 ha and the number and total size of sanctuaries in the estuary is limited to five, totaling approximately 800 ha. During the hunting season, large numbers of snow geese are confined during the day to these sites because of hunting pressure elsewhere. Consequently, Giroux and Bédard (1987) and Bélanger and Bédard (in press) showed that grazing by greater snow geese greatly lowered plant production within these sanctuaries. A reduction in feeding time due to disturbances, combined with a depleted food resource within these sanctuaries, could severely affect the distribution and the length of the staging period of greater snow geese along the St. Lawrence estuary during the fall hunting season (see Maisonneuve and Bédard 1992). Moreover, high levels of disturbance may have deleterious energetic consequences for staging birds. We have shown that 2 disturbances/hr may cause an energy deficit that no behavioral compensatory mechanism, including night feeding, can counterbalance.

A primary goal for managers in improving and maintaining the carrying capacity of staging habitats of greater snow geese along the St. Lawrence River should be to reduce disturbance to a rate ≤1/hr. Since geese are slow to habituate to such disturbance, aircraft flights over sanctuaries should be strictly regulated. Although further study is needed, we suggest that flights below 500 m should be prohibited (sensu Owens 1977). Sanctuaries should also be large enough (e.g., >200 ha) so that geese could fly away once but promptly return to the foraging site and resume feeding (Response A rather than B). Very small contiguous sanctuaries could be connected by no-hunting corridors at least during days of high predictable disturbance rates such as during weekends. Finally, since hunting-related activities are responsible for about 30% of all disturbance flights during fall, new management plans should be put in place as recommended by Giroux and Bédard (1986). These new sanctuary or refuge designs should allow birds to be less vulnerable to gun shots and movements of hunters in the marsh by providing screened buffer zones and escape corridors. The deleterious effects of disturbance caused by outdoor recreational activities during the hunting season could be mitigated by the establishment of some kind of spatial or temporal buffer zone (see Chapter 20).

Knowledge Gaps

Few studies have addressed the impact of disturbance on staging waterfowl (see Dahlgren and Korschgen 1992 for a review). Most studies have shown moderate to pronounced impacts on birds, mostly in terms of distribution, site use, and interruption in feeding activities. Many of these studies, however, have been conducted at one study site and considered only the daylight period in their assessment of bird response to disturbance (short-term impacts on birds at local levels). Our evaluation is also very limited in space and time. Researchers need to compare different sites to get a better idea of the long-term impacts of disturbance on waterfowl at the population level and on a regional scale. Finally, response of birds to disturbance also likely depends on environmental conditions. As was shown in our study, the nature of disturbance and several biotic (flock size) and abiotic factors (date, time, tidal level) affected responses of flocking birds to disturbance. Weather conditions also are believed to influence sensitivity of birds to disturbance. Finally, it is possible that repetitive disturbance may have a cumulative impact on geese. Indeed, for various reasons, waterfowl could become more tolerant over a short period of time to particular disturbance types (for instance, the slow passage of a ferry boat in this study). An understanding of how external factors shape birds' responses to variations in disturbance intensity would enable managers to develop more effective management strategies to minimize disturbances to waterfowl.

The consequences of disturbance during the hunting season on energy budgets of ducks and geese needs to be better understood, as energetic constrains may be as serious as hunting mortality or emigration of birds. So far, the question of the impacts of disturbance on bird energetics has received little attention. Most studies have been conducted on captive or semi-captive birds and have concentrated on basal activities (Owen 1970; Burton et al. 1979). Responses of migratory birds to disturbances and adequate bioenergetic measurements are difficult to obtain. Studies using a telemetry approach with marked birds should be conducted. Researchers should also consider that individual birds show different responses, depending on their breeding status, physiological condition, and so on. More information is also needed on distribution, activities, and energetics of waterfowl at night to determine if night feeding may compensate for loss of daytime feeding opportunities due to disturbance.

Managers should be aware of the problems from recreation-induced disturbance and should develop land management plans that increase public appreciation of wildlands without affecting bird activities and energetic requirements. Indeed, different management options should be compared and experimentally tested. Except for some recent works (e.g., Klein 1993), most studies have used an observational rather than an experimental sampling approach (Gutzwiller 1991). New and improved research designs and methods are needed. This will greatly help to establish causal relations and more effectively measure the recreational impacts of disturbance on wildlife (Gutzwiller 1991).

Literature Cited

Anderson, D.W. and J.O. Keith. 1980. The human influence on seabird nesting success: conservation implications. *Biological Conservation* 18:65–80.

Bartelt, G.A. 1987. Effects of disturbance and hunting on the behavior of Canada goose family groups in eastcentral Wisconsin. *Journal of Wildlife Management* 51:517–522.

Batten, L.A. 1977. Sailing on reservoirs and its effects on water birds. *Biological Conservation* 11:49–58.

Bélanger, L. and J. Bédard. 1989. Responses of staging greater snow geese to disturbance. *Journal of Wildlife Management* 53:713–719.

Bélanger, L. and J. Bédard. 1990. Energetic cost of man-induced disturbance to staging snow geese (*Chen caerulescens atlantica*). *Journal of Wildlife Management* 54:36–41.

Bélanger, L. and J. Bédard. Role of ice scouring and goose grubbing in marsh plant dynamics. *Journal of Ecology*, in press.

Bell, D.V. and L.W. Austin. 1985. The game-fishing season and its effects on overwintering wildfowl. *Biological Conservation* 33:65–80.

Burger, J. 1981. The effect of human activity on birds at a coastal bay. *Biological Conservation* 21:231–241.

Burton, B.A. and R.J. Hudson. 1978. Activity budgets of lesser snow geese wintering on the Fraser estuary, British Columbia. *Wildfowl* 29:111–117.

Dahlgren, R.B. and C.E. Korschgen. 1992. Human disturbance of waterfowl: an annotated bibliography. U.S. Department of the Interior, Fish and Wildlife Service, Reserve Publication 188, 62 pp.

Drent, R., B. Ebbinge, and B. Weijand. 1979. Balancing the energy budgets of arctic-breeding geese throughout the annual cycle: a progress report. In *Proceedings of the Symposium on the Feeding Ecology of Waterfowl. Verhandlungen der Ornithologischen Gesellschaft in Bayern* 23:239–264.

Frederick, R.B., W.R. Clark, and E.E. Klaas. 1987. Behavior, energetics and management of refuging waterfowl: a simulation model. *Wildlife Monograph* 96:1–35.

Gauthier, G., Y. Bédard, and J. Bédard. 1988. Habitat use and activity budgets of greater snow geese in spring. *Journal of Wildlife Management* 52:191–201.

Giroux, J.F. and J. Bédard. 1986. Sex-specific hunting mortality of greater snow geese along firing lines in Québec. *Journal of Wildlife Management* 50:416–419.

Giroux, J.F. and J. Bédard. 1987. The effects of grazing by greater snow geese on the vegetation of tidal marshes in the St. Lawrence estuary. *Journal of Applied Ecology* 24:773–788.

Giroux, J.F. and J. Bédard. 1988. Use of bulrush marshes by greater snow geese during the spring and fall staging periods. *Journal of Wildlife Management* 52:415–420.

Gutzwiller, K.J. 1991. Assessing recreational impacts on wildlife: the value and design of experiments. In *Transactions of the North American Wildlife and Natural Resources Conference* 56:233–237.

Havera, S.P., L.R. Boens, M.M. Georgi, and R.T. Shealy. 1992. Human disturbance of waterfowl on Keokuk pool, Mississippi River. *Wildlife Society Bulletin* 20:290–298.

Henson, P. and T.A. Grant. 1991. The effects of human disturbance on trumpeter swan breeding behavior. *Wildlife Society Bulletin* 19:248–257.

Hume, R.A. 1976. Reactions of goldeneyes to boating. *British Birds* 69:178–179.

Johnson, S.R., D.R. Herter, and M.S.W. Bradstreet. 1987. Habitat use and reproductive success of Pacific eiders *Somateria mollissima v-nigra* during a period of industrial activity. *Biological Conservation* 41:77–89.

Kahl, R. 1991. Boating disturbance of canvasbacks during migration at Lake Poygan, Wisconsin. *Wildlife Society Bulletin* 19:242–248.

Klein, M.L. 1993. Waterbird behavioral responses to human disturbances. *Wildlife Society Bulletin* 21:31–39.

Korschgen, C.E. and R.B. Dahlgren. 1992. Human disturbances of waterfowl: causes, effects and management. In Waterfowl management handbook, Fish and Wildlife leaflet 13.2.15, 8 pp.

Korschgen, C.E., L.S. George, and W.L. Green. 1985. Disturbance of diving ducks by boaters on a migrational staging area. *Wildlife Society Bulletin* 13:290–296.

Madsen, J. 1985. Impact of disturbance on field utilization of pink-footed geese in West Jutland, Denmark. *Biological Conservation* 33:53–63.

Maisonneuve, C. and J. Bédard. 1992. Chronology of autumn migration by greater snow geese. *Journal of Wildlife Management* 56:55–62.

Owen, R.B. 1970. The bioenergetics of captive blue-winged teal under controlled and outdoor conditions. *Condor* 72:153–163.

Owens, N.W. 1977. Responses of wintering brent geese to human disturbance. *Wildfowl* 28:5–14.

Pedroli, J.C. 1982. Activity and time-budget of tufted ducks on Swiss lakes during winter. *Wildfowl* 33:105–112.

Pfister, C., B.A. Harrington, and M. Lavine. 1992. The impact of human disturbance on shorebirds at a migration staging area. *Biological Conservation* 60:115–126.

Raveling, D.G. 1979. The annual cycle of body composition of Canada geese with special reference to control of reproduction. *Auk* 96:234–252.

Thomas, V.G. 1983. Spring migration: the prelude to goose reproduction and a review of its implication. In *Fourth Western Hemisphere Waterfowl and Waterbird Symposium,* ed., H. Boyd, 73–81. Ottawa, Canada: Canadian Wildlife Service.

Thornburg, D.D. 1973. Diving duck movements on Keokuk pool, Mississippi River. *Journal of Wildlife Management* 37:382–389.

Tuite, C.H., M. Owen, and D. Paynter. 1983. Interaction between wildfowl and recreation at Llangorse Lake and Talybont Reservoir, South Wales. *Wildfowl* 34:48–63.

White-Robinson, R. 1982. Inland and saltmarsh feeding of wintering brent geese in Essex. *Wildfowl* 33:113–118.

CHAPTER 15

Balancing Wildlife Viewing with Wildlife Impacts: A Case Study

Richard A. Larson

Each year since 1980, approximately 16% of the U.S. population (16 years and older) took trips away from home for the primary purpose of observing, photographing, and feeding wildlife (USFWS 1992). This represents more than twice the number of people who hunt and underscores the role nonconsumptive wildlife use plays in American life.

Similarly, a strong demand for wildlife viewing opportunities was found among people in Denver, Colorado (Manfredo et al. 1992). Six out of ten Denver metro residents currently participate in wildlife viewing recreation on a regular basis, and nine in ten have some degree of interest in pursuing this activity. As Colorado prepares for the next century, more people are paying attention to impacts that a growing population will have on Colorado's natural resources. By 2020, Colorado's population is expected to increase by approximately 1.3 million people, bringing the state's population to 4.4 million. Most people will settle along the Front Range, a congested area already suffering from fragmentation and loss of wildlife habitat.

Colorado's population growth, coupled with an expanding interest in wildlife viewing, has begun to stress natural resources in unprecedented ways. Traditionally, observing, feeding, and photographing wildlife were considered to be "nonconsumptive" activities because removal of animals from their natural habitats did not occur (Shaw and Mangun 1984). In the past, nonconsumptive wildlife recreation was considered relatively benign in terms of its effects on wildlife; today, however, there is a growing recognition that wildlife-viewing recreation can have serious negative impacts on wildlife. This does not mean that recreation opportunities should be avoided; rather, management strategies must be implemented to avoid deleterious and detrimental effects. As Lucas (1979) pointed out, "Any recreational use of wildlands will produce *some* environmental change. 'No change' or 'unmodified natural conditions' may sound noble, but neither can be achieved in areas visited by recreationists and therefore both are unrealistic management objectives." If we assume that change will occur over time, managers and researchers need to identify impacts and acceptable levels of those impacts. To this end, a variety of models have been developed to classify and describe impacts to wildlife from wildlife-related recreation (Schoenfeld and Hendee 1978; Stankey 1982; Pomerantz et al. 1988; Kuss et al. 1990).

To deal with the growing pressures on natural resources, managers and researchers must begin to think in new terms. We can provide wildlife-viewing recreation without causing detrimental impacts to wildlife, if the relationships between these two are better understood.

User classification models can guide the development and management of wildlife-viewing areas (Manfredo and Larson 1993). This approach divides user types according to their preferred viewing experiences, allowing managers to manipulate impacts while alleviating user conflicts and meeting user preferences. Wildlife viewing is a management and research-based program designed to provide an array of wildlife recreation opportunities (wildlife observation, feeding, photography, etc.) and educational and interpretive activities that encourage awareness and responsible action toward wildlife and the environment. The emphasis on experience-based recreation management provides a framework for effectively meeting the preferences of recreational viewers. By emphasizing that education, management, and research are all necessary to mitigate negative impacts of recreationists on wildlife, the concept that viewers do not impact wildlife is diminished.

Chatfield State Park Wildlife Viewing Area

In Colorado, the Division of Wildlife (DOW) and the Division of Parks and Outdoor Recreation (DPOR) both are mandated to provide the greatest possible variety of wildlife-related opportunities while preserving, protecting, enhancing, and managing Colorado's diverse wildlife heritage. To support this directive and to foster agency partnerships, the DOW and the DPOR initiated the development of a Wildlife Viewing Area (WVA) at Chatfield State Park in September 1992. Development of such a facility required a comprehensive management plan (Larson and Melcher 1992) that outlined strategies for achieving three goals: (1) to provide the public with opportunities to view the nesting activities of colonial waterbirds; (2) to enhance visitor experiences by providing educational opportunities that highlight human-wildlife interactions and the natural history of colonial waterbirds; and (3) to protect the waterbird colony by promoting an understanding of human impacts on the environment.

Chatfield State Park is located in Jefferson County, Colorado, approximately 16 km southwest of Denver. Within the next few years, the park anticipates hosting 1.5 million visitors annually. Depending on fluctuating water levels, the park encompasses an average of 1660 ha of land and 607 ha of water. The waterbird nesting colony, consisting of great blue herons and double-crested cormorants, covers 60 ha of water. The WVA was constructed on a bluff overlooking the colony, where the South Platte River flows into Chatfield Reservoir. The area's vegetation consists of live and dead cottonwoods, which are inundated frequently during spring run-off or when reservoir levels rise.

Prior to construction of the Chatfield Dam in 1973, there were more than 200 nests in the cottonwood trees that grew along the north bank of the South Platte River. Once the reservoir was filled, however, nesting trees were subjected to mechanical erosion by wind, wave, and ice action, ultimately leading to deterioration and decline of the trees. By 1988, most of the trees had fallen down and the number of birds using the colony had decreased substantially (Fig. 15.1). In February 1990, DOW wildlife managers placed nesting platforms and heron decoys in a stand of cottonwood trees 300–400 m south of the old nesting site. It was hoped that herons would be encouraged to re-establish the colony at this new site. In 1991, herons and cormorants began to

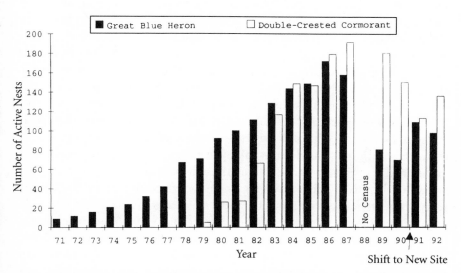

Figure 15.1   Population data for great blue herons and double-crested cormorants at Chatfield State Park waterbird colony, 1971–1992 (American Birds 1971–1986; H. Kingery pers. comm. 1987–1992; S. Skagen pers. comm. 1992).

use the new site (Rucks 1991). During the 1992 breeding season, the new colony consisted of 97 active great blue heron nests and 135 active double-crested cormorant nests (Fig. 15.1).

Recreational Influences

The scientific literature is mixed with respect to the effects of human disturbance on great blue herons and double-crested cormorants. It is clear, however, that nesting waterbirds are vulnerable to a variety of disturbances simply by virtue of their clumped-nesting habits: entire populations of these birds can be affected by single disturbance events (e.g., Parnell et al. 1988). Changes in behavior, redistribution, population declines, and colony abandonment are among the possible consequences of disturbance (Buckley and Buckley 1976). When adult herons are flushed from their nests, they can cause egg breakage or push chicks out of the nest (e.g., Nordstrom 1980). Nestlings may be exposed to excessive solar radiation (Blus et al. 1980), lethal cold temperatures (Kury and Gochfeld 1975), or predation (e.g., Krebs 1974) when adult herons are flushed from nests. In addition, breeding and foraging behaviors may be disrupted by human activities (Parnell et al. 1988).

Conversely, some studies have shown that human activities that pose no direct threat to colonial waterbirds have little to no effect on the birds. Under certain circumstances, colonial waterbirds will tolerate human activities, and some colonies may even increase in the presence of human activities (Ryder et al. 1979). Burger and Gochfeld (1981) concluded that many species of colonial birds should habituate to human activity if the birds are able to distinguish whether or not the human intrusion presents an actual threat to them. Taylor et al. (1982) found that, despite the close proximity of human activities (175 m), great blue herons at a colony in Indiana had excellent reproductive success; the birds appeared habituated to repetitive activities that posed no threat to them. Similarly, Webb and Forbes (1982) expressed "cautious optimism" about the continuation of a heron colony in a high human activity area where herons had become habituated to human encroachment.

Part of the mixed information regarding the effects of human disturbance stems from attempts to extrapolate from one species to another or from one set of circumstances to another. "The adaptive characteristics of wildlife, the recreationists' behavior, and the context of the disturbance all seem to be important" (Roggenbuck 1992). Erwin (1989) noted that colonies of nesting wading birds did not respond to human intrusion at distances greater than 150 m. Graul (1981) found that human activities beyond 100 m seldom disturbed great blue herons, and Vos et al. (1985) found that 67% of the great blue herons studied in north-central Colorado did not flush from their nests at all as a result of experimental human intrusions. Erwin (1989) also suggested that previous exposure to human activity, i.e., habituation, is a major factor in determining flushing distances.

Management Options for Coexistence

A goal in developing wildlife-viewing areas should be coexistence between recreationists and wildlife, with emphasis on management options that are based on scientific research. The Chatfield WVA (Figs. 15.2, 15.3, 15.4, 15.5, 15.6) programs and facilities were designed to minimize visitor intrusions and disturbances to the birds while still providing a variety of recreation opportunities. Specific goals and strategies, based on expert opinion and scientific literature, were developed to manage human visitors and to diminish disturbances to the birds. The main strategies are: (1) spatial and temporal zoning of human activities; (2) visitor education and interpretive programs; and (3) law enforcement.

Zoning is the primary strategy for protecting avifauna during sensitive times and consists of controlling visitor access to the WVA. Access to the WVA can be gained from the nearby handicap parking lot or by walking (20 minutes) from the main parking area; bicycle riders may approach the WVA only as far as the viewing shelter, which is 150 m from the closest nesting tree. In addition, a recreation access schedule (Fig. 15.7) defines timing and types of human activities allowed within 150 m of the waterbird colony (Larson and Melcher 1992). From 1 March through 30 April, when the birds return to their nesting colony, undergo courtship, build nests, lay eggs, and begin incubation, the risk of human disturbance-related abandonment of the colony is high. Therefore, access beyond the viewing shelter is prohibited during that time period. Although nesting activities continue into June, birds are less likely to abandon the colony during that time. As a result, pre-arranged, guided groups will be allowed on the viewing deck between 1 May and 15 May. Normal public access throughout the entire WVA will be allowed from mid-May through February.

Wildlife-viewing recreation may not be the only source of disturbance at a WVA and should not be considered in isolation; rather, all recreational activities, as well as the aggregate effects, should be taken into account. For example, Chatfield's WVA is situated on a bluff, 7.5 m above the mean water level, which restricts people from going closer to the nesting trees. There are several points from which people could approach the colony by water; therefore, buoys and signs are placed around the colony at a distance of 240 m to restrict

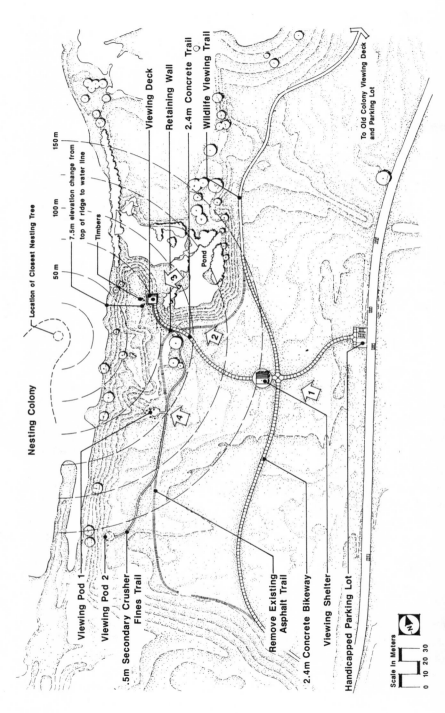

Figure 15.2 Chatfield State Park Wildlife Viewing Area. (Figures 15.2–15.6 prepared by Shalkey Walker Associates, Inc., Denver, Colorado.)

Nesting Colony

Location of Closest Nesting Tree

7.5m elevation change from top of ridge to water line

Timbers

Pond

50 m
100 m
150 m

To Old Colony Viewing Deck and Parking Lot

Viewing Pod 1
Viewing Pod 2
.5m Secondary Crusher Fines Trail

Remove Existing Asphalt Trail
2.4m Concrete Bikeway
Viewing Shelter
Handicapped Parking Lot

Viewing Deck
Retaining Wall
2.4m Concrete Trail
Wildlife Viewing Trail

Scale In Meters
0 10 20 30

N

Figure 15.3   Chatfield State Park Wildlife Viewing Area, viewing shelter.

Figure 15.4   Chatfield State Park Wildlife Viewing Area, trail to viewing deck.

Figure 15.5   Chatfield State Park Wildlife Viewing Area, viewing deck.

Figure 15.6   Chatfield State Park Wildlife Viewing Area, viewing pod #1.

boating, fishing, and other water-based activities to areas outside the buoy line. A variety of other avoidance prompts (e.g., signs, barriers) are erected in strategic locations to prohibit inappropriate approaches to the colony by hikers, equestrians, bikers, boaters, or anglers. A bicycle trail within 75 m of the closest nesting tree was removed. These zoning strategies are essential if successful reproduction and continued growth of the colony are to occur.

Wildlife viewing areas present opportunities to educate visitors about wildlife protection. The use of educational and interpretive methods that modify human behaviors and diminish visitor impacts on wildlife should be encouraged (Buckley and Buckley 1976; Parnell 1988). Educational techniques that enhance visitor awareness of the impacts of their activities are critical for motivating users to adopt low-impact behaviors (Lucas 1979). Furthermore, Roggenbuck (1992) described how persuasive communication may reduce resource impacts caused by some types of recreationists if they can perceive their impacts.

A vital persuasion technique at the Chatfield WVA is the recruitment of uniformed (to enhance their credibility and authority) volunteers to work during high-use times to model appropriate behaviors for visitors and to conduct educational programs. Additionally, 10 full-color, interpretive panels and associated trail signs are placed throughout the WVA. The panels were designed to meet the needs of a variety of users and to highlight the values of natural systems and the ecology of colonial waterbirds.

Persuasive communication strategies are used repeatedly at Chatfield WVA to emphasize both viewing ethics and the detrimental impacts of human dis-

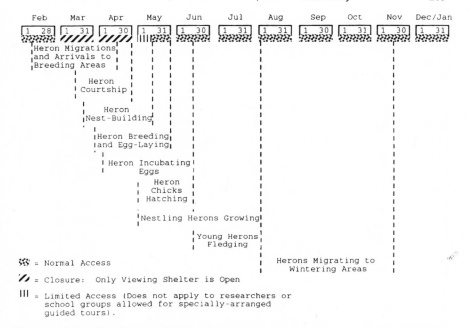

Figure 15.7   Chatfield Wildlife Viewing Area recreation access schedule (Larson and Melcher 1992).

turbance. These strategies are evident in the illustration and text of an interpretive panel (21″ × 35″) at the viewing shelter:

You Are Being Watched

As a visitor to Chatfield, you have a unique opportunity to get a close look at the wildlife that lives here. Yet, even "watching" wild animals can stress them. To minimize your disturbance, please act responsibly while observing wildlife.

Great blue herons and other waterbirds are shy and wary. Human disturbance, even unintentional, can cause adult birds to flush from their nests, leaving eggs or young vulnerable to heat, cold or predators.

One way to watch animals is to follow their examples. For instance, try to imitate a great blue heron as it stalks its prey. Walk slowly and quietly. Be patient.

This text gives visitors information about the consequences of their behaviors, suggests appropriate behaviors, and actively attempts to persuade the visitor to behave in a certain manner by using reasoned arguments.

The third strategy for managing human activities at Chatfield WVA is law enforcement. It is the most direct method of influencing human behaviors. Park rangers and DOW wildlife managers have the authority to enforce regulations and policies both within the park and at the WVA. Funding is ongoing to enhance law enforcement.

A variety of facility design elements (Fig. 15.2) are incorporated into the Chatfield WVA to further minimize human-caused disturbances to herons and cormorants. There is evidence that tangential approaches to colonial birds elicit minimal responses when compared with direct approaches (Burger and Gochfeld 1981). Therefore, trails leading to the viewing deck and pods are indirect to the colony to reduce perceived threats by approaching visitors. In addition, a number of other design features are incorporated, including: (1) timbers (vertical positions) of various heights have been installed along the trail to the viewing deck and to viewing pod number 1 to disrupt human profiles; (2) the viewing deck trail was built below ground level, minimizing the probability that herons will see approaching humans; (3) leaving existing vegetation to block views of approaching people; and (4) positioning the viewing deck so that people on the deck will be obscured by an embankment.

Historically, colonial waterbirds at Chatfield have been subjected to considerable recreational and industrial disturbances (Kingery pers. comm. 1992). The general area is urban, and the park is used extensively by boaters, anglers, hikers, bikers, and picnickers. In addition, dredging activities occurred within 100 m of the colony at one point. Despite this human activity and the decline in the number of cottonwood trees, heron and cormorant populations increased between 1971 and 1987 (Fig. 15.1). It is important that long-term monitoring of waterbird behavior, nest attentiveness, and productivity be undertaken to assess the effects of human disturbance associated with the WVA. Therefore, the final element of management at Chatfield WVA is research, both before and after construction of the viewing facility.

To assess possible impacts of the Chatfield WVA on the waterbird colony, the DOW and the DPOR initiated a four-year field study. Phase I of the study (pre-construction, control) took place during the breeding seasons of 1992 and 1993. The results will be the basis of comparison for determining whether or not herons and cormorants exhibit negative responses to the WVA and associated human activity. Phase II (post-construction, treatment) will take place during the breeding seasons of 1994 and 1995.

It was hypothesized that, if human activities at the Chatfield WVA disturb herons and cormorants, a shift in nest distribution away from the WVA would be detected. Furthermore, if human activities detrimentally affect the birds, it is expected that abnormal breeding behaviors and/or declines in productivity would be detected, as well. To test these ideas, several kinds of data are being collected, including: (1) nest locations, (2) time/activity budgets, and (3) productivity.

Mapping nest locations will permit researchers to detect whether birds redistribute away from the WVA. Quantifying the birds' time/activity budgets and describing their behaviors at different distances from the viewing areas will allow insights regarding disturbance-related stress that may not be evident otherwise (e.g., increased vocalizations, altered parental behaviors, and temporary or permanent nest abandonment). Because determination of productivity and population trends are necessary for management decisions, a long-term monitoring program will be implemented.

All of these data are essential to determine both short- and long-term impacts of the WVA. Replication of this protocol will provide managers with new information that will permit general predictions of potential impacts at other WVAs.

Knowledge Gaps

Of 166 papers containing original data on wildlife recreation impacts, only 27 dealt with the effects of people observing and photographing wildlife (Boyle and Samson 1985). To address the dual goal of protecting natural resources and providing an array of recreation opportunities, greater emphasis is needed on developing biological research in concert with public involvement programs. To date, public involvement programs have increased dramatically while biological research continues to lag. As a result, the gaps in our knowledge of how to protect natural resources, while simultaneously providing wildlife viewing opportunities, are widening. To build effective public involvement programs, biological research must be integrated with human dimensions research. This would allow an understanding of recreational impacts as well as of what types of experiences recreationists desire. At Chatfield WVA, research will be in-depth and long-term, covering both biological and human dimensions to meet this challenge.

Past research in the area of human impacts on wildlife has been relatively sparse and fragmented. Currently, it is difficult to predict impacts of recreational activities in most situations. Rather, results of studies conducted under different circumstances must be extrapolated to a particular situation of interest. Given the variations of interspecific and intraspecific responses to disturbance, the array of recreation preferences, the cumulative effects of those activities, and other factors that influence responses to disturbance, a synthesis of existing knowledge coupled with the collection of new information in well-designed experiments is imperative. There is need for future research on mitigation of human impacts and for the development of guidelines and standards for managers.

Wildlife-viewing recreation is now recognized to have detrimental impacts on wildlife and their associated habitats. While balancing wildlife-viewing recreation with impacts on wildlife, it is important for managers to understand that strong political support for wildlife stems from the public's ability and opportunity to experience wildlife. In this light, it is critical for managers to continue to develop wildlife-viewing recreation opportunities while simultaneously protecting wildlife for present and future generations. Similarly, there is a critical need for researchers to work hand-in-hand with managers to assess the impacts of recreationists enjoying wildlife. Research results should be applied to program development and to the development of mitigation strategies for recreation impacts. As recreation demands increase and habitat fragmentation continues, managers and researchers are faced with few alternatives but to balance wildlife-viewing recreation with wildlife impacts.

Literature Cited

Blus, L.J., C.J. Henny, and T.E. Kaiser. 1980. Pollution ecology of breeding great blue herons in the Columbia Basin, Oregon and Washington. *Murrelet* 61:63–71.

Boyle, S.A. and F.B. Samson. 1985. Effects of nonconsumptive recreation on wildlife: a review. *Wildlife Society Bulletin* 13:110–116.

Buckley, P.A. and F.G. Buckley. 1976. Guidelines for the protection and management of colonially nesting waterbirds. Boston, Massachusetts: National Park Service.

Burger, J. and M. Gochfeld. 1981. Discrimination of direct versus tangential approach to the nest by incubating herring and great black-backed gulls. *Journal of Comparative and Physiological Psychology* 95:676–684.

Erwin, R.M. 1989. Responses to human intruders by birds nesting in colonies: experimental results and management guidelines. *Colonial Waterbirds* 12:104–108.

Graul, W.D. 1981. Population surveys of selected bird and mammal species in Colorado, Part I. Job Program Report, Wildlife Resource Report. Colorado Division of Wildlife, Denver, Colorado. 19 May 1992.

Krebs, J.R. 1974. Colonial nesting and social feeding as strategies for exploiting food resources in the great blue heron (*Ardea herodias*). *Behavior* 51:99–134.

Kury, C.R. and M. Gochfeld. 1975. Human interference and gull predation in cormorant colonies. *Biological Conservation* 8:23–24.

Kuss, F.R., A.R. Graefe, and J.J. Vaske. 1990. Impacts to the recreation experience. In *Visitor Impact Management*, vol. 1, 187–229. Washington, D.C.: National Parks Conservation Association.

Larson, R.A. and C. Melcher. 1992. Chatfield Wildlife Viewing Area: wildlife and recreation management plan. Colorado Division of Wildlife, Denver, Colorado: Colorado Division of Parks and Outdoor Recreation.

Lucas, R.C. 1979. Perceptions of nonmotorized recreational impacts: a review of research findings. In *Recreational Impact on Wildlands Proceedings* R-6-001-1979, eds., R. Ittner, D.R. Potter, J.K. Agee, and S. Anschell, 24–31. Seattle, Washington: United States Forest Service and National Park Service.

Manfredo, J.J., A. Bright, and M. Stephenson. 1992. Public preferences for nonconsumptive wildlife recreation in the Denver area. In *The Rocky Mountain Arsenal Wildlife Viewing Recreation: Analysis and Recommendations*, ed., R.A. Larson, 129–224. Denver, Colorado: Colorado Division of Wildlife.

Manfredo, M.J. and R.A. Larson. 1993. Managing for wildlife viewing experiences: an application in Colorado. *Wildlife Society Bulletin* 21:226–236.

Nordstrom, W.R. 1980. Colonial waterbird protection program. Resource Assessment and Management Section, Alberta Department of Recreation and Parks. Alberta, Canada: Parks Division. 125 pp.

Parnell, J.F., D.G. Ainley, H. Blokpoel, B. Cain, T.W. Custer, J.L. Dusi, S. Kress, J.A. Kushlan, W.E. Southern, L.E. Stenzell, and B.E. Thompson. 1988. Colonial waterbird management in North America. *Colonial Waterbirds* 11:129–168.

Pomerantz, G.A., D.J. Decker, G.R. Goff, and K.G. Purdy. 1988. Assessing impact of recreation on wildlife: a classification scheme. *Wildlife Society Bulletin* 16:58–61.

Roggenbuck, J.W. 1992. Use of persuasion to reduce resource impacts and visitor conflicts. In *Influencing Human Behavior*, ed., M.J. Manfredo, 149–208. Champaign, Illinois: Sagamore Publishing Inc.

Rucks, J.A. 1991. A home for herons. *Colorado Outdoors* 40:12–13.

Ryder, R.A., W.D. Graul, and G.C. Miller. 1979. Status, distribution, and movements of ciconiiformes in Colorado. *Proceedings of 1979 Conference of Colonial Waterbird Group* 3:49–58.

Schoenfeld, C.A. and J.C Hendee. 1978. *Wildlife Management in Wilderness*. Pacific Grove, California: The Boxwood Press.

Shaw, W.W. and W.R. Mangun. 1984. Nonconsumptive use of wildlife in the United States. Washington, D.C.: United States Fish and Wildlife Service Resource Publication 154.

Stankey, G.H. 1982. Carrying capacity, impact management, and the recreation opportunity spectrum. *Australian Parks and Recreation* May:24–30.

Taylor, T.M., M. Reshkin, and K.J. Brock. 1982. Recreation land use adjacent to an active heron rookery: a management study. *Proceedings of the Indiana Academy of Sciences* 91:226–236.

U.S. Fish and Wildlife Service. 1992. *1991 Preliminary Findings of the National Survey of Fishing, Hunting, and Wildlife-Associated Recreation.* Washington, D.C.: United States Fish and Wildlife Service.

Vos, D.K., R.A. Ryder, and W.D. Graul. 1985. Response of breeding great blue herons to human disturbance in northcentral Colorado. *Colonial Waterbirds* 8:13–22.

Webb, R.S. and L.S. Forbes. 1982. Colony establishment in an urban site by great blue herons. *Murrelet* 63:91–92.

CHAPTER 16

Hawk Mountain Sanctuary: A Case Study of Birder Visitation and Birding Economics

Paul Kerlinger and Jim Brett

Bird watching is one of the most popular forms of nonconsumptive, wildlife-associated recreation (Kellert 1985; Kerlinger 1993). The ranks of birders has grown dramatically in the past three decades with estimates ranging as high as 60 million participants in the United States (Hall and O'Leary 1989), although the number of "committed" birders is actually smaller. Although there may be as many, if not more, birders than hunters or anglers, birders and their interests are relatively unrepresented in federal and state wildlife agencies. The reason for this is that birders historically have not been considered wildlife users, nor have they shared the cost of supporting state wildlife programs. In many cases, the needs of birders and nongame birds have been regarded by wildlife agency personnel as antithetical to, or not compatible with, management of game species.

Birders tend to focus on areas where large numbers of species can be seen or where birds are abundant. Usually this is during the migration or non-breeding season, although breeding areas can also attract large numbers of birders. These locations are often rare ecosystems and, therefore, biologically interesting. They are also centers of biodiversity and support endangered, threatened, or unusual species. Hundreds of these sites exist across North America and are well known to the birding community. These sites serve as magnets, sometimes attracting more than a hundred thousand birders annually, or many thousands during a particular season. Birders come with professional tour groups, bird clubs, or by themselves, bringing millions of dollars to the areas surrounding these birding hotspots (Kerlinger 1993).

Studies of the economic impact of birders on communities near birding hotspots, and demographic studies of the birders who visit those areas, are becoming important conservation tools to counter arguments that wildlife refuges are an economic burden to local communities (Kerlinger 1993). It has long been known that hunters and anglers provide substantial revenue to wildlife management, but until recently we have had little information about the economic impacts of birding. Here we present results of a one-year case study of birding economics and birder demographics at Hawk Mountain Sanctuary, Pennsylvania. We were particularly interested in characterizing the birders who visited this private, nonprofit sanctuary and determining the economic impact of this user group on neighboring communities. Information from our study is being used as a management tool for preventing habitat degradation by visitors and as a public-relations tool to promote open-space conservation in the area surrounding the sanctuary.

Recreational Effects

In the early 1930s the shooting of migrating hawks was a popular activity along the ridges of Pennsylvania. In 1934, after witnessing the autumn carnage at Hawk Mountain, a group of conservationists purchased 550+ ha of the Kittatinny Ridge in eastern Pennsylvania to protect migrating hawks. Hawk Mountain Sanctuary became the first sanctuary created specifically for birds of prey. A warden was hired to patrol the ridgetop sanctuary during the hawk migration season, and the Hawk Mountain Sanctuary Association (HMSA) was chartered in 1938. Since its establishment, the HMSA has grown steadily with membership totaling 2,500 in 1954 and 4,500 in 1971. Currently, HMSA has about 15,000 members, 15 staff, and 200 volunteers. Initially, the association was unpopular among the locals, and the early history of community relations was poor. After several projects initiated by HMSA in the 1970s, the local community began to realize birds of prey were an integral part of natural ecosystems, and the Hawk Mountain Sanctuary (HMS) was an important contributor to the local economy.

Since the 1930s HMSA has been a leader in the field of hawk and owl protection. Because of the large number of hawks, and the ease with which they could be seen, HMS has become a mecca for bird watchers. Today HMS encompasses nearly 1,000 ha and stands as one of the most popular birding sites in the world. In addition, it provides educational opportunities for primary and secondary school groups, college interns, graduate students, and others. The HMS also provides quality recreational opportunities including hiking, photography, botanizing, birding, and cross-country skiing. Trails and facilities are open year-round, and visitors use the site at all times.

To characterize the people who visit the sanctuary and determine their economic impact on the local community, we queried 1,350 people during June 1990 to May 1991. Each respondent was asked to complete a questionnaire of 25 items. The time required of each respondent was about 5 to 10 minutes, and most visitors were glad to participate. Because we were interested in the economic impact of avitourists on the communities surrounding HMS, we defined the study area as a 48-km radius centered at the sanctuary. These data were also used to compare HMS with several other popular birding areas that have been studied.

Visitation

The number of people who passed through the gates of HMS during the study year was 53,853, similar to the numbers counted in the preceding five years. Visitation is highly seasonal, with most (67.4%) coming during the hawk migration season (1 September–30 November). As many as 3,000 birders have arrived at the sanctuary on one day.

Forty-five percent of questionnaire respondents were first-time visitors to HMS, suggesting that there is increasing interest in this form of recreation. Only 23% of all visitors were members of the HMSA. A majority of our sample were residents of Pennsylvania (69%), although respondents came from more than 32 states and eight foreign countries. Respondents from New Jersey, New York, and Maryland, all contiguous states, accounted for 15% of the sample. Visitors from adjacent counties totaled 32.9% of the sample.

Demographics of Visitors

As with other studies of avitourism, birders who visited HMS were mostly male (58.8%), middle-aged (average = 38 years), well educated, and with above-average incomes. Only 6.9% of visitors were retirees and 11.8% listed themselves as students. Most visitors (59.5%) were in the 31–50-year range, but ages ranged from <10 to more than 80 years. Although the proportions of the sexes found at HMS were similar to those reported in other studies, HMS visitors were slightly younger than avitourists studied at High Island, Texas (Eubanks et al. 1993); Cape May, New Jersey (Kerlinger and Wiedner 1991); and elsewhere (Payne 1991; Wauer 1991; and Wiedner and Kerlinger 1990). Average age in these studies ranged from 44 to 54.8 years. We conclude from these results that hawk viewing attracts a wider age-class of participants.

The income of respondents was greater than the national average, with 63.2% earning more than $25,000 per year, including retirees. Nearly 16% had incomes in excess of $50,000. HMS visitors were also highly educated. More than 62% reported at least a four-year college degree, and another 14.5% reported having at least two years of college. These results are similar to those found in the previously cited studies.

An interesting measure of disposable income is the type and price of optical equipment used by birders, and their membership in conservation/birding organizations. About 7% of visitors to HMSA owned binoculars that retail for more than $1,000 (Zeiss, Leitz/Leica, and Bausch and Lomb). Although 54.4% of respondents reported that they belonged to no conservation organizations, the average number of memberships reported by the remainder of the sample was greater than two memberships. In comparison, the average number of organizations to which active birders belonged was 3.2 (Wiedner and Kerlinger 1990), suggesting that visitors to HMS are a mix of conservationists. There was an interesting variety of organizations listed by respondents, totaling 157, several of which (e.g., National Association of Environmental Professionals, People for the Ethical Treatment of Animals) were not strictly conservation groups. These findings on binoculars and memberships indicated that birders are willing to commit resources to their pastime.

Economic Impact on Local Communities

Income to local communities as a result of birder visitation to HMS includes expenditures for lodging, food, gasoline, and income to HMS (Table 16.1). Accommodations accounted for the largest expenditure by visitors, followed closely by eating establishments. Forty-three motels, hotels, and guesthouses, along with 23 camping facilities, were listed by respondents. Visitors supported a minimum of 150 restaurants (including delis and food stores), more than 65 different lodgings (motels, campsites, and bed-and-breakfasts), and many gas stations. This means that HMS attracts people who contribute to over 200 businesses and accompanying jobs in the area.

Perhaps the most important component of the economic impact of HMS on the community is the HMSA budget. Although some HMSA members live in nearby communities, most live outside the area. Membership support through contributions and purchases in the gift shop by members and non-members is revenue to the community. The total budget for HMSA in 1992 was about $800,000, much of which reaches the community through staff wages and purchases, or goods and services from local vendors. Staff wages are then spent on such diverse things as property taxes, clothing, automobiles, appliances, rent, and real estate.

Thus, about $1.5 million dollars, in total, came into the communities surrounding HMS as a result of HMSA activities during the study year. An estimate of the total economic impact of HMSA on the surrounding communities

Table 16.1

Summary of Economic Impacts of Birders on the Communities
Adjacent to Hawk Mountain Sanctuary, June 1990 to May 1991

Expenditure	Calculation[a]	Amount
Lodging		
Motels	13.7% × $40 per night × 1 night	$295,120
Campsites	9.2% × $15 per night × 1 night	$74,317
Other	8.7% × no estimate made	
Meals	39.8% × $5 per meal × 2.01 meals	$215,400
Gas	29.5% × $ 10 per tank	$158,870
Hawk Mountain Sanctuary Association budget		$800,000
TOTAL		$1,543,707

[a] Calculations are based on 53,853 visitors to Hawk Mountain Sanctuary Association during the study year.

is difficult. Most economists employ multipliers to determine the actual impact of expenditures on the communities involved. Standard economic multipliers range from 1.6 (i.e., Texas Department of Commerce for tourism) to greater than 2.4 (the national average). Application of these multipliers to the $1.5 million estimated as direct input, yields estimates of between $2.5 and $3.7 million per year. Because visitation at HMS increased to nearly 70,000 per year in 1992, the range of economic impact given above is conservative.

How does the magnitude of this economic impact of avitourists on the communities surrounding HMS compare with the impacts found in other studies? The numbers reported in Table 16.1 are similar to those found in studies in New Jersey, Nebraska, Ontario, and Texas (Table 16.2). At all of these sites, avitourism is a multi-million dollar industry.

Because the economic impact of birders on the communities surrounding HMSA is diffuse, many business people may not realize the magnitude of spending by avitourists in the region. Several small businesses in communities like Orwigsburg, located nearest HMS, do benefit greatly and realize that avitourists are an important source of their annual income (unpublished data). It is the motels and restaurants closest to the mountain that benefit most during the autumn migration season. This is especially true because the immediate area has few other attractions for tourists during autumn.

Avitourism at HMSA is not without negative environmental impacts. With a maximum of 3,000+ birders visiting in a given day and hundreds visiting on most weekends during autumn, traffic jams, parking shortages, crowded

Table 16.2

Comparison of Visitation and Economic Impact at
Five Major Birding Sites in the United States and Canada[a]

Site	Visitors per year	Annual economic impact	Reference
Cape May, New Jersey	100,000[b]	$10 million[b]	Kerlinger and Wiedner 1991
Hawk Mountain, Pennsylvania	53,000	$2.4 million	This study
High Island, Texas	6,000+	$2.5 million	Eubanks, Kerlinger, and Payne 1993
Grande Isle, Nebraska	80,000	$40 million	Lingle 1991
Point Pelee, Ontario	56,000	$3.2 million	Hvenegaard et al. 1988

[a]Comparisons given in table reflect differing methodologies, study area sizes, seasonal differences, and differences in the application of economic multipliers. They are given to provide the reader with an idea as to the magnitude of economic benefit accrued by local communities.

[b]Numbers reflect recalculation of economic impact in 1992, four years after the original study was published, based on increased visitorship.

trails, and a crowded lookout are not unusual. To alleviate some of the problems caused by large numbers of avitourists during the migration season, a comprehensive strategy was devised. On days when too many people arrive at the sanctuary, people are asked to come back at a later time. This is to prevent trail deterioration and to ensure a high quality birding experience. In 1990 to 1991, the visitor center was enlarged and new parking areas were placed in the forest in a way that an unbroken canopy was retained. Parking is also allowed along some of the roads leading to the sanctuary, which impedes the flow of traffic. Volunteer "wardens" and well marked trails minimize destruction of the forest. With nearly 70,000 visits during 1992 (J. Brett unpublished data), automobile traffic has become the most important negative environmental impact on the area. Exhaust fumes, noise, parking, and collisions with wildlife are now the concern of many locals.

Management Options for Coexistence

The primary implication of our case study is that businesses and wildlife benefit by the development of a world class refuge. HMS is a proven attractant of a specific type of tourist who visits during a season when few other tourists are present. That tourist is a well educated, middle-aged, affluent individual who spends money locally and has little negative impact on the community. When local communities realize that wildlife viewing areas support local economies, and at the same time, protect open space, they become supporters of such activities. In effect, the HMSA has been responsible for a shift in public opinions, both locally and nationally, in which birds of prey are now viewed as integral parts of our environment.

Although many nonprofit conservation organizations are exempt from property taxes, HMSA has been taxed at the same rate as other property owners for more than 55 years. This means that the township has not lost tax ratables (ratables are revenue units derived from taxes on property), other than what might have been gained if HMSA had been subdivided and developed as single-family dwellings. In fact, the sanctuary may have saved taxpayers by preventing single-family home development, which is now known to be a tax drain through the cost of services (schools, fire, police, and infrastructure) to residents (Real Estate Research Corporation 1974; American Farmland Trust 1986). Because of the international reputation of this sanctuary, the value of adjoining and nearby properties has increased dramatically, raising valuation of property and, therefore, taxes paid by property owners. Thus, avitourism provides significant revenues without burdening the majority of taxpayers in the region. In fact, HMSA pays nearly $8,000 per year to the township in amusement taxes, which is 10% of entrance fees paid by trail visitors. Without open-space conservation, the economic benefits provided by avitourism could not be realized.

In addition to bringing revenues to the community, HMSA provides environmental education and recreational opportunities to residents at little cost. Nearby residents are now supportive of the sanctuary and perceive it as an asset. An ecotourism study (Estes et al. 1992) done by the University of Maryland revealed that a resounding 86% of respondents to their survey were in favor of preserving Hawk Mountain, although 71.4% did not want to see more visitors. Their response was based on the increase in traffic during the hawk migration season and on the potential for environmental degradation. Most interesting was the finding that more than one-half of the respondents who owned businesses derived 10% to 25% of their revenues from Hawk Mountain visitors. This percentage translates into a significant income for these local businesses. The growth of home businesses in the area adjacent to HMS is also evident. Because of the increase of tourism, the number of prospective buyers of home-made goods has increased.

The HMSA case study is an important public-relations tool for open space conservation and could prove to be an incentive to development of ecotourism in other areas. With economic and environmental planning on a regional scale, ecotourism can provide much needed revenue, especially in economically depressed areas (see Chapter 11). Nonprofit organizations must play a role in offsetting the loss of ratables in areas where tax exempt, nonprofit organizations are now purchasing land.

In Cape May, New Jersey, local businesses were made aware of the positive economic impact of birders following a study similar to ours (Kerlinger and Wiedner 1991) (Table 16.2). Many made membership contributions to New Jersey Audubon Society's Cape May Bird Observatory and wanted to know how their businesses could attract birder dollars. Subsequent interactions between businesses and the Cape May Bird Observatory have been financially beneficial to both, and more business owners now realize the value of open space for attracting avitourists. Studies done at sites in Texas (Eubanks et al. 1993), Nebraska (Lingle 1991), New Jersey (Kerlinger and Wiedner 1991), and Ontario (Hvenegaard et al. 1988) showed that birding attractions brought large numbers of people into these areas, and that their spending was significant. What is most interesting about these studies is that the avitourist dollars come into these communities outside the traditional tourist season (Cape May, Texas, and Ontario), or they bring revenues to communities that have no other tourist attractions (Nebraska). In Cape May, High Island (Texas), and Point Pelee (Ontario), birders come mostly during migration season, which occurs in spring or autumn when beach goers or other tourists are not numerous.

Knowledge Gaps

Hawk Mountain Sanctuary is one of hundreds of birding sites in North America. Together, these sites attract millions of people annually. For example, J. N. "Ding" Darling National Wildlife Refuge attracts nearly three-quarters of a million people per year, most of whom come to view birds (Louis Hinds pers. comm.). Other preserves like the Nature Conservancy's Ramsey Canyon Preserve in Arizona draw over 30,000 visitors annually (T. Wood pers. comm.), attracted by the diversity of hummingbirds and vagrant Mexican bird species that frequent the preserve. The difference in attendance is large, but it illustrates the variation in popularity and accessibility of birding sites. To date, case studies of avitourist demographics and birding economics similar to the HMSA study have been conducted at fewer than five sites in the world. This leaves a wide gap in our understanding of one of the most popular forms of nonconsumptive wildlife use and its corresponding user-group.

Because there is no central clearinghouse for monitoring birding sites in North America, there is no means of calculating the magnitude of this aspect of birding economics. A fertile area for research would be to trace the revenues that come into the communities surrounding a birding hotspot. Such a study at HMS could trace where dollars go once they come into the area and how many and what types of jobs are affected.

Our study did not examine the expenses incurred by visitors on travel to and from the HMS area. Although we did ask if visitors had purchased gasoline locally, we did not inquire as to how much they spent on their round-trip to and from the sanctuary. Based on the geographic origin of most visitors, it is likely that they arrive in their own automobiles. It is unknown how many flew on commercial aircraft and rented cars to visit the sanctuary. Travel expenses incurred by birders can be substantial. Eubanks et al.(1993) found that 35% of the avitourists who visit High Island, Texas, to observe the spring migration of songbirds arrived via airplane. Most of them rented automobiles. Approximately 50% of the cost of ecotourism is related to transportation. Similarly, Wiedner and Kerlinger (1990) found that "active" birders spend about $1,850 per year on their avocation, with more than half of that for travel.

Because there are so few studies of birding economics at either publicly or privately owned refuges, research is needed at an array of birding locales. Site-specific case studies like this one should also be conducted outside of North America, wherever biodiversity attracts ecotourists. These studies will provide a more complete picture of birders and their spending, and will promote better management of the resource as well as provide insight into how birders may be transformed into a source of funding for wildlife programs that manage for biodiversity.

Acknowledgments

This chapter is HMSA Publication 11. Our study was funded by a grant from the National Fish and Wildlife Foundation and New England Biolabs Foundation, and by contributions of the members of New Jersey Audubon Society, Cape May Bird Observatory, and Hawk Mountain Sanctuary Association. We thank June Trexler-Kuhns for computer services.

Literature Cited

American Farmland Trust. 1986. Density related public costs, Washington, D.C.

Estes, E., C. Huang, C. Revenga, C. Romano, and J. Touval. 1992. Ecotourism destination report: Hawk Mountain, Pennsylvania. Unpublished report. College Park, Maryland: University of Maryland.

Eubanks, T., P. Kerlinger, and R.H. Payne. 1993. High Island, Texas: a case study in avitourism. *Birding* 25:415–420.

Hall, D.A. and J.T. O'Leary. 1989. Highlights of trends in birding from 1980 and 1985 national surveys of nonconsumptive wildlife-associated recreation. *Human Dimensions in Wildlife Newsletter* 8(2):23–24.

Hvenegaard, G.T., J.R. Butler, and D.K. Krystofiak. 1988. Economic values of bird watching at Point Pelee Park, Canada. *Wildlife Society Bulletin* 17:526–531.

Kellert, S.R. 1985. Birdwatching in American society. *Leisure Science* 7:343–360.

Kerlinger, P. 1993. Birding economics and birder demographic studies as conservation tools. In *Proceedings of Status and Management of Neotropical Migratory Birds,* eds., D. Finch and P. Stangel. Fort Collins, Colorado: Rocky Mountain Forest and Range Experimental Station, General Technical Report RM-229, 32–38.

Kerlinger, P. and D.S. Wiedner. 1991. The economics of birding at Cape May, New Jersey. In *Ecotourism and Resource Conservation,* vol. 1, ed., J.A. Kusler, 324–334. Miami, Florida: Second International Symposium: Ecotourism and Resource Conservation.

Lingle, G.R. 1991. History and economic impact of crane-watching in central Nebraska. *Proceedings of the North American Crane Workshop* 6:25–29.

Payne, R.H. 1991. Potential economic and political impacts of ecotourism: a research note. *Texas Journal of Political Studies* 13:65–77.

Real Estate Research Corporation. 1974. The costs of sprawl: detailed cost analysis. Washington, D.C.: U.S. Government Printing Office.

Wauer, R. 1991. Profile of an ABA birder. *Birding* 23:146–154.

Wiedner, D.S. and P. Kerlinger. 1990. Economics of birding: a national survey of active birders. *American Birds* 44:209–213.

CHAPTER 17

Beach Recreation and Nesting Birds

Joanna Burger

Coastal habitats are exposed to continued threats from natural processes such as tides, storms, shore erosion, and accretion, as well as to encroachment by humans. The impacts of human populations have become heightened along the Atlantic Coast as a result of an increase in numbers and concentrations along the Northeast corridor. Human growth along coasts has been possible because of engineering techniques that made coastal development economically feasible, and mosquito control that made coastal living enjoyable.

The zone between terrestrial and oceanic, however, is also prime nesting habitat for a variety of birds. Many of these species feed in the land/ocean interface, as well as inland or out into the ocean. The high density of people and birds provides the potential for conflicts.

My overall objective is to examine how people and birds interact on beaches and associated coastal habitats, and how coexistence can be encouraged and fostered. Although I emphasize New Jersey, the relationships exist along both coasts of North America and for many other beaches of the world. To this end I will examine: (1) beach and coastal habitats and their avifauna; (2) human use of beaches and associated habitats; (3) effects of human recreation on nesting birds in these habitats; (4) ways to manage and mitigate these beach habitats for coexistence of birds and people; and lastly (5) necessary research efforts to understand and manage these habitats for improved use by both people and birds in the future.

The combining of recreation and wildlife conservation can have three outcomes: (1) they can come in direct conflict (usually with negative impacts for the wildlife); (2) they can exist in separate locations (by law in some cases); or (3) they can be organized so that each derives benefits (or at least a lack of negative impacts). An additional objective of my chapter is to examine the degree of overlap, the need for separation, and the potential for positive coexistence.

Avian and Human Use of Beaches

The coastal regions of New Jersey, and the mid-Atlantic in general, involve a complex of barrier beaches, bays and estuaries, salt marsh islands, sandy shoals, and the mainland (see Westergaard 1987; Burger 1991; Burger and Gochfeld 1990, 1991a). These coastal landscapes differ from many other habitats, making them attractive to both birds and recreationists (Burger and Gochfeld 1991b). They are the interface between the open ocean and the continents; they are longitudinal in nature (in some places very narrow), and they are a mosaic of patches of land of various sizes interspersed with bodies of water of various sizes. Moreover, coastal habitats are ephemeral, because the dynamic nature of the regular tides and storms shift sand and rocks, creating and destroying sand spits, barrier islands, and sandy islands alike.

The whole complex, from open ocean through the barrier islands, bays, and salt marsh islands to the mainland is a mosaic of habitats where many species of birds nest and feed. One important aspect of the spatial use of coastal habitats by birds is their nesting dispersion (Burger 1981). Some birds nest solitarily, separated by a hundred or more meters, whereas others nest in colonies of a few individuals to many hundreds. These nesting patterns affect their vulnerability to recreationists.

Birds, however, do not use the habitats equally at all times of the year. Table 17.1 indicates the temporal use of coastal habitats for breeding and feeding by the four major groups of coast-nesting birds discussed in this chapter: herons and egrets, gulls, terns, and shorebirds. Breeding can occur from late February through mid-August. Some breeding birds are present all year (i.e., herring gulls, ducks), however, and so these habitats are used for foraging throughout the year.

People use beaches and coastal habitats for a variety of activities including fishing and clamming, hunting (ducks), boating (including water skiing), swimming, sunbathing, picnicking, jogging and walking, photography, and bird-watching (Burger 1986). These activities have spatial and temporal patterns (Fig. 17.1). Other activities, such as photography and bird-watching occur along the entire gradient from the surf to the mainland. There is human use of nearly all of the habitats, although the greatest use is on the beaches, and the lowest use is on the salt marshes. Salt marshes are usually avoided when there is no sandy beach or upland sand areas, and because they have high populations of mosquitos and other insects.

Temporal patterns of human use vary daily and seasonally. Generally, beachgoers do not arrive until 0900 hr or so, and numbers build up until about 1600 hr when people begin to leave. Exceptions are anglers, joggers, and

Table 17.1

Potential Overlap of Major Avian Groups and Recreationists on Northeast Coastal Beaches and Coastal Habitats[a]

	Jan.	Feb.	March	April	May	June	July	Aug.	Sept.	Oct.	Nov.	Dec.
Avian use												
Breeding												
Herons and egrets												
Gulls												
Terns												
Shorebirds												
Feeding by nesters												
Herons and egrets												
Gulls												
Terns												
Shorebirds												
Recreation												
Consumptive												
Fishing												
Clamming												
Hunting												
Nonconsumptive												
Boating												
Sun-bathing/swimming												
Jogging												
Photography												
Bird watching												

[a]Bar for recreation indicates peak of activity for New Jersey and New York areas, and dots for avian use indicate low usage.

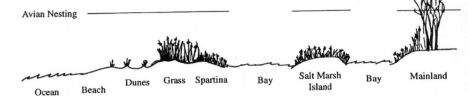

Figure 17.1   Spatial overlap of birds and people in coastal zone.

walkers who often frequent beaches in the early morning or late afternoon. In New Jersey, there are considerable numbers of anglers at night on the beaches (often with off-road vehicles).

Human use also varies seasonally (Table 17.1). Hunting is a highly restricted seasonal activity. Some activities are restricted both legally (clamming), as well as by presence of the resource (fishing). Others occur all year, but use is sporadic in the winter months and is heaviest from May until September (boating, swimming, jogging, and walking).

The relative relationship of human use varies with the specific habitat (beach, bay, islands) and time of year. Because beaches provide the greatest area of temporal and spatial overlap, data are presented for these habitats.

The pattern of human activity varies seasonally, both in types of recreation and intensity of use. In March, nearly 100% of the people on beaches are jogging or walking, whereas from June through August people are swimming or sunbathing (60%), walking or jogging (30%), or engaged in other activities. Overall human use varies seasonally at bathing beaches that are also used by birds for breeding (Fig. 17.2). The number of recreationists increases dramatically during the summer months.

Recreational Influences

Effects of recreationists on birds can be classified as indirect (habitat loss, increased predators) and direct (death, displacement, and reduced reproductive success).

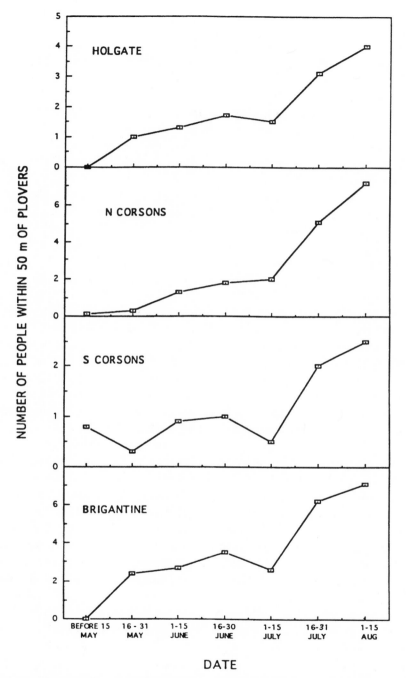

Figure 17.2   Human use at four beaches used by nesting shorebirds and colonially nesting birds as a function of date (after Burger 1991). Given are number of people per 100 m of beach per census.

Habitat Loss and Predators

Habitat loss is a major factor in the decline of beach-nesting birds (Burger 1987, 1989). Such declines not only eliminate nesting and foraging habitats, but concentrate birds and their predators into smaller areas. Habitats are rendered less useful to birds because of excessive human activities during some parts of the year. Birds can accommodate or adapt to some human presence depending on the intensity, level, and type of use (see Chapter 5). The effect of recreationists or other people can be examined by directly observing the effects of people on birds; however, this fails to examine the level of recreation that forces birds to avoid using an area.

The presence of recreationists is detrimental because it restricts the available foraging habitat for nesting birds, may eliminate the abundance of some types of nesting habitats, and may reduce the abundance of some types of prey. For example, piping plovers may suffer both nesting and foraging habitat loss when tide lines are unavailable due to recreationists.

Death, Displacement, and Reduced Reproductive Success

In addition to habitat loss and increased numbers of predators, beach-nesting birds are vulnerable to death, daily disturbances, premature departure from the colony, and lowered reproductive success. Birds nesting on the open beach primarily include shorebirds (piping plover, oystercatcher), terns, and black skimmers. These species also nest in the area of highest human recreational use.

The response of beach-nesting birds to recreationists depends on species' characteristics, habituation, exposure, seasonality, and nesting dispersion (Burger and Gochfeld 1990, 1991a). With increased human exposure, many species habituate (see Chapter 6). Some species such as the black skimmer, however, are less able to habituate to people. Most birds show increased response to people when their young are most vulnerable (i.e., at hatching).

Colonially nesting birds are also vulnerable to disturbances by people. When adults defend a colony against an intruder, eggs or chicks are unattended. Such disturbances can injure eggs or chicks, and expose them to increased heat stress or predation. Direct intrusion into least tern colonies has resulted in colony failures over several years (Burger 1984; Fig. 17.3). The percentage of colonies in New Jersey that have completely failed has decreased since the early 1980s, due mainly to protection and increased public awareness.

Solitary-nesting birds are in some ways more vulnerable because they have only themselves to look for intruders and defend against them. The piping plover is a solitary nesting species and illustrates impacts recreationists may have. Effects of people are varied and include: (1) People walking or jogging on beaches step on eggs or chicks, scare incubating parents from nests or from

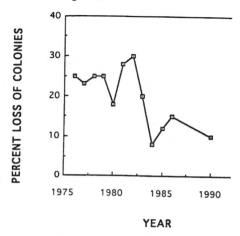

Figure 17.3   Percentage of all least tern colonies in New Jersey that failed (no reproduction) due to use by recreationists on their beach habitats (after Burger 1989).

guarding their brood, or scatter foraging adults and young. (2) Photographers and bird-watchers often get too close to nesting, brooding, or foraging plover, forcing them to shift habitats or abandon nests. (3) Sunbathers often lay too close to nests, forcing adults to leave eggs or young chicks to die in the sun. (4) Swimmers move in and out of the waves, causing nesting adults and young to stop foraging or even to leave suitable foraging sites.

In an ongoing study of the interactions of people and piping plovers at several New Jersey beaches, we found that the plovers lost valuable foraging time to the presence of recreationists and often foraged at night to obtain enough food (Burger 1991; Staine and Burger 1994). At several beaches heavily used by people, parents and young lost considerable foraging time because of people (Table 17.2). They devoted nearly half of their time to watching for or avoiding people.

The problem of foraging birds is particularly severe because parents spend much of their time trying to keep their brood together. Every time a jogger or walker passes, the chicks scatter (Burger 1991). This continues until the parent eventually has only one or two chicks remaining. Only then can the parents remain with these chicks and devote sufficient time to feeding. Thus, reproductive success may be reduced considerably.

Not only is it possible to show a direct, daily effect of recreationists on the foraging behavior of piping plover, it is also possible to compare overall foraging behavior of nesting plovers with numbers of people on the beach (Fig. 17.4). Clearly, when more people are present, the plovers have to spend more

Table 17.2

Effects of Recreationists on Foraging Piping
Plover Nesting along Several New Jersey Beaches[a]

	Incubation	Early chick phase	Late chick phase
Number of people within 50 m	2.1 ± 1.2	5.6 ± 7.5	4.8 ± 1.5
Number of disturbances per 2 min	1.0 ± 1.2	2.1 ± 1.8	1.8 ± 1.5
Seconds feeding plover can actually feed per 2 min sample	54 ± 11	48 ± 25	62 ± 11
Seconds devoted to watching for or avoiding people per 2 min	48 ± 27	54 ± 21	46 ± 17

Source: After Burger (1991).
[a]Values given are means plus or minus standard deviation.

time being alert. The number of people on the beaches in 1988 and 1989 was lower due to excessive human waste on the beaches. This illegal offshore dumping was eliminated after 1990, and recreation use went back up to the 1985–1987 levels.

Management Options for Coexistence

The extensive use of coastal marshes seems destined to continue, and some areas will experience increased human use. It is imperative to develop reasonable means of coexistence so that beach-nesting birds and human populations both can prosper.

Protecting Colonies of Nesting Birds

Colonies of nesting birds are very vulnerable because they are often large, conspicuous, and noisy. However, the presence of dense, clearly defined colonies allows for their protection through public education, posting and fencing, wardening, and legal measures (Burger 1991). Signs and information brochures can be placed at the boundary of colonies, enlisting public support in their protection.

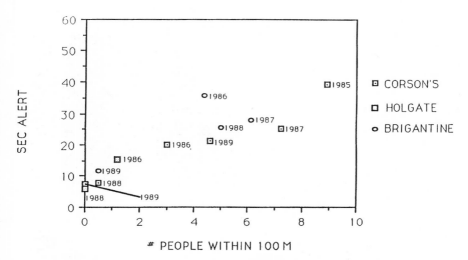

Figure 17.4   Nesting-season relationship between number of seconds that foraging plover devote to vigilance (in response to people) and the number of people within 100 m (after Burger 1991 and continuing research).

Many colonies of vulnerable species, such as least terns, are fenced and posted when the birds first arrive to breed. The number of people visiting the West End and Cedar Beach tern colonies increased from 1.2 in 1964 to 25.4 in 1978 (Burger and Gochfeld 1991a); however, use decreased thereafter near the colony due to sign-posting, wardening, and public awareness. Often, conservation groups and other volunteers regularly take part in the posting and fencing of colonies, and by these actions become active participants in their protection. Enlisting local people to fence "their" colony has been particularly effective. Towns and individuals also can be enlisted to adopt a colony.

Protection of colonies by fences and wardens can markedly decrease reproductive losses. For example, in New Jersey, least terns were declared endangered in 1978, due to habitat loss and increased human disturbance (primarily from recreationists) at breeding colonies. The Endangered and Non-Game Species Project instituted a management plan that involved the use of fences and wardens. With these procedures, both population levels and reproductive success rose (Fig. 17.5).

We found that the least terns learned that people would stay away from the colony, and they continued to use sites even though they were surrounded by hordes of sunbathers and swimmers.

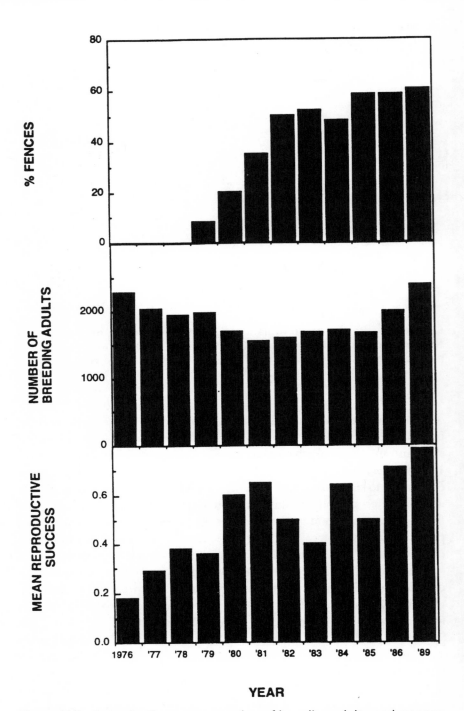

Figure 17.5   Reproductive success, number of breeding adults, and management practices (percentage of colonies with fences and/or wardens) for least terns nesting on beaches (1976–1989).

Additional methods that have proven useful include leash laws for dogs or the complete exclusion of dogs from beaches. It is imperative to keep dogs off the beaches, for the terns respond to them as predators. Beach closures during the sensitive breeding period have been very effective in increasing tern and plover breeding success where beach use is uncontrollable. Beach closures can eliminate recreationists in off-road vehicles as well as those on foot. Beach closure should be used sparingly, only with particularly vulnerable species, and should be limited to periods in the breeding season when adults and young are most susceptible to impact.

A more radical procedure for protecting colonially nesting birds is to create beach habitats in areas removed from people or from heavy human use. In New Jersey, the creation or rehabilitation of sandy shoals or small inlet islands has been particularly useful in attracting least terns and black skimmers. These species prefer the early successional stages of bare sand with little vegetation. However, these islands must be immediately posted, or boaters searching for picnic sites will use them. We have found that posting only half of such an island, while leaving half for recreationists, has increased breeding success of the birds. Most people respect the nesting space of the birds if they are provided suitable space for picnics and other beach activities.

Protecting Solitary-Nesting Birds

Solitary-nesting birds are more difficult to protect because much larger areas of beach must be protected, managed, or wardened. The primary solitary, beach-nesting species in the East that has suffered because of recreationists is the piping plover. Because it nests on the low dunes or open beach and feeds along the surf, it occupies just the area preferred by a wide range of beach recreationists. It is impossible to draw the plover to other, less disturbed nesting sites because they prefer to feed along the surf.

The management options for ensuring the coexistence of people and plovers (or other solitary-nesting shorebirds) include public education, active protection (small fences around nests, signs, wardens, predator control), legal measures (leash laws, enforcing protection laws), and beach closures. All of these measures have been used in New Jersey, with the result that populations and reproduction have increased.

On many New Jersey and New York beaches, piping plovers and people have coexisted by having wardens that educate and guide people around nests or vulnerable chicks. It has also been necessary to limit the use of off-road vehicles, even at night, because chicks can be run over or fall in the tire tracks (and are unable to climb out). In this case it helps to provide alternate routes to the beach for anglers who may have engaged in these activities for many years. Anglers often use the back or bay parts of beaches, and these may be reached without driving along the beachfront.

Knowledge Gaps

Research on the impacts of people on beach-nesting birds and on possible mitigation measures has been conducted for nearly 25 years. The realization that human use and development of beaches has escalated, that the number of natural undisturbed beaches is dwindling, and that birds on beaches face special survival and reproductive problems has prompted action. Moreover, it is clear that people want to continue using beaches and want birds to continue breeding and feeding there.

Our major gaps in knowledge about how people affect birds include: (1) defining disturbance; (2) determining the effect of specific recreational activities rather than combining all effects; (3) determining the seasonal and daily differences in effects of people on birds; (4) determining the effects of night activities by people on reproductive success of birds; (5) determining the effects of different sizes and time of beach closures on birds and people.

Defining Disturbance

Much of the literature on the effects of recreationists and avian activity has measured human presence and activities, and has failed to distinguish presence from negative impact. For example, sometimes anglers and foraging birds can exist side by side with no negative impact on either. Yet, this coexistence may be recorded as "disturbance." It is essential for researchers to document when human presence does not cause an effect as well as when it does. The presence of all people (with their associated activities) should be recorded in conjunction with the response of birds.

Determining the Effects of Specific Recreational Activities

Most studies of human disturbance and birds combine all human activities or focus on only hunting or fishing. There is a need for more studies that document the specific effect of other types of human activities.

Table 17.3 lists some variables that might be recorded to determine the effects of particular human activities on beach-nesting birds. This type of information is critical for management because it will help managers to make decisions about the types of activities that are compatible with avian use, and it may aid in educating recreationists on how to reduce their impact. For example, if walking slowly reduces adverse effects, then people can be encouraged to walk slowly around nesting and foraging shorebirds, and then to continue their brisk walk or jog.

Determining the Seasonal and Daily Differences in Effects of People on Birds

We know relatively little about how birds vary in their response as a function of time of day or season. It is well documented that birds respond differently

Table 17.3

Variables to Be Recorded for Study of Effects
of Human Activities on Beach Birds

Physical factors
Weather
Habitat
Subhabitat (i.e., surf, wet beach, dry beach,
    dunes)
Distance from water, dunes, vegetation

Human activity
Type (bird watcher, sunbather, jogger, swimmer,
    kite person)
Age
Group size (both number and area covered by
    group)
Movement speed (in one place, slow walk, run)
Associated object (pets, kites, sticks, balls)
Duration of activity

Avian response
Total birds present (by species)
Original activity (by percentage engaged in each
    activity)
Response
    Percentage of birds responding
    Variations in response
    Duration of response
    Time to resume original behavior
Distance between people and birds at which par-
    ticular responses are elicited

at different times of the season to predators (Burger and Gochfeld 1990, 1991a). Yet we have not examined this factor for people. It may be that birds are differentially vulnerable to various human activities at different points in their breeding cycle.

The factors listed in Table 17.3 are also applicable to this research need. Additional data to record would include time of day, time of year, and stage in the breeding cycle (percentage of birds in each stage). This information would be useful for public information and to manage human use during critical time periods.

Determining the Effects of Night Activities by People on Beach Birds

Only recently have we realized that some birds may be active at night. For the most part, we have assumed that our nocturnal beach activities (parties, swimming, jogging, angling) could not conflict with those of birds because they are asleep or roosting on distant marshes or the open ocean.

It is becoming increasingly clear that nesting birds may feed at night on beaches, and that these activities may be critical to breeding (Robert and Mc-Neil 1989; Staine and Burger 1994). Because the visibility of birds is reduced at night, it may be essential to restrict or limit human activities on beaches for critical time periods. But without knowing the extent of avian and human use of beaches at night, it is impossible to manage human activities.

Protocols should include the data from Table 17.3, as well as information on relative darkness, moon phase, distance at which birds appear to perceive people, the importance of sound cues, and the time that elapses before avian behavior returns to normal.

Determining the Effect of Different Sizes and Time of Beach Closures on People and Birds

What is the optimal time of the year and length of time period for closure? What is the optimal size to be closed? What size exclosure leads to increased reproductive success? What size exclosure can people tolerate, and what size will allow them to engage in their preferred, recreational activities? What is the relative effect of limits on the number of people? Would limiting the number of people eliminate the need for closure?

Beaches and coastal habitats provide an excellent example of the successful coexistence of birds and recreationists. This coexistence, however, requires the continued vigilance of management and conservation personnel, and the cooperation of a variety of recreationists. Without these, nesting populations will quickly decrease.

Acknowledgments

Over the years many people have accompanied me on the beaches of New York and New Jersey, helped in innumerable ways with logistics and field assistance, and discussed ideas with me. I thank them now: Michael Gochfeld, Carl Safina, Fred Lesser, Larry Niles, Jim Jones, Betsy Jones, Brook Lauro, Kevin Staine, Dave Jenkins, Debbie Gochfeld, and David Gochfeld. My work on beaches has been supported by the New Jersey Endangered and Non-Game Species Project, the Environmental Protection Agency, and the U.S. Fish and Wildlife Service.

Literature Cited

Burger, J. 1981. The effect of human activity on birds at a coastal bay. *Biological Conservation* 21:231–241.

Burger, J. 1984. Colony stability in least terns. *Condor* 86:61–67.

Burger, J. 1986. The effect of human activity on shorebirds in two coastal bays in the northeastern United States. *Environmental Conservation* 13:123–130.

Burger, J. 1987. Physical and social determinants of nest site selection in piping plover (*Charadrius melodus*) in New Jersey. *Condor* 89:811–818.

Burger, J. 1989. Least tern populations in coastal New Jersey: monitoring and management of a regionally-endangered species. *Journal of Coastal Research* 5:801–811.

Burger, J. 1991. Foraging behavior and the effect of human disturbance on the piping plover (*Charadrius melodus*). *Journal of Coastal Research* 7:39–52.

Burger, J. and M. Gochfeld. 1990. *The Black Skimmer: Social Dynamics of a Colonial Species*. New York: Columbia University Press.

Burger, J. and M. Gochfeld. 1991a. *The Common Tern: Its Breeding Biology and Social Behavior*. New York: Columbia University Press.

Burger, J. and M. Gochfeld. 1991b. Human activity influence and diurnal and nocturnal foraging of sanderlings (*Calidris alba*). *Condor* 93:259–265.

Robert, M. and R. McNeil. 1989. Comparative day and night feeding strategies of shorebird species in a tropical environment. *Ibis* 131:69–79.

Staine, K.J. and J. Burger. 1994. The nocturnal foraging behavior of breeding piper plover *Charadrius melodus* in New Jersey. *Auk*, in press.

Westergaard, B. 1987. *New Jersey: A Guide to the State*. Piscataway, New Jersey: Rutgers University.

CHAPTER 18

Waterborne Recreation and the Florida Manatee

Thomas J. O'Shea

A member of the small mammalian order Sirenia, the Florida manatee (Fig. 18.1) is a subspecies of the West Indian manatee. The latter occurs outside the United States in fragmented and declining populations in coastal areas and large rivers of the Caribbean, Central America, and northern South America. Sirenians occupy the unique niche of large mammalian aquatic herbivores. They have never been diverse in the fossil record, and have been largely tropical or subtropical throughout their evolutionary history. Florida and Georgia are at the northern limit of the range of the West Indian manatee, which is circumscribed by an intolerance of cool temperatures. Unlike other marine mammals that consume more energy-rich food, manatees have low metabolic rates, lose heat rapidly in cool water, and do not dive deeply. In Florida, their intolerance of cool water leads them to aggregate at warm water sources in winter, which include a few natural springs and several industrial effluents. Manatees occur close to shore where aquatic vegetation is available, primarily in coastal lagoons, estuaries, large rivers, and canals. Shallow inshore waters are also favored places for recreational activities.

Recreational Influences

Recreational influences on Florida manatees involve boating and related activities such as sport fishing and waterskiing, and diving and snorkeling at Crystal River, a major winter aggregation site.

Boating

Florida is one of the nation's leading recreational boating states. Boating occurs year-round, and Florida has more than 12,800 km of saltwater tidal

Figure 18.1   Florida manatees. Note healed propeller scars on the topmost individual. Photo courtesy of National Biological Survey (J. Reid).

coastline, 1,200,000 ha of lakes, and 19,000 km of rivers and streams. Its coastline is longer than that of California, Washington, and Oregon combined. Florida waterways are plied by 1 million boats each year, a number projected to grow to 1.6 million by the year 2000 (Florida Department of Natural Resources 1989). Boating activity will continue to expand in Florida, where the population has grown at a rate of 800 new residents daily, increasing from about 8 million in the mid-1970s to over 13 million today. Growth is largely due to immigration from other states, in part based on attraction by waterborne recreational opportunities. Florida also hosts over 40 million visitors a year, including some that cruise from northern areas in large pleasure craft each winter. Although it will be possible to build more roadways to accommodate vehicles associated with growth, the number and area of waterways will remain constant.

Enter into this scenario the unusual manatee, and the situation grows in complexity. Manatees can persist in an intricate interface with human activities, but are vulnerable to accidental strikes by boats. They have captivated public interest and are protected by several laws that include: (1) the U.S. Endangered Species Act of 1973 (ESA); (2) the U.S. Marine Mammal Protection Act of 1972 (MMPA); (3) the Florida Endangered and Threatened Species Act of 1977; and (4) the Florida Manatee Sanctuary Act of 1978. Protective laws demand actions to promote manatee recovery, but the task must be carried out against both the background of increasing pressure from recreational boating and a lack of certainty on key aspects of manatee population biology.

DIRECT MORTALITY

The major conflict between recreation and manatee conservation is direct mortality from accidental boat strikes. Historical accounts of propeller scars on living manatees indicating such contacts began in the 1940s and 1950s, but detailed documentation awaited organization of a carcass salvage network in the 1970s. Subsequent to passage of the MMPA and ESA, the U.S. Fish and Wildlife Service (USFWS) and cooperators instituted a program to retrieve every manatee carcass reported. A toll-free "hotline" was maintained by the Florida Department of Natural Resources (FDNR, recently reorganized as the Department of Environmental Protection) to encourage public reporting of dead manatees. The hotline was advertised by posters, news releases, and public service announcements. Necropsies were performed on each carcass to determine the cause of death.

During the late 1970s and early 1980s, the carcass recovery program clearly demonstrated that the major identifiable causes of death were human-related, and that accidental strikes by boats led the list (O'Shea et al. 1985). Deaths resulted from deep wounds from propellers (Fig. 18.2) and blows from boat hulls without involvement of propeller blades. Manatees were more likely to survive collisions with smaller outboard-driven boats (less than about 7.3 m in length) than larger craft (Beck et al. 1982). During the remainder of the 1980s and through the early 1990s, the number of manatees killed by boats increased at an average rate of about 9% per year, and was correlated with a large rise in the number of registered boats, which increased from about 440,000 in 1975 to 715,000 in 1992 (Ackerman et al. in press) (Fig. 18.3). Design of boat motors also changed over this period, allowing boats to traverse shallower depths at higher speeds than previously possible (Wright et al. in press).

The effects of deaths due to collisions with boats on manatee population dynamics are difficult to assess, but it generally is considered to be additive to natural mortality (see Chapter 9). Over 2,000 carcasses dead from all sources were examined between 1974 and 1993. Predation was never observed, and the incidence of death due to disease was low (O'Shea et al. 1985; Ackerman et al. in press). Mortality due to boat strikes affects all age classes and occurs year-round over extensive areas (O'Shea et al. 1985; Ackerman et al. in press). In January, 1992, at least 1,850 manatees were counted in winter aerial surveys; in the previous year 174 were found dead, at least 53 of these killed by boats (Ackerman et al. in press). Due to sampling difficulty, total population size has not been estimated with statistical precision or accuracy; moreover, the proportion of true mortality represented by the number of carcasses found

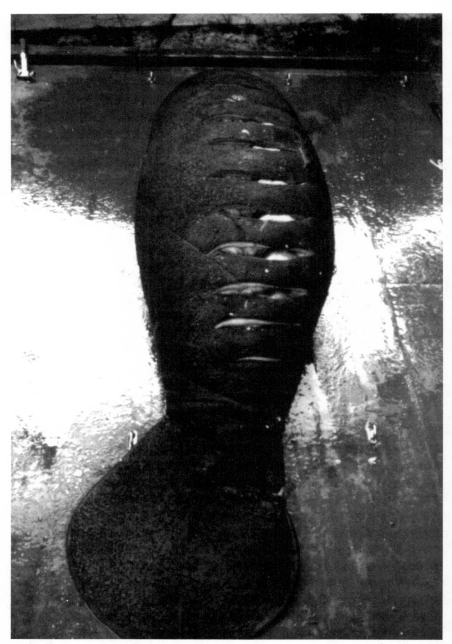

Figure 18.2   A Florida manatee killed in a collision with a boat. Note the series of deep lacerations from propeller blades. Many manatees are also killed by the force of impact of the hull without contact by propellers. Photo courtesy of National Biological Survey (C. Beck).

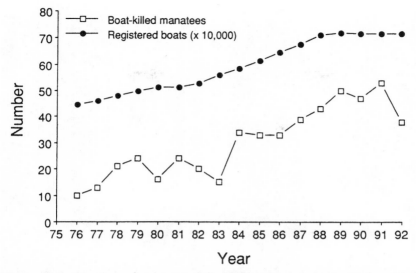

Figure 18.3   Number of manatees killed by boats and numbers of registered boats in Florida, 1976–1992. Data courtesy of the U.S. Fish and Wildlife Service, Florida Department of Environmental Protection (Marine Research Institute and Bureau of Vessel Registration and Titling), and cooperators.

is also unknown. However, given low reproductive rates, it is unlikely that the population can sustain recent annual mortality and the increasing incidence of boat kills. It is also unlikely that population declines can be detected by current estimation techniques unless large numbers were lost, a situation that could require decades to recover to previous levels. Given the increases in boats and numbers of boat-killed manatees (Fig. 18.3), managers have taken a prudent, aggressive approach despite uncertainty on key aspects of population biology.

INJURY AND INDIRECT EFFECTS OF BOATING

Boating has several potentially serious indirect effects on manatees. Prominent among these are injuries to survivors of collisions with boats. Scars from such injuries are extremely common on manatees. Many suffer from multiple past injuries, some of which appear debilitating (Fig. 18.4). Such injuries may have negative effects on the population through impairment of reproductive output. Boats also produce intense underwater noise pollution within the range of manatee hearing. It is not known if this noise causes impairment of the auditory system, which might also result in greater susceptibility to boat strikes. Boat traffic can disturb manatee behavior patterns, such as feeding, resting and nursing, and thereby disrupt activity budgets and energy balances.

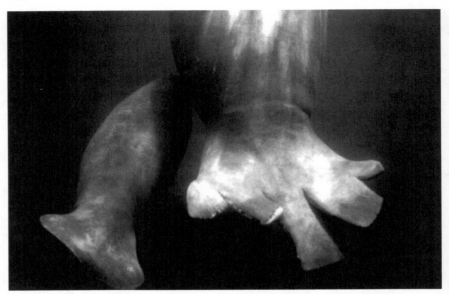

Figure 18.4   Healed tail wounds in a Florida manatee that survived a propeller strike. Many manatees suffer gruesome wounds but continue to reproduce (note the young calf). Photo courtesy of National Biological Survey (J.
Reid).

Of increasing concern is the advent of personal watercraft that traverse, with
notable disturbance, the nearshore waters used by manatees.

Indirect impacts of boating can involve adverse affects on habitat quality.
Large scale weed control activities to allow freer movement of recreational
boats take place in freshwater systems used by manatees. Aquatic plant control
can reduce available forage, and certain herbicides may be harmful if ingested
in large quantities. In marine areas, boats scour bottoms in seagrass meadows,
and wake action may raise turbidity, possibly affecting growth of forage.

SPORT FISHING

Disturbance of manatees can occur when fishing takes place in resting and
feeding areas. People commonly fish among winter aggregations of manatees
attempting to rest. Lures have been seen imbedded in manatee skin in these
aggregations. However, the primary problems are related to discarded
monofilament line. Manatees ingest line during feeding. Based on the carcass
salvage program, Beck and Barros (1991) reported monofilament line in the
gastrointestinal tracts of 11% of 439 manatees sampled from 1978 through
1986. Line caused two deaths from intestinal intussusception and impaction
(Beck and Barros 1991). A few of the manatees also had ingested fish hooks.

Figure 18.5   Florida manatee with partially amputated right flipper. Entanglement and binding by monofilament line can cause loss of limb elements. In severe cases ensuing septicemia causes death. Photo courtesy of National Biological Survey.

Besides problems caused by ingestion, some manatees have suffered amputation, injury, and death (from septicemia) due to entanglement of flippers in monofilament (Fig. 18.5). About 4% of living manatees have missing or scarred flippers due to entanglement, which also results in expensive rescue and rehabilitation efforts.

Diving, Snorkeling, and Viewing at Crystal River

Most recreational diving in Florida takes place in marine waters and inland springs where manatees infrequently occur. A major exception occurs at the headwaters of the Crystal River, near the city of Crystal River in northwestern peninsular Florida. Historical records indicate little use of the area by manatees during the first half of this century, but numbers seen in aerial counts in winter have increased steadily over the last three decades, reaching about 300 per year (Rathbun et al. 1990). This increase has been due to internal recruitment from reproduction, immigration from elsewhere in Florida, and high adult survival with low human-related mortality. The manatee increase has been paralleled by an increase in exotic aquatic vegetation used as winter food, increased protection, and a surge in human interest. The area provides memorable experiences for divers and snorkelers who view manatees, for many the closest large-mammal encounters of their lifetime. Up until the late 1970s, there was no effort to protect manatees or to regulate human activities at Crystal River. The situation has changed rapidly as manatee numbers, diving and snorkeling, and human interest have increased.

Conflicts stem from the harassment, both intentional and unintentional, that arises from the sheer numbers of recreationists involved. The number of divers at Crystal River in winter 1990–91 was estimated at 60,000 to 80,000, double that in 1980, and is likely to continue to increase (U.S. Fish and Wildlife Service 1993). The ESA defines harassment as "an intentional or negligent act or omission which creates the likelihood of injury to wildlife by annoying it to such an extent as to significantly disrupt normal behavior patterns which include, but are not limited to, feeding or sheltering." Manatees seek out Crystal River headwaters to avoid cold temperatures. In winter they rest near the main sources of warmer artesian water during cool morning hours and after the passage of cold fronts, but during most of the year they disperse from the area (Rathbun et al. 1990). Manatees are sensitive to cool temperatures because of low metabolic rates, and prolonged exposure to cold weather may be fatal, particularly to subadults and those with low energy reserves (O'Shea et al. 1985). Many manatees have become accustomed to humans at Crystal River and ignore their presence. Some approach people to solicit rubbing and petting. This behavior is sought by divers and snorkelers, who unknowingly approach intolerant manatees with expectations of such encounters. Animals unaccustomed to humans are thereby interrupted while resting or feeding, and sometimes flee to colder water. The implications for manatee time and energy budgets make such actions clearly fit the legal definitions of harassment. Although nearly 90% of the visitors to Crystal River stated that they had been informed of manatee protection regulations, 37% had seen illegal harassment of manatees, including chasing, crowding, and violations of no-entry sanctuaries (Buckingham 1989).

Management Options for Coexistence

Management to promote coexistence of manatees and recreation has been evolving over a number of years. Options include creation of sanctuaries and speed limit zones, consideration of manatee needs in planning of boating-related development, law enforcement, and public education.

Boating
Recognition of boats as manatee mortality agents prompted the adoption of seasonal slow-speed zones at winter aggregation sites beginning in 1979. This

approach was based on (1) the fact that boats travelling at slow rates of speed will collide with manatees at lower impact forces, and (2) the premise that boat operators and manatees will have more time to avoid each other. These zones were created primarily under the Florida Manatee Sanctuary Act of 1978, but may also be instituted under authority of federal regulations.

Although it is obviously important to slow boat speeds in aggregation areas, zones established in 1979 covered only 87 km in a state with many thousands of km of waterways inhabited by manatees. Mortality records showed that boat-caused deaths occurred outside these zones and were not seasonally restricted (O'Shea et al. 1985). Broad regions with higher numbers killed by boats included key counties on the east coast, the St. Johns River, and the Caloosahatchee River and its estuary in southwestern Florida (O'Shea et al. 1985; Ackerman et al. in press).

Concern over the finding of more and more boat-killed manatees reached a peak in 1989, when the Florida governor and cabinet directed the FDNR to recommend specific actions to protect manatees and increase boater safety. Recommended actions included instituting operator licenses, statewide nighttime speed limits for all waters and daytime speed limits for channels, increased law enforcement, and mandatory boater safety education with manatee awareness components. Recommendations for manatee protection targeted 13 key counties where over 80% of mortality had occurred or where important manatee travel corridors exist. Shoreline slow-speed zones were suggested for all inland waterways accessible to manatees in these counties, as was an interim boating facility expansion policy that allowed only one new slip per 30 m of shoreline owned or controlled by the applicant. Also recommended were provisions to: designate preserves; accelerate development of a geographic information system (GIS) for analysis of data on manatee distribution, boating facility locations, and other spatial information; distribute manatee awareness information to the public; and increase sign posting and maintenance (Florida Department of Natural Resources 1989).

Many of the recommendations on boater licensing and safety require legislative approval and have not yet been passed into law. Development of regulations for manatee protection in the 13 key counties, however, was authorized by the governor and cabinet in October 1989. The FDNR then developed comprehensive protection zones through a process involving coordination with local governments and special interest groups; review of manatee distribution, mortality, and boat-related activity information using GIS; navigational and economic impact considerations; and detailed review and comment. Fortunately, by the time this process had begun considerable data were available on distribution of manatees from radio and satellite telemetry studies of movements, aerial surveys, and mortality locations. Eleven of the 13 key counties had manatee protection speed-zone rules approved by the governor and cabinet by mid-1993.

Speed-limit zones for manatees have the support of the public and may have positive economic impacts. A 1989 survey by Florida State University showed that 95% of Florida boat owners supported then-existing regulations, 91% supported designating no-entry areas during parts of the year, and 76% supported restricting boat access to some areas completely. In addition, an analysis of the economic impact of the Manatee Sanctuary Act on recreation-related business was carried out, in 1993, as a result of business concerns in one of the affected counties. The analysis included estimates of the contingent valuation of the "public good" of recreational boating and manatee protection, based on a survey of the amount the average adult Florida resident would be willing to pay each year for manatee protection, and the amount each resident boat owner would be willing to pay for unregulated boat speeds. The results were biased towards costs, and excluded measures of indirect benefits of speed regulations (improved boating safety, and shoreline, dock, pier and estuary preservation) as well as national benefits and contingent values. Despite bias, the economic benefits of manatee preservation outweighed the estimated costs. Although more than 70% of the boaters polled lived in counties with speed-limit restrictions, only 10% indicated that speed limits reduced their ability to enjoy boating. The average resident was willing to pay nearly four times as much for manatee protection as the average boater was willing to pay for unlimited speeds. Although the net economic impacts were beneficial, some individual businesses could suffer adverse effects in a few locations.

Enforcement of boat speed-limit zones is difficult. Few officers must cover very extensive areas. In addition, the appearance of violations can be subjective, particularly when speeds are defined as minimum wake or slow speed, and most boats lack speedometers. In busy areas officers have large digital read-out electronic signs coupled to radar detectors to inform passing vessels of their speeds, but such means are unwieldy over large areas. An alternative not yet implemented would be to waive speed limits for boats powered by small (less than 10-horsepower) motors, or to not allow boats into protected areas unless powered by such motors. Small motors are unlikely to cause manatee mortality (Beck et al. 1982; Wright et al. in press) and are easily recognized. Many boats already carry such small engines as "kickers," and compliance with their use in designated areas would be unambiguous.

In addition to speed zones, the State of Florida has another mechanism to enhance long-term manatee protection. The Florida Local Government Comprehensive Planning and Land Development Regulation Act was enacted in 1985 as a result of pressures from population growth and land and waterfront development. It requires county governments to produce comprehensive plans for development, including locations for housing, roadways, and other infrastructure necessary to support growth. Endangered species elements are part of planning considerations, and manatee protection plans will be developed as part of the comprehensive plans of 33 counties. Protection plans will detail locations of sanctuaries, speed zones, navigational channels, and marina and dock development. Zoning for waterskiing will be included, because skiing is incompatible with manatee coexistence. Plans will also address education, and water and habitat quality. These are important steps that could serve as planning models for other regions as the world's shrinking biota becomes more strongly interfaced with a burgeoning human population.

Other means for curtailing manatee mortality from boats include detailed review of permit applications for building or expanding marinas and boat docks, designation of reserves as sanctuaries from waterborne activities, and acquisition of wetlands and surrounding uplands to prevent deterioration of vital habitat by boat traffic or other factors. Review of permit applications by the USFWS has resulted in more jeopardy opinions (a legal category leading to project denial or modification under the ESA) due to potential impacts on manatees or their habitat than for all other U.S. endangered species combined. Land acquisition has been carried out to expand national wildlife refuges, state parks and other areas. Crystal River National Wildlife Refuge, for example, was an outgrowth of the purchase of several small islands from private owners by The Nature Conservancy. Among other actions taken in October 1989, the Florida governor and cabinet also directed FDNR to recommend acquisition of critical manatee areas under the State's Conservation and Recreation Lands purchasing program and to strengthen aquatic preserve management to ensure seagrass protection. Efforts to reduce the impact of sport fishing have not been strong, but could entail excluding fishing from aggregation sites, which would eliminate disturbance and reduce the presence of monofilament fishing line, and increasing efforts at educating the fishing public to reduce disturbance and discarding of line.

Diving, Snorkeling, and Viewing at Crystal River

Partly in response to manatee harassment, the Crystal River National Wildlife Refuge was created in 1983. However, the 18 ha refuge consists of just nine small islands, adjacent submerged lands, and a few scattered lots on the mainland. Three permanent seasonal (15 November to 31 March) manatee sanctuaries were established in 1980. All waterborne activities are prohibited in these areas, which total 4.3 ha, and are clearly marked by floating lines and buoys with warning signs above and below the surface. Manatees have learned to use these zones to avoid disturbance, and use them disproportionately when human activity is high. The USFWS recently proposed the establishment of

three additional sanctuaries and extending an existing sanctuary to increase
the total protected area to 16 ha (U.S. Fish and Wildlife Service 1993). These
areas were selected to accommodate the increase in the number of manatees
using the area and to offset harassment from increasing public use while
avoiding the exclusion of divers from favorite sites. Future options for coexistence include (1) expanding sanctuaries; (2) increasing public education
through measures such as addition of an interpretive center and an information broadcasting radio station; (3) increasing enforcement; (4) developing a
permit system to regulate the numbers of divers and boaters allowed in critical
areas; and (5) closing key areas during the most stressful (coldest) times of the
day and year.

Public Education and Information
Education is a cornerstone in reducing conflicts between manatee conservation and recreationists. Numerous public pamphlets emphasize the need for
coexistence in what is one of the largest endangered species education efforts
in the nation. The Florida Power and Light Company has issued hundreds of
thousands of brochures and maps of manatee speed zones and has conducted
manatee awareness workshops. Guidelines for diving with manatees at Crystal
River are distributed at dive shops and elsewhere. Volunteers also promote
manatee awareness at Crystal River. Millions of residents and visitors are educated at oceanaria and theme parks in Florida that display captive manatees.
The Save-The-Manatee Club, with approximately 30,000 dues-paying members, holds workshops for public-school teachers and distributes information
to a wide range of outlets. The State of Florida also provides several avenues
for informing the public, including a specialty manatee license plate for motor
vehicles. More than 130,000 plates were sold the first year of the program, net-
ting nearly $400,000 for manatee conservation. Public service announce-
ments, television documentaries, buttons, posters, dolls, t-shirts, bumper and
boat stickers featuring manatees abound in Florida.

Knowledge Gaps

Efforts to reduce the numbers of manatees killed by boats have strengthened
considerably in the last few years. It will be difficult, however, to evaluate
short-term progress toward achieving mortality reduction goals. For example,
numbers of manatees killed by boats dropped in 1992 in comparison with previous years (Fig. 18.3) and were also lower through mid-1993. Perhaps efforts
begun in late 1989 have had an effect (considering lags in 1990-1991 as due to
delays in speed-zone implementation, sign posting, etc.). However, it is also

plausible that although numbers of boats increased, actual boat usage has declined, perhaps for economic reasons, resulting in fewer manatee deaths. Indices of boating activity, such as amount of marine fuels sold or estimates of traffic levels and patterns, should be included in the design of analyses of trends in numbers of manatees found killed by boats. Unfortunately, ambiguous knowledge of manatee population size or trends and lack of estimates of true numbers killed hamper assessments of progress. Continued research on manatee population dynamics must be emphasized to fully evaluate the impact of protection measures.

Other avenues to reduce manatee mortality and injury include determining the potential utility of propeller guards and warning devices (U.S. Fish and Wildlife Service 1989). However, enforcing blanket requirements for propeller guards would not only be an enormous task, considering the numbers of boats involved, but propeller cuts cause fewer manatee deaths than the force of impact of the hull alone. Cuts are involved in 39% of watercraft-related deaths, whereas impacts cause 55%, with the remainder attributable to both or not easily categorized (Wright et al. in press). However, slash sizes on those killed by propellers suggest that fatal wounds are inflicted by medium-sized or larger vessels, whereas scars on living manatees indicate that collisions with smaller boats (less than 7.3 m long) are less likely to be fatal (Beck et al. 1982; Wright et al. in press). Further analysis is warranted to determine if guards should be tested experimentally in specific areas where mortality from cuts implicating certain size classes of boats is highest. Alternatively, voluntary use of guards could be promoted on the principle that some reduction in injuries is better than none, and that prevention of human deaths and injuries would also be likely. However, implementation of such actions should be accompanied by a monitoring design that allows evaluation of confounding variables such as changes in manatee usage, boating activity, and compliance.

The suggestion to require boats to carry acoustic warning devices would also face logistic problems related to implementation. Such an idea is also questionable for other reasons. Boat motors themselves produce loud sounds within the range of manatee hearing and can be detected at great distances underwater. Careful determination must be made of the likelihood that such devices would improve the ability of manatees to detect and avoid oncoming boats without frightening animals away from important habitat. It is more likely they would merely add to the underwater sound pollution that already exists, and be subject to sensory habituation by manatees. Along these lines, additional research can be designed to judge the impacts of boat noise on manatee hearing by measuring and modeling underwater sound energy in motor noise in relation to potential hearing damage. Histopathological study of manatee hearing organs might be conducted from specimens found dead, incorporating analysis of elements related to cause of death, manatee age, and boat traffic in the region of occurrence. Knowledge of the impacts of boat noise and its physical behavior over distance might also be used to plan the sizes and locations of sanctuaries from acoustic disturbance. Other research needs in manatee biology include developing knowledge of the susceptibility of manatees to diseases of humans and companion animals, an overlooked consideration related to recreational swimming and viewing.

The bioenergetic effects of harassment should be estimated. Modifications of physiological techniques used with other large mammals need to be applied to determine the impact of disturbance on energy and time budgets, especially when manatees are disrupted from resting in waters of preferred temperatures. Field and laboratory components of such studies should be carried out, the latter making use of captive animals held at oceanaria. Quantification of the energetic effects of disturbance could assist managers in deciding the need and timing for restrictions of human activities. Currently there are no data to show when or how much harassment and disruption are either trivial or serious.

In addition to biological research, more study is needed to determine details of vessel traffic patterns in regions of contrasting levels of manatee mortality, degree of compliance with diving and boating regulations, and demographic, temporal, and geographic patterns of recreational boating and diving. Social-science research is needed to judge and track public perceptions and values regarding conflicts between manatee conservation and recreation (see Chapters 2 and 3). Initial surveys and economic analyses suggest there is strong demand for both; however, further socioeconomic study and monitoring will be required, particularly as increasing pressures from recreation and development apply greater stress on this flagship species symbolic of ecosystem health in Florida.

Acknowledgments

The following friends and colleagues reviewed manuscript drafts: Bruce Ackerman, Pat Rose, and Scott Wright, Florida Department of Environmental Protection; David Laist, U.S. Marine Mammal Commission; Lynn Lefebvre, National Biological Survey; and Bob Turner, U.S. Fish and Wildlife Service. Figures were provided courtesy of the Sirenia Project and Florida Marine Research Institute.

Literature Cited

Ackerman, B.B., S.D. Wright, R.K. Bonde, D.K. Odell, and D.J. Banowetz. Trends and patterns in manatee mortality in Florida, 1974–1991. In *Popu-*

lation Biology of the Florida Manatee, eds., T.J. O'Shea, B.B. Ackerman, and H.F. Percival. Washington, D.C.: National Biological Survey Biological Report, in press.

Beck, C.A. and N.B. Barros. 1991. The impact of debris on the Florida manatee. *Marine Pollution Bulletin* 22:508–510.

Beck, C.A., R.K. Bonde, and G.B. Rathbun. 1982. Analyses of propeller wounds on manatees in Florida. *Journal of Wildlife Management* 46:531–535.

Buckingham, C.A. 1989. Crystal River National Wildlife Refuge, public use survey report. Technical Report 37. Gainesville: Florida Cooperative Fish and Wildlife Research Unit, University of Florida.

Florida Department of Natural Resources. 1989. Recommendations to improve boating safety and manatee protection for Florida waterways. Final report to the Governor and Cabinet. Tallahassee, Florida.

O'Shea, T.J., C.A. Beck, R.K. Bonde, H.I. Kochman, and D.K. Odell. 1985. An analysis of manatee mortality patterns in Florida, 1976–1981. *Journal of Wildlife Management* 49:1–11.

Rathbun, G.B., J.P. Reid, and G. Carowan. 1990. Distribution and movement patterns of manatees (*Trichechus manatus*) in northwestern Peninsular Florida. *Florida Marine Research Publications* 48:1–33.

U.S. Fish and Wildlife Service. 1989. Florida manatee (*Trichechus manatus latirostris*) recovery plan. Prepared by the Florida Manatee Recovery Team for the U.S. Fish and Wildlife Service, Atlanta, Georgia. 98 pp.

U.S. Fish and Wildlife Service. 1993. Endangered and threatened wildlife and plants; proposed rule to establish additional manatee protection areas in Kings Bay, Crystal River, Florida. *Federal Register* 58(91): 28381–28385.

Wright, S.D., B.B. Ackerman, R.K. Bonde, C.A. Beck, and D.J. Banowetz. Analysis of watercraft-related mortality of manatees in Florida, 1979–1991. In *Population Biology of the Florida Manatee*, eds., T.J. O'Shea, B.B. Ackerman, and H.F. Percival. Washington, D.C.: National Biological Survey Biological Report, in press.

CHAPTER 19

Rattlesnake Round-ups

P. C. Arena, C. Warwick, and D. Duvall

In at least 10 U.S. states, but particularly Texas and Oklahoma, large numbers of rattlesnakes are removed annually from wild populations for two reasons. One is for inclusion in events most commonly known as "rattlesnake round-ups," wherein snakes are used for diverse public entertainment and often converted into various fashion, curio, and culinary items (Black 1981; Seippel 1988; Campbell et al. 1989; Warwick 1990; Warwick et al. 1991). The other purpose is for the commercial trade sector (Warwick et al. 1991). Both of these collection drives involve recreational elements, although for the wholly commercial interests the recreation component is almost negligible. This chapter will focus primarily, therefore, on the largely recreational round-ups.

For decades, rattlesnake round-ups, occasionally also called "rattlesnake rodeos" and "rattlesnake hunts," have attracted a great deal of criticism from biologists, conservationists, and those concerned with animal welfare. Major concerns associated with round-ups include: species and environmental degradation (Speake and Mount 1973; American Association of Zoological Parks and Aquaria 1974; Lueck 1975; Black 1981; Mulvany 1987; Seippel 1988; Warwick 1990; Warwick et al. 1991); poor public health and safety (American Association of Zoological Parks and Aquaria 1974; Lueck 1975; Mulvany 1987); miseducation (Speake and Mount 1973; Seippel 1988); negative effects on rattlesnake populations (Pisani and Fitch 1992); and inhumane treatment of snakes (American Association of Zoological Parks and Aquaria 1974; Black 1981; Mulvany 1987; Seippel 1988; Warwick 1989, 1991).

Comparatively little investigation has been made into the implications of round-ups and their unregulated and indiscriminate harvesting, as well as their disregard for humane treatment. While venomous snakes might be perceived negatively by the public, increasing numbers of people recognize their

313

ecological role and vulnerabilities to human disturbance (Webber 1951; Duvall et al. 1985; Gibbons 1988; Anonymous 1990; Williams 1990). It is therefore surprising that, considering the actual and potential damage to rattlesnake populations and ecological consequences, round-ups have attracted so little scientific study. Although animal welfare considerations are arguably as important as conservation and ecological issues, the relatively low interest in humane factors has meant that, of the eight published and three unpublished scientific reports addressing round-ups, there appears to be only one that addresses primarily animal-welfare considerations (Warwick 1991).

Round-ups are one of the most extreme examples of wildlife exploitation in the United States (Warwick et al. 1991). In some areas, an effort has been made to remove all rattlesnakes from areas habitable by humans (Lawler 1977; Weir 1992). Rattlesnake round-ups, therefore, may be a globally unique and especially destructive human-wildlife interaction.

History

Rattlesnake hunts have existed for about 300 years (Dodd 1987). However, the earliest "ancestors" of today's round-ups seem to have appeared around 1930 to 1940, when cattle ranchers organized hunts to remove rattlesnakes from their land, intending to minimize livestock casualties due to snake bite (Black 1981). Snakes were killed and often cooked to provide meals for local participants. However, the more formalized current annual round-ups, held in the spring and early summer, appear to have emerged with the Okeene, Oklahoma, round-up, circa 1930; other established events in Oklahoma and Texas have a similar history (Shelton 1981; Fitch and Pisani 1988; Warwick 1989; Adams et al. 1991; Warwick et al. 1991). The most widely publicized of these round-ups, and probably the largest, occurs at Sweetwater, Texas, where approximately 90 tons of rattlesnakes have been accounted for between its inception in 1958 and the round-up of 1991 (Weir 1992). The majority of rattlesnake round-ups conducted in other states are modeled on those in Oklahoma and Texas (Lawler 1975).

Recreational Influences

The recreational collection of rattlesnakes has several direct negative ecological influences on the target species, including adverse effects on rattlesnake populations and their environments. Undesirable consequences also occur for nontarget species that are incidentally removed or coincidentally suffer be-

cause of habitat degradation. In addition, there are numerous indirect negative ecological influences. For example, poor collection techniques and
housing practices result in high mortalities and associated excess takes of
snakes (Warwick 1991). A rattlesnake round-up is not simply hunting for
sport. This recreational activity involves snake collection, and in many cases
their mistreatment. In fact, very little of the "round-up phenomenon" is
straightforward when compared to other recreation-wildlife conflicts and related management principles.

Removal of Target Species

The majority of round-ups have targeted the western diamondback rattlesnake, although some target with the eastern diamondback rattlesnake and
prairie rattlesnake (Campbell et al. 1989; Warwick 1990; Warwick et al. 1991).

Removal of Coexisting Species

Rattlesnake crevices and den sites accommodate numerous vertebrates and
invertebrates. Very little information is available on other reptiles collected
during the round-ups, but these are known to include additional crotalines
and several nonvenomous colubrids (Campbell et al. 1989) and lizards (Warwick 1989, 1991; Warwick et al. 1991). The list of secondary targets and incidental captures of snake species includes other western rattlesnake subspecies,
canebrake rattlesnakes, massasaugas, racers, coachwhips, glossy snakes, and
western bull or gopher snakes (Campbell et al. 1989). Many of the nonvenomous (and occasionally certain venomous) species are taken to round-ups
and sold to participants; they may then serve as targets for shooting practice or
may die of dehydration and are then discarded. Anecdotal accounts suggest
that a large number of a wide variety of species are collected both as secondary
targets and incidentally, when rattlesnake captures are less successful than anticipated. It has been estimated that up to 40 nonvenomous snakes may be incidentally destroyed for every 100 "targeted" rattlesnakes (Weir 1992).

Hunters and Harvests

Two basic categories of "hunter" typically are associated with round-ups. The
first and most prominent is the so-called "professional," who collects snakes
seasonally or year-round exclusively for profit. The second category includes
members of the public who participate in organized hunts as part of the usual
weekend festivities, and are recreational collectors.

Probably more than 50 round-ups occur throughout at least 10 American
states, although few data are available on the impact of large-scale collection of
rattlesnakes from these round-ups (Warwick et al. 1991). An extrapolation
based on reported collections of snakes at specific round-ups multiplied by

the probable number of round-ups suggests that between 60,600 and 101,000 snakes may be harvested annually (Warwick et al. 1991). The actual number of snakes lost from nature may be increased by factors such as mortality in the field during collection and in certain storage conditions. These figures do not include the wholly commercial trade. Recent anecdotal accounts indicate that since the study of Warwick et al. (1991), there has been an unquantifiable decrease in the number of round-ups and number of snakes collected. Reasons for this possible reduction are unclear, but increasingly poor harvests and public disdain have been suggested. In a study of rattlesnake round-ups in Texas, the majority of hunters and attendees did not perceive round-ups to have a major negative impact on rattlesnake populations (Adams et al. 1991); however, participants and residents in both Texas and Oklahoma concede that the numbers of rattlesnakes are decreasing (Warwick et al. 1991).

Populations may not be able to sustain present levels of harvesting (Black 1981; Lueck 1975; Warwick 1990; Clark 1991; Warwick et al. 1991). Round-up data that imply harvests are constant and sustainable must be questioned, as many of the snakes registered at round-ups are brought in from other areas, including other states (Campbell et al. 1989; Clark 1991; D. Duvall, unpub. obs.). A study of the Sweetwater rattlesnake round-up indicated that the total number of snakes collected annually has increased significantly over its 28 years. However, the total number of snakes collected per hunter has remained relatively constant as a result of corresponding increases in the number of hunters (Campbell et al. 1989). Furthermore, apart from nonconfirmable anecdotes, we have no means of measuring the intensity of hunting or hunting effort.

As prizes are awarded for the longest, smallest, and greatest quantity of rattlesnakes (among other declared goals), both mature breeding adults and juvenile snakes are actively sought. The more snakes a hunter registers, the greater his chance of winning prizes (Seippel 1988). Not only are rattlesnake populations affected at critical age classes, but there is added incentive for hunters to collect as many snakes as possible. Estimated loss from collection of potentially reproductive females of the western diamondback rattlesnake suggests negative long-term implications for the survival of this species in these areas (Campbell et al. 1989; Weir 1992).

Snake Collection: Species and Environmental Considerations

Various methods are employed to catch rattlesnakes. Some collectors use the relatively benign "road-riding" approach and take animals found crossing selected roads. Others use explosives to unearth snake dens. Probably the most popular method of capturing rattlesnakes is the introduction of toxic substances (e.g., gasoline) into crevices, dens, or any place that rattlesnakes may use for shelter or hibernacula. Insecticide spraying devices are commonly used, with enough attached tubing to reach deep into the dens. This process is known as "gassing," and hunters, particularly the inexperienced, gas holes indiscriminately (Lawler 1975; Seippel 1988). Rattlesnakes are driven out of their refuges and collected. If the snakes are unable to emerge, they succumb to the fumigant and perish below ground. If the hunter has reason to believe that a "prime" specimen has been located, the site may be excavated.

Gasoline drastically reduces the lifespan of those rattlesnakes exposed to it (Speake and Mount 1973; Lawler 1975; Speake and McGlincy 1981; Speake 1986; Campbell et al. 1989); it also irritates mucous membranes (including the vent and associated copulatory structures), respiratory tissue, and the skin (Lueck 1975). In addition, gasoline may impair the sense of smell and affect sight, depending on the form of exposure/contact, as well as mask or destroy chemical cues important to the snakes in their biology and behavior.

Collectors in some states, such as Oklahoma, rarely use fumigants. Instead, snakes are collected through manual searches of ground cover, crevices, and deep dens. Both professional and amateur hunters employ this method, but even this nontoxic approach has detrimental consequences on the habitat. In the process of investigating ground cover, items are overturned and otherwise misplaced, degrading shelter size and quality (for example, incurring loss of valuable moisture); searched areas are often left uninhabitable for snakes. Although disrupted rocks inevitably create some new cover, more shelter is lost than is gained (Warwick 1990; Warwick et al. 1991). Microhabitat disturbance of this kind is one reason that the Arizona Game and Fish Commission recently enacted strict bag limits for collection of selected amphibians and reptiles by recreationists (Burkhart 1993; J. Howland, M. Sredl, and M. Goode pers. comm.). Unfortunately, the new bag limits do not apply to the two most heavily hunted rattlesnakes in Arizona, the western diamondback and the Mojave.

Warwick (1990) and Warwick et al.(1991) have suggested that the displaced populations of rattlesnakes driven into rocky areas by hunting-related habitat alteration, as well as general land development, provide false impressions of abundance because of forced aggregation. Although hunters report repeated harvests of large numbers of rattlesnakes from the same dens each year, this simply may indicate that the snakes have few choices as shelter becomes increasingly limited, and as many of the smaller sites are disturbed continually by hunters or rendered uninhabitable by the use of gasoline.

Several factors—indiscriminate collection methods, habitat alteration, number of snakes collected, indiscriminate size, sex, and seasonal targeting, compromised reproductive potentials, reduced population recruitment, and forced aggregation—suggest two major points of concern. First, perceived sustainability of current harvests based on ongoing "sufficient" collections provides an incorrect impression of snake abundance; the true status may be one of severe damage to population integrity. Second, further declines in remaining populations may occur and, if they do, may be sudden and dramatic.

Welfare Considerations

Little attention has been given to the acts of brutality and cruelty associated with rattlesnake round-ups. From the point of initial capture procedures to their death, the rattlesnakes' welfare is a matter of indifference and disregard. Where gassing and excavation prove unsuccessful, some form of grappling hook is employed to pull snakes from their dens. This frequently results in physical injury; in some cases the skin and underlying tissues are torn and in many cases the spinal cord and ribs are damaged as the animals are lifted without adequate support for the weight of their bodies.

Once the snakes are removed, they often are secured with grab-sticks and deposited into containers that may be contaminated with oil, gasoline, pesticides, and herbicides. Alternatively, wooden and wire crates are used (Lawler 1975, 1977; Warwick 1991). Snakes may remain so "stored" for up to eight months prior to the round-ups. No food or water is provided and hygiene provisions are ignored; not surprisingly, many snakes perish during storage. Typical causes of fatalities are effects of fumigants, asphyxiation as a result of overcrowding and animal size disparities, crushing injuries, dehydration and starvation, thermal extremes, injuries from capture, handling abuse, co-occupant bites, unsanitary conditions, and disease (Speake and Mount 1973; Lawler 1975, 1977; Lueck 1975; Black 1981; Fitch and Pisani 1988; Seippel 1988; Campbell et al. 1989; Warwick 1989, 1990, 1991; Warwick et al. 1991; Pisani and Fitch 1992).

In the "demonstration pits" that are common to the round-up festivities, snakes are provoked into striking targets such as balloons, grasped to expose fangs, have venom "milked" (Seippel 1988), are thrown or stepped on (in some round-ups this is incorporated into "stomping competitions" where bagged snakes are trodden to death for prizes), are injured through improper use of snake hooks, and are dropped from heights sufficient to cause physical injury (Lawler 1975; Mulvany 1987). While alive, they may also have beer or formaldehyde forced down their throats, their rattles cut or pulled off, their jaws defanged, and their mouths sutured shut to enable human portraits to be taken with the snakes (Black 1981; Fitch and Pisani 1988; Warwick 1989; Pisani and Fitch 1992). Furthermore, during each stage where snakes are transferred between containers, physical abuse often occurs.

At round-ups the usual fate of rattlesnakes is decapitation, which may be performed by handlers or in the course of a contest where the aim is to sever the heads from as many snakes as possible in a set time period. As a consequence of low metabolic rate and a resilience of the nervous system to conditions of anoxia, decapitation of a reptile does not result in neurogenic shock and immediate death (Cooper et al. 1989; Warwick 1991). Warwick (1991) reported that the heads of decapitated rattlesnakes displayed signs of life in coordinated response to stimuli (such as gaping mouths, fang erection, and pupillary actions) during periods of 27 to 65 minutes. The period of time varied depending on snake size and ambient (and presumed body) temperature.

Management Options for Coexistence

Considering the paucity of sound data on problems inherent in round-ups, these animals should, at the very least, be granted total protection until a proper assessment of status and damage to natural populations is completed. A moratorium on round-ups is urgently needed. A good sign here is that informal discussions, which may lead to formal programs of conservation research, appear to be becoming more frequent and more serious at North American professional herpetological conferences (D. Duvall, pers. obs.). Gradual abatement of round-ups as they are currently structured is, we believe, too casual and unpredictable. In our view, the only feasible means of managing rattlesnake round-ups, at least in the short run, is total abolition of the use of snakes in these events. Round-ups are entrenched in the culture and economy of many towns, so this will be difficult to achieve (Black 1981; C. Painter and L. Fitzgerald pers. comm.). Without the necessary scientific data, it will be difficult to stop, control, or even manage these events.

Pisani and Fitch (1992) proposed the following guidelines for the state of Oklahoma: spring only round-ups and accompanying hunting; game animal status for the western diamondback (the main target species in this state); unspecified bag, possession, and snake size limits; the release of surplus animals back into nature; a ban on gassing; moderated destruction of snake dens; increased basic research on rattlesnake biology; and increased education of the public. We support some, but not all, of these recommendations. For example, uncertain geographical origin and poor physical condition of many captive snakes may mean that snakes rounded up and then released could introduce disease into natural populations. Yet another weakness with these recommendations is that the approach depends upon the efficacious cooperation of round-up organizers and participants, as well as all relevant enforcement agencies. Efforts to date primarily have been the dissemination of advisory notes suggesting that organizers "clean up their act," and that indicate the management value of hunting seasons and bag limits. There has been no evidence of a positive response to this reasonable and practical guidance.

Many believe that rattlesnakes should be declared a game animal (e.g., Lawler 1975; Pisani and Fitch 1992). The hope is that such a designation would lend rattlesnakes some or all of the benefits derived from increased knowledge among recreationists and from hard data on demographic impacts of hunting and collecting. However, rattlesnakes are currently classified as game species in some states, with as yet no apparent benefit. One potentially positive example, though, exists in Colorado, where rattlesnakes have been granted game status and bag limits are enforced (R. L. Knight pers. comm.). We can only hope that such policies will lead to rattlesnake population sustainability and wise use of this resource by all concerned. Colorado should now follow up this progressive start with basic studies of rattlesnake population status and demographics in both hunted and nonhunted populations.

In Pennsylvania, local populations of timber rattlesnakes have been decimated by overhunting to levels below those considered viable, despite the imposition of licenses and bag limits by state authorities (Brown 1992; H. Reinert pers. comm.). Comparable enforcement problems are known to occur in other states holding round-ups (c.f. Warwick 1989; Warwick 1990; Warwick et al. 1991). More than enough is known, however, about environmental and species conservation and humane effects of round-ups to warrant serious concern and probably alarm.

Where state laws exist that effectively prohibit the exploitation of rattlesnakes for the purposes of round-ups, such legislation is often not enforced (Mulvany 1987). In Oklahoma, for example, the staging of a round-up and associated activities breaches at least six state laws, from illegal methods of collection, storage, and transportation, to prohibited acts of cruelty; yet proper action is not taken. Wildlife officers should be granted the powers necessary to enforce the law.

If applied in the true spirit of animal and environmental protective regulations, *available* legislation would essentially end the characteristic destructiveness of most rattlesnake round-ups. However, greater centralization of both action, such as at the federal level, may be required. Round-ups are notorious for their interstate trade activities, and the breadth of their effects exceeds state considerations. The federal Lacey Act is already available for prosecuting violations involving transport across state lines. We encourage federal agencies to apply these regulations and spearhead the control of round-ups.

Knowledge Gaps

Because there are so few accurate data about the impacts of rattlesnake round-ups on the viability of natural populations, it is difficult to determine

the detailed effects of small- or large-scale harvesting of rattlesnake populations. Little is known of the life cycles and life histories of rattlesnakes (Ernst
1992). Even less is known about the ecology of undisturbed rattlesnake populations, including population dynamics and densities, as well as demographic
data (Campbell et al. 1989; Weir 1992).

Some data on snake populations and life-history traits (e.g., using body
size, estimated age, and reproductive states) have been collated by Fitch and
Pisani (1988, 1993) from snakes exhibited and used at selected round-ups.
However, because the snakes studied were derived from highly varied, selective regimes of collection, storage, and distribution, the data are very difficult
to interpret. As noted above, large numbers of these snakes are transported in
from both distant and proximate areas. What such data do reflect, however,
are the kinds of snakes and related products in circulation among recreationists at any given time. Use of these data for the assessment of status of natural
populations is extremely inappropriate.

We know almost nothing of the demography, recruitment, age-specific reproduction and mortality, reproduction and environment, habitat needs, and
behavior of most rattlesnakes, especially the western diamondback, the focus
of most round-ups (but see Tinkle 1962). Historical and current rattlesnake
population status and trends, possible harvest levels, appropriate target populations and taxa, general reproduction, accurate information on hunter/
recreationist activity, and many other areas of interest could profit from
formal investigations.

The model research that points the way to the kinds of data needed to understand and manage rattlesnakes, and, in turn, rattlesnake hunters, is the
study by Brown (1991, 1992) of the population, reproductive, and habitat
ecology, as well as conservation, of the timber rattlesnake in New York State.
Brown found, for example, that overall female lifetime reproductive potential
is very low and that up to eight or nine years of growth and development are
needed for females to reach sexual maturity. These timber rattlesnake populations probably could not withstand the intensive collecting characteristic of
the typical Oklahoma or Texas round-up. In addition, we need long-term
studies that focus on the western diamondback, the prairie rattlesnake, and all
other pit viper populations that experience intensive hunting.

In this regard, we suggest conducting a long-term study (20 years) of the
population and habitat ecology of hunted and nonhunted populations of
western diamondbacks. Both kinds of populations are known to exist, in East
Texas for example, in some cases in close proximity to each other (D. Duvall
pers. obs.). In such a study it would be important to gather comparable data
for the two populations on (1) growth, development, and reproductive condition of all individuals; (2) population age structure; (3) age-specific

reproduction and mortality; (4) litter sizes and female reproductive cyclicity; (5) population recruitment; (6) female reproductive rates; (7) the local mating system; (8) seasonal and daily movement patterns; and (9) habitat use and needs. As information accrued from such a study, it would be possible to make defensible recommendations for that area regarding possible harvest levels.

Acknowledgments

We thank Ms. Catrina Steedman and Ms. Clairly Lance for reading through the manuscript. A grant to D. Duvall from ASU West provided financial support for a portion of P. Arena's work on this project.

Literature Cited

Adams, C.E., K.J. Strnadel, S.L. Lester, and J.K. Thomas. 1991. Texas rattlesnake round-ups. Contract report, number 370-0516. Austin, Texas: Texas Parks and Wildlife Department.

American Association of Zoological Parks and Aquaria. 1974. Notice. Association of Zoological Parks and Aquaria, 17 October 1974. Fort Worth, Texas: Fort Worth Zoological Park.

Anonymous. 1990. Human nature. *British Broadcasting Corporation Wildlife Magazine* 8:122.

Black, J.H. 1981. Oklahoma rattlesnake hunts—1981. *Bulletin of the Oklahoma Herpetological Society* 6:39–43.

Brown, W.S. 1991. Female reproductive ecology in northern populations of timber rattlesnake *Crotalus horridus. Herpetologica* 47:101–113.

Brown, W.S. 1992. Biology, status and management of the timber rattlesnake (*Crotalus horridus*): a guide for conservation. *Society for the Study of Amphibians and Reptiles, Herpetological Circular* 22:1–78.

Burkhart, B. 1993. Amphibians, reptiles subject of new rules. *Arizona Republic*, 24 October 1993.

Campbell, J.A., D.R. Formaniwicz, Jr., and E.D. Brodie, Jr. 1989. Potential impact of rattlesnake round-ups on natural populations. *Texas Journal of Science* 41:301–317.

Clark, V.S. 1991. A report on rattlesnake roundups in Oklahoma and other parts of the United States. *Herpetology* 21:12–15.

Cooper, J.E., R.E. Ewbank, C. Platt, and C. Warwick. 1989. Euthanasia of amphibians and reptiles: report of a joint working party. Potters Bar and

London: Universities Federation for Animal Welfare/World Society for the Protection of Animals.

Dodd, C.K., Jr. 1987. Status, conservation, and management. In *Snakes: Ecology and Evolutionary Biology*, eds., R.A. Seigel, J.T. Collins, and S.S. Novak, 478–513. New York: Macmillan.

Duvall, D., M. King, and K. Gutzwiller. 1985. Reconstructing the rattlesnake. *British Broadcasting Corporation Wildlife Magazine* 3:80–82.

Ernst, C.H. 1992. *Venomous Reptiles of North America*. Washington D.C.: Smithsonian Institution Press.

Fitch, H.S. and G.R. Pisani. 1988. Report to Oklahoma Department of Wildlife Conservation on the 1988 rattlesnake round-ups.

Fitch H.S. and G.R. Pisani. 1993. Life history traits of the western diamondback rattlesnake (*Crotalus atrox*) studied from roundup samples in Oklahoma. *Occasional Papers, Museum of Natural History, The University of Kansas, Lawrence* 156:1–24.

Gibbons, J.W. 1988. The management of amphibians, reptiles and small mammals in North America: the need for an environmental attitude adjustment. In *Management of Amphibians, Reptiles, and Small Mammals in North America*, eds., R.C. Szaro, K.E. Severson, and D.R. Patton, 4–10, USDA Forest Service Gen. Tech. Report Rm-166.

Lawler, H.E. 1975. A zoo perspective on rattlesnake round-ups in the southeastern U.S. Paper presented to the American Association of Zoological Parks and Aquaria. Southern Zoo Workshop, Knoxville, Tennessee, 27–29 April.

Lawler, H.E. 1977. Rattlesnake roundups revisited. *American Association of Zoological Parks and Aquaria. Regional Workshop Proceedings* 1977–78:376–381.

Lueck, B. 1975. Rattlesnake round-ups: inhumane and dangerous. *National Humane Review* November:6–9.

Mulvany, P.S. 1987. Dealing with Oklahoma's rattlesnake hunts. *Newsletter of the Oklahoma Herpetological Society* 2:1–7.

Pisani, G.R. and H.S. Fitch. 1992. A survey of Oklahoma's rattlesnake round-ups. *Kansas Herpetological Society Newsletter* 92:7–15.

Seippel, A.J. 1988. Rattlesnake round-ups. *Newsletter of the Oklahoma Herpetological Society* 3:1–6.

Shelton, H. 1981. A history of the Sweetwater Jaycees rattlesnake roundup. In *Rattlesnakes in America*, eds., J. Kilmon and H. Shelton, 95–234. Sweetwater, Texas: Shelton Press.

Speake, D.W. 1986. *Vertebrate Animals in Need of Special Attention*, ed., R.H. Mount, 48–49. Auburn, Alabama: Alabama Agricultural Experimental Station, Auburn University.

Speake, D.W. and J.A. McGlincy. 1981. Response of indigo snakes to gassing their dens. *Proceedings of the Annual Conference of Southeastern Association of Fish and Wildlife Agencies* 35:135–138.

Speake, D.W. and R.H. Mount. 1973. Some possible ecological effects of "rattlesnake roundups" in the southeastern coastal plain. *Proceedings of the 27th Annual Conference of Southeastern Association of Game and Fish Commissioners* 1973:267–277.

Tinkle, D.W. 1962. Reproductive potential and cycles in female *Crotalus atrox* from northwestern Texas. *Copeia* 1962:306–313.

Warwick, C. 1989. Atrocity Texas. *British Broadcasting Corporation Wildlife Magazine* 7:508–513.

Warwick, C. 1990. Disturbance of natural habitats arising from rattlesnake round-ups. *Environmental Conservation* 17:172–174.

Warwick, C. 1991. Observations on collection, handling, storage and slaughter of western diamondback rattlesnakes (*Crotalus atrox*). *Herpetopathologia* 2:31–37.

Warwick, C., C. Steedman, and T. Holford. 1991. Rattlesnake drives—their implications for species and environmental conservation. *Oryx* 25:39–44.

Webber, J.P. 1951. Snake facts and fiction. *Texas Game and Fish* 9:12–13.

Weir, J. 1992. The Sweetwater rattlesnake round-up: a case study in environmental ethics. *Conservation Biology* 6:116–127.

Williams, T. 1990. Driving out the dread serpent. *Audubon* September:26–32.

PART IV

Ethics and Answers

CHAPTER 20

Wildlife and Recreationists: Coexistence through Management

Richard L. Knight and Stanley A. Temple

It is expected that outdoor recreational activity will continue to increase, while the amount of wild land where wildlife may seek refuge from disturbance will decrease. This dilemma presents natural resource managers with difficult choices: protect biological diversity by denying humans the opportunity to enjoy outdoor recreation, or allow unrestricted use of wildlands and harm wildlife. Clearly, the answers fall within these two extremes of the land-management spectrum. Because of the necessity for balancing the increasing pressures of recreational activities with the needs of wildlife, chapter authors in Parts II and III of this book were asked to devote a portion of their writing to "Management Options for Coexistence." These sections present in detail ideas that resource managers may find useful in various situations. In this penultimate chapter, we intend to present a general overview of management approaches land stewards may consider when faced with wildlife-recreational conflicts.

Two general approaches are currently used to minimize the effects of recreational disturbance on wildlife. One is to deny human access to sensitive areas by instituting closures. This approach is generally only used in cases where critically endangered or sensitive species occur. Because species that fall in this category often have large spatial requirements, closures are difficult to enforce and unpopular with recreationists.

Closures or refuges exclude recreationists from areas so wildlife can exist unmolested. In Denmark, "game-pockets" are established where recreationists are prohibited and where roe deer can seek shelter during orienteering events. In the United States, refuges also may exist where game animals can retreat during the hunting season (Root et al. 1988). The National Park Service closed campgrounds near osprey nests once it was demonstrated that camping lowered reproductive success (Swenson 1979). Tuttle (1979) has suggested that caves used by bats for hibernating and rearing young should be closed to all human activity. In Sweden, certain islands were established as sanctuaries for Arctic loons because nesting success on unprotected islands was low (Göt-mark et al. 1989). Waterfowl were able to use a London reservoir despite an intensification of sailing activities only after a portion of the reservoir was made inaccessible to boating activity (Batten 1977). Reserves may unintentionally exist where wildlife can escape from disturbing activities. Because of topography, tree-stand density, or lack of roads, Maine deer had access to areas where snowmobiles could not go (Richens and Lavigne 1978).

The second approach is to devise management schemes that allow recreationists and wildlife to coexist. This is perhaps the most realistic approach in that it ensures that wildlife is still accessible to people, one of the primary reasons that recreationists visit wildlands. This approach requires detailed knowledge of specific populations. Managers must know how recreational activities affect particular species and at what intensities and times during the species' annual cycle such activities are harmful (see Chapters 4 and 5).

Restrictions That Promote Coexistence

There are four categories of restrictions that may facilitate the management of recreationists and wildlife for coexistence: spatial, temporal, behavioral, and visual.

Spatial

Spatial restrictions are perhaps the most common management technique used to control recreational disturbance. Spatial delineations most closely resemble the original approach used to control harmful disturbance, by using closures or refuges to deny human access. In this approach, recreationists and wildlife are spatially separated by buffer zones, spatial areas within which managers temporarily isolate wildlife from factors that would disrupt normal life-history processes.

Closures and refuges differ from buffer zones in that the former are permanently set aside, whereas buffer zones are temporary. Because buffer zones are designed to allow wildlife to function in a normal state, they center around areas that are crucial to the survival and reproduction of wildlife such as feeding, breeding, roosting, and nursery areas.

Flushing responses and flight distances are often used to determine the spatial distances necessary for buffer zones. The use of a single buffer-zone distance for a species throughout its range will not necessarily be adequate, as there is considerable variation between populations in their responses to the same type of disturbance (e.g., Fraser et al. 1985; Götmark et al. 1989). When establishing buffer zones, it is equally important to decide what proportion of the population using an area will be protected (see Chapter 13). A buffer zone of 200 m may protect 50% of a population, whereas a buffer zone of 500 m may be necessary to protect an entire population (Stalmaster and Newman 1978; Stalmaster 1980; Knight and Knight 1984).

Variations in flushing response and flight distance among different populations have caused considerable frustration for natural resource managers who seek uniform regulations for species over wide areas. One site may have wide buffer-zone requirements because individuals in that population are persecuted, whereas a population that is not persecuted may allow narrower limits. For example, white-fronted and bean geese decreased their flight distances from 500 m to 200 m following a ban on shooting. Understanding the root of this variation within species and populations will allow managers to both better manage wildlife and explain to recreationists why restrictions may vary (see Chapter 6).

In addition to spatial variation, there are temporal differences in flight distances within a population. Flocks of brant geese flew, on average, at a distance of 211 m in September, but by October the distance had increased to 367 m (Madsen 1985). To protect this population from disturbance would require knowing the greatest distance at which individuals took flight during the entire hunting season. The flight distance of roe deer when disturbed was noticeably shorter at night than during the day; spatial restrictions could be different during these two time periods (Jeppesen 1987).

Temporal

The time frame during which disturbance occurs is of critical importance in shaping wildlife responses (see Chapter 5). Therefore, temporal restrictions are an appropriate management tool for time periods when wildlife use critical resources. For example, in the Pacific Northwest, wintering populations of bald eagles feed primarily in the morning hours (Knight et al. 1991; Skagen et al. 1991). Restricting the timing of human activities between 0800 and 1200 from October to March would enable eagles to feed undisturbed (Stalmaster 1980).

Temporal restrictions are traditionally counted in weeks or months, but in some cases other units of time may be more useful. For example, many ducks feed at night, or at dawn and dusk, so that daytime limitations may not be necessary (Tuite et al. 1983). Van der Zande et al.(1984) determined that human activity on weekends had greater impacts on birds than disturbance during weekdays. Depending on the particular levels of disturbance and sensitivity of the species involved, temporal restrictions may therefore be tailored to the day of the week.

Behavioral

Although spatial and temporal restrictions on human activities are the most often used management techniques, alteration of human behavior is also a viable management approach (see Chapters 2 and 3). Because such things as noise, speed, and type of recreational activity elicit different responses from wildlife, aspects of these categories could be altered to minimize the impacts of recreationists (Klein 1993). For example, waterfowl are wary species that seek refuge from most forms of disturbance, particularly if the activities are associated with loud noise and rapid movement. If noise and movement of recreationists could be lessened, there would be an increased likelihood of coexistence and easing of restrictions. A paucity of information on how human behavior affects wildlife has kept the usefulness of this coexistence strategy from being used to its full potential.

Visual

Researchers have noted that wildlife are often less affected when visually shielded from human activities. The flight response of white-tailed deer to snowmobiles depended on whether or not animals were in areas of dense vegetative cover. Timber allowed deer to observe an approaching snowmobile without being exposed; in these cases deer would remain standing rather than flee (Richens and Lavigne 1978). Likewise, mountain goats were more likely to cross roads when they were shielded from humans by forest cover (Singer 1978). When roe deer in Denmark were disturbed, they ran into dense vegetative cover and remained there until the disturbance had passed (Jeppesen 1987).

The role of visual buffers is an important concept as it can result in reduced spatial restrictions separating critical wildlife-use areas from disturbances. For example, chamois in Austria were put to flight by vehicles at distances of 300–500 m. However, when in forests, chamois could be approached by a vehicle to within 30 m (Hamr 1988).

Components of visual screening that can influence wildlife response to disturbance include the juxtaposition of the animals, the location of the vegetation, and the location of potential danger. Feeding bald eagles and other avian scavengers prefer to be far away from vegetation, which may hinder their detection of potential danger (Skagen et al. 1991). Likewise, geese are more reluctant to feed in fields when their views are obstructed (Owens 1977; Madsen 1985). However, when screening vegetation is near the source of disturbance (as opposed to near the animals), it may allow animals to use areas closer than usual to the disturbance (Batten 1977).

Design Mechanisms That Allow for Coexistence

Although largely ignored in the past, resource managers are increasingly recognizing the importance of designing recreational facilities to minimize negative impacts on wildlife (see Chapter 15).

Design of Facilities in Wild Areas

To the degree that we understand how disturbance affects wildlife and wildlife responses to various human activities, we should be able to design facilities that minimize wildlife disturbance. For example, because raptors and other cliff-nesting birds are more sensitive to disturbance from the top of a cliff than from its base, overlooks could be situated so that they minimize wildlife disturbance (Hooper 1977). Campsites are often placed in riparian areas, which may also be important nesting sites for birds that are susceptible to disturbance (e.g., Ream 1976; Swenson 1979). Campsites could be designed or placed to ensure adequate spatial and visual restrictions that would allow sensitive wildlife to exist nearby. Such sites could be situated so patches or strips of vegetation lie between them and necessary wildlife habitat. Existing understory vegetation could be maintained, and both horizontal and vertical heterogeneity of vegetation could be increased, allowing for greater species diversity as well as minimizing overt effects of disturbance (Blakesley and Reese 1988).

Because of variations both within and between species, and between-area variation in wildlife responses to disturbance, natural resource managers will have to design management plans on a site-by-site basis. For this approach to be successful, land-management entities need many more carefully designed studies on wildlife responses to recreationists. At present, there does not exist an adequate data base from which generalizations can be made and from which management plans can be prepared. A collection of studies that addresses the responses of wildlife is needed, representing the spectrum of life-history strategies (e.g., long-lived versus short-lived species), to a variety of different types, frequencies, and intensities of recreational activities (e.g., Klein 1993; Holmes et al. 1993). The success of any management approach designed to minimize harmful effects of disturbance will hinge on how well the manager understands: (1) the specific wildlife species and population being managed; (2) the environment the population inhabits; and (3) the recreational activities affecting that population.

Literature Cited

Batten, L.A. 1977. Sailing on reservoirs and its effects on water birds. *Biological Conservation* 11:49–58.

Blakesley, J.A. and K.P. Reese. 1988. Avian use of campground and noncampground sites in riparian zones. *Journal of Wildlife Management* 52:399–402.

Fraser, J.D., L.D. Frenzel, and J.E. Mathisen. 1985. The impact of human activities on breeding bald eagles in north-central Minnesota. *Journal of Wildlife Management* 49:585–592.

Götmark, F., R. Neergaard, and M. Ählund. 1989. Nesting ecology and management of the Arctic loon in Sweden. *Journal of Wildlife Management* 53:1025–1031.

Hamr, J. 1988. Disturbance behaviour of chamois in an alpine tourist area of Austria. *Mountain Research and Development* 8:65–73.

Holmes, T.L., R.L. Knight, L. Stegall, and G.R. Craig. 1993. Responses of wintering grassland raptors to human disturbance. *Wildlife Society Bulletin* 21:461–468.

Hooper, R.G. 1977. Nesting habitat of common ravens in Virginia. *Wilson Bulletin* 89:233–242.

Jeppesen, J.L. 1987. The disturbing effects of orienteering and hunting on roe deer (*Capreolus capreolus*). *Danish Review of Game Biology* 13:1–24.

Klein, M.L. 1993. Waterbird behavioral responses to human disturbances. *Wildlife Society Bulletin* 21:31–39.

Knight, R.L. and S.K. Knight. 1984. Responses of wintering bald eagles to boating activity. *Journal of Wildlife Management* 48:999–1004.

Knight, R.L., D.P. Anderson, and N.V. Marr. 1991. Responses of an avian scavenging guild to anglers. *Biological Conservation* 56:195–205.

Madsen, J. 1985. Impact of disturbance on field utilization of pink-footed geese in West Jutland, Denmark. *Biological Conservation* 33:53–63.

Owens, N.W. 1977. Responses of wintering brent geese to human disturbance. *Wildfowl* 28:5–14.

Ream, C.H. 1976. Loon productivity, human disturbance, and pesticide residues in northern Minnesota. *Wilson Bulletin* 88:427–432.

Richens, V.B. and G.R. Lavigne. 1978. Response of white-tailed deer to snowmobiles and snowmobile trails in Maine. *Canadian Field-Naturalist* 92:334–344.

Root, B.G., E.K. Fritzell, and N.F. Giessman. 1988. Effects of intensive hunting on white-tailed deer movement. *Wildlife Society Bulletin* 16:145–151.

Singer, F.J. 1978. Behavior of mountain goats in relation to U.S. Highway 2, Glacier National Park, Montana. *Journal of Wildlife Management* 42:591–597.

Skagen, S.K., R.L. Knight, and G.H. Orians. 1991. Human disturbance of an avian scavenging guild. *Ecological Applications* 1:215–225.

Stalmaster, M.V. 1980. Management strategies for wintering bald eagles in the Pacific Northwest. In *Proceedings of the Washington Bald Eagle Symposium*, eds., R.L. Knight, G.T. Allen, M.V. Stalmaster, and C.W. Servheen, 49–67. Seattle, Washington: The Nature Conservancy.

Stalmaster, M.V. and J.R. Newman. 1978. Behavioral responses of wintering bald eagles to human activity. *Journal of Wildlife Management* 42:506–513.

Swenson, J.E. 1979. Factors affecting status and reproduction of ospreys in Yellowstone National Park. *Journal of Wildlife Management* 43:595–601.

Tuite, C.H., M. Owen, and D. Paynter. 1983. Interaction between wildfowl and recreation at Llangorse Lake and Talybont Reservoir, South Wales. *Wildfowl* 34:48–63.

Tuttle, M.D. 1979. Status, causes of decline, and management of endangered gray bats. *Journal of Wildlife Management* 43:1–17.

van der Zande, A.N., J.C. Berkhuizen, H.C. van Latesteijn, W.J. ter Keurs, and A.J. Poppelaars. 1984. Impact of outdoor recreation on the density of a number of breeding bird species in woods adjacent to urban residential areas. *Biological Conservation* 30:1–39.

CHAPTER 21

Taking the Land Ethic Outdoors: Its Implications for Recreation

Max Oelschlaeger

Aldo Leopold's *land ethic* is the best known environmental ethic of our time, although he did not use the word "environmental" as a modifier of the term "ethic." "Environmental" carries a connotation that Leopold attempted to avoid by conjoining land with ethic. The difference is this: the land ethic encourages humans to behave as members of the "land community," rather than as rational agents who have found reasons to treat the environment prudentially. I will argue that recreation, recontextualized through the land ethic, leads to something quite different than a "recreational ethic," which advises recreationists to travel lightly upon, avoid harm to, and leave no trace of human presence in the environment.

The genius is not in extending Leopold's land ethic to recreation, but in his work. It is likely that he would acknowledge the land community as his teacher (Leopold 1949; Flader 1974). In truth, there would be no call for this paper if society lived in accord with Leopold's norms, otherwise put, nature's way—however problematic that notion seems to some of my contemporaries (Botkin 1990). If the land ethic was culturally institutionalized, recreation would be more akin to going to church rather than an escape from daily routines. I hasten to add that Leopold did not think of the land ethic as a recreation ethic *per se,* even though he explicitly addressed questions concerning the kind of behavior we call recreational. His asked how one could live on the land without spoiling it. He found that ethics guided by utilitarian goals concerned only with human-use value led to the mindless exploitation and either potential or actual ruin of the land (Leopold 1949).

Let me begin with a preliminary exploration of the land ethic and its ties to recreation. In the preface to *Sand County Almanac* (1949), Leopold asserts without argument that, "We abuse the land because we regard it as a commodity belonging to us. When we see land as a community to which we belong, we may begin to use it with love and respect. There is no other way for land to survive the impact of mechanized man, nor for us to reap from it the aesthetic harvest it is capable, under science, of contributing to culture." Notwithstanding the false generic "man," Leopold's words admirably frame what recreation is and what it could be. Recreation is an industry that manufactures commodities for consumers. Therefore, those hoping to locate themselves in the outdoors can rarely find their place as members of a land community. Recreation might yield what Leopold calls an aesthetic harvest, a term for which "imaginative experience," "reflective encounter," and "opportunity to cross the boundary between civilization and wildness" are virtually synonymous. Conceptualized as such, recreation is more *re-creation*, an experience that puts one in a land community by enhancing awareness of ecological relationships as primary affiliations, which precede those that are merely civil.

Since Leopold died in 1948, he probably did not fully anticipate the explosive growth in population or the percentage of people involved in some form of recreation. To him, the growth of the recreation industry itself would be surprising, including the method of training for recreation managers and researchers, the manufacture of an endless stream of recreation-related products, the development of resort facilities, and the so-called windshield and eco-tourist industry. Nevertheless, he speculated early in his career about the consequences of the increasing humanization of the earth. In "The River of the Mother of God" (Leopold 1991), he anticipates many of the negative aspects of the post-World War II recreation boom, including the problem of "loving the wilderness to death." In his maturity, he explicitly called into question the implications and consequences of his own presence in wilderness ecosystems (Leopold 1949).

Partly through his own recreational experience, Leopold's attitudes toward the land changed remarkably. Curt Meine (1989), Leopold's latest biographer, persuasively argues that in 1917, Leopold saw the wilderness as a place to hunt, fish, and camp. Consistent with this perspective, he saw his profession as management of the natural world. Accordingly, he believed that predator elimination would benefit deer hunting. By 1947, the wilderness was an idea or "cerebral entity, an alternative to which civilization could turn to assess not only its ecological health, but even its social and psychological well-being" (Meine 1989:504–505). Leopold no longer regarded himself as the lord and master of the outdoors, but rather a cautious steward guided foremost by respect for the integrity, stability, and beauty of the land. Human agendas, like visitor numbers and amount of game harvested, were secondary. Although Leopold was not a Romantic, he came to see outdoor experience, much as

William Wordsworth, as an opening of the gates of perception. By the end of his life, Leopold no longer considered wilderness recreation as a mere diversion but as a necessity for the human animal. Through constant reflection upon his own experience, he realized that wilderness was a context in which culture could be evaluated. By providing a perspective outside normal channels of socialization and education, wilderness becomes more than a playground. So conceptualized, recreation is more than a vacation or escape from daily routine; it is an opportunity to call into question the transitory and contingent nature of culture.

Sand County Almanac (1949) represents the mature expression of Leopold's thought, but *Game Management* (1933) expressed a concern for wild nature, later amplified in the land ethic, that went beyond mere efficiency and utility.

> We of the industrial age boast of our control over nature. Plant or animal, star or atom, wind or river—there is no force in earth or sky which we will not shortly harness to build "the good life" for ourselves.
>
> But what is the good life? Is all this glut of power to be used for only bread-and-butter ends? Man cannot live by bread, or Fords, alone. Are we too poor in purse or spirit to apply some of it to keep the land pleasant to see, and good to live in? (Leopold 1933)

Leopold implies that recreation is characteristically governed by the same anthropocentric orientation held for all of nature. The recreation industry too often encourages consumers to seek fun, excitement, trophies, adventure, and scenery as the primary focus of their experience in the great outdoors. Yet, as Leopold matured, he grasped the notion that recreation could be the portal through which humans entered a more balanced relationship with the land.

While the expansion of the recreation market might surprise Leopold, he did anticipate the impulse to commercialize the outdoors. While living in the southwestern United States, he discovered that human greed is voracious, and that the entrepreneur (personified as Mr. Babbitt) viewed nature as a "cash cow." The growth of the recreation industry is in part a consequence of the Babbittian impulse, turning wild America into a playground, a diversion for hordes of recreationists. Investors had anticipated the exponential growth of leisure time and discretionary income following World War II. The world, on the other hand, had never seen anything like it. Whatever the impulses and motivations that lead recreationists into the woods, most are now consuming a product that has been, appearances aside, carefully packaged and marketed.

Increasingly, it appears that recreation is governed by the domination and exploitation of nature for profit and entertainment, concepts prevalent in Western culture. More importantly though, through the lens of the land ethic, it is possible to discover something that has been concealed from the recreationist by commercialization of the experience: the primal address of place (Bugbee 1974). Whatever recreation is now, it could be a part of a larger process by which we re-create ourselves and our culture, perhaps pushing us onto a road toward sustainability, where humans might recognize that they are only one group of many in the land community.

Recreation and the Evolution of the Land Ethic: Leopold's Childhood Experiences

I will begin with Leopold's own recreation, since these experiences, including hiking, camping, and hunting, helped him attain his land ethic. He grew up in Burlington, Iowa, living in proximity to the Mississippi River and the rolling lands of southeastern Iowa. Meine's (1989) stunning biography emphasizes the importance of outdoor experience in forming heart and soul of the young man. As Leopold matured, his outdoor experiences became less a matter of entertainment and diversion and more something that allowed him to discover his place on earth. Even as a youngster, he reflected on and wrote about his interaction with the natural world. Leopold's wilderness odysseys, characterized as Thoreauvian journeys on a modest scale, gave him occasion to slip outside the bounds of convention and cross over to "the green world" on the other side of culture.

The importance of recreation in forming the human psyche is typically disregarded in favor of theories that valorize early parental conditioning, later socialization (with emphasis on the peer group), religious indoctrination, and formal education. A few thinkers, though, such as Paul Shepard (1978), Edith Cobb (1977), and Morris Berman (1989), have escaped present dogma. They discuss in detail the psychological ramifications of wilderness recreation for children, both its nurturing presence and the stultifying effects of its absence. I will not dwell on their theories, so the briefest of characterizations must suffice. These theorists argue that civilization is a thin skin pulled over a human nature fashioned during time immemorial, outside of our concept of history; ironically, this second nature is disproportionately responsible for human behavior. Thus, humans raised exclusively in the hothouse of culture never encounter nature as other, that is, as life forms with patterns and structures independent of human artifice and manipulation. Yet, given opportunity to encounter such a nature, the human psyche is shaped so that individuals can see themselves both in their cultural artificiality (as products of society) and as the basic, big-brained mammals we are (products of nature). Shepard, Cobb, and Berman argue in one fashion or another that the malaise, the angst, and the violence of culture reflect the fact that we are embedded in an illusion that deprives the psyche of the rich store of information and pattern inherent in natural ecosystems. Thus, humans do not achieve maturity, being totally en-framed by the artifice of culture.

Unlike Thoreau, who reveled in walking, or Muir, who loved to climb mountains, Leopold's preferred form of recreation as an adult was hunting (primarily bowhunting for deer). Although Thoreau and Muir, as well as many of our contemporaries, looked with great skepticism on hunting, for Leopold it fostered a sense of the dynamic interrelation of hunter and quarry, an intuition that uncovered deeper questions about the relation between cul-ture and nature. Leopold was keenly aware of anti-hunting sentiment, but he effectively met such criticism by raising difficult questions posed at the core of human nature. Through hunting, Leopold disclosed something concealed by culture (our second nature): namely, the reality of a first nature conditioned by tens of thousands of years of evolutionary history. He came to see that in-dustrial culture with its mores, laws, customs, religions, and philosophies were all contingencies.

> Hunting for sport is an improvement over hunting for food, in that there has been added to the test of skill an ethical code, which the hunter formulates for himself, and must live up to without the moral support of bystanders. That the code of one hunter is more advanced than that of another is merely proof that the process of sublimation, in this as in other atavisms, is still advancing.
>
> The hope is sometimes expressed that all these instincts will be "outgrown". This attitude seems to overlook the fact that the resulting vacuum will fill up with something, and not necessarily with something better. It somehow overlooks the biological basis of human nature, the difference between historical and evolutionary time scales. We can redefine our manner of exercising the hunting instinct, but we shall do well to persist as a species at the end of the time it would take to outgrow it (Leopold 1933: 391).

Leopold's hypothesis of a hunting instinct is dubious, but his point is well taken: human beings have a genetic structure that is essentially fixed when viewed against a cultural time line. This nature has, regardless of any illusions that deny our grounding in nature, profound implications for human be-havior—and recreation.

What Is Recreation Today?

The land ethic is simply stated: an action is right when it preserves the beauty, integrity, and stability of the ecosystem (Leopold 1948). Clearly, the land ethic counters the idea that nature is nothing more than a resource to be exploited for profit and pleasure, or both, in the case of recreation. It allows us to re-describe recreation as re-creation, to shake it free of the entangling vines of civilization. When recreation is dominated by and appropriated for the purposes of profit, it furthers the domestication of ourselves and the land community. Nature recedes, concealed by activities that create the illusion of wildness.

By calling on us to heed the integrity, stability, and beauty of the land community, the land ethic reminds recreationists of the opportunity to act as members of a land community. It also reveals that recreation today is an industry, a business carefully packaged to give the illusion of an authentic outdoor experience. Recreationists who bracket the recreation market can "cross the fence" between wilderness and civilization, to momentarily see that they are domesticated, over-civilized. Typically, though, as critics point out, an outdoor experience occurs within a cultural framework that is indistinguishable from "raw" nature (Birch 1990; Baudrillard 1988).

Recreationists are, ironically, destroying the very thing they love: the blooming buzzing confusion of nature. Ersatz replaces genuine in the natural economy of value as manufacturers, guides, and resorts provide the products and services that recreationists turned consumers have been conditioned to desire. Professionals manage recreational resources increasingly akin to Disneyland: predictable, uniform, and guaranteed to please. Insofar as economic motives drive outdoor recreation, and entrepreneurs, manufacturers, and states compete in the market for recreation consumers, adverse effects on ecosystems, flora, and fauna are to be expected. Recreation is, after all, big business, and competition for the tourist dollar is vigorous. With few exceptions (such as guides marketing genuine "wilderness experiences"), there can never be too many tourists, since every tourist signifies more money for the local economy.

The conundrum is inescapable (at least at a philosophical level): more people mean more profits and less wildness. Just as scientists cannot observe the atom without leaving the trace of human presence, people cannot recreate without leaving their tracks. The consequence is that the primordial address of place is concealed beneath the commodification and regularization of the market. The recreation industry deserves to be listed on the same page with interests that are cutting the last of the old-growth forests, washing fertile topsoils into the sea, and pouring billions of tons of greenhouse gases into the atmosphere. Through commercialization, the outdoors has been converted into a standing reserve existing for profitable exploitation by entrepreneurs and consumer entertainers.

Edgar Anderson (1967) suggests a useful metaphor: humans are weeds, opportunistic invaders who colonize the land. We can trace the spread of humankind across the face of the globe through fossilized pollens and seeds. Alfred Crosby (1986) suggests the term "ecological imperialism" to describe the actual course of events by which Europeans colonized the new world. As the twenty-first century nears, the same cultural dynamic that lead to ecological imperialism is now taking a new form of recreational imperialism. National parks and even wilderness areas are now only approximations of nature, testimony to human arrogance as "the other" is controlled, so that nothing is beyond civilization. Edward Abbey (1988) observed that we are "coming so close to the end [of the process of humanization] that we can easily foresee an American state, inhabited by our children, in which swamp and forest, desert, seashore, and mountain are nothing but recreational parks for organized tourism." In short, recreational imperialism is leading toward the displacement of the wilderness by hyperreality, a condition contrived to entertain, amuse, and otherwise provide a spectacle for consumers.

Recreation as Hyperreality: The Factory System

Hyperreality is the collapse of reality into an image, where that which is represented is taken as more real than that which is signified (Rosenau 1992). Baudrillard (1988:118) suggests that "All societies end up wearing masks." The world lies subject to appropriation through signification. He fears the dominant-cultural frames that determine signification. Nothing escapes the cultural dialectic, not even agglomerations of stone and rock. Reflecting on Monument Valley, he observes that this geological, even cosmological heap of signs, was "destined to become like all that is cultivated—like all culture—natural parks" (Baudrillard 1988:4). Perhaps America, as Baudrillard claims, epitomizes a culture that is increasingly embedded in hyperreality. In any case, there is no reason to think that recreation is an exception.

Recreational imperialism overwhelms national parks, forests, and even so-called wilderness areas insofar as they are prepackaged for those who seek an "outdoor experience." Recreational consumers view a world framed through their windshields as they drive along Rocky Mountain National Park's Trail Ridge Road, or any of hundreds of others. They attend scheduled events, such as the eruption of Yellowstone's geysers or nature interpretation tours. They fish, catching factory produced "meat on the fin" that often goes from the hatchery to the skillet within days. They hunt "wild" game, enjoying not only the kill but the performing of a valuable duty in herd management. They hike and camp, trekking to their wilderness destination on carefully constructed

342 IV Ethics and Answers

trails that lead to carefully constructed campsites with levelled surfaces and occasionally fire pits. Ah, the great outdoors! What a life!

Even the best of us, such as John Muir, are deceived by hyperreality. Michael P. Cohen (1992) notes in a wry commentary that Muir did not know that as much as 50% of the California flora and fauna he observed (c. 1870s and 1880s) had been introduced by Euroculture. Even "real" features of the terrain can deceive the recreationist, because they are sometimes the consequence of human modification. The balds of the southeastern forests are another consequence of human intrusion, just as the coverts that grouse hunters celebrate in literature are areas cut over by settlers and farmers, which allow grouse to exist in numbers beyond their natural levels.

Throughout this collection of chapters, many examples have been used to illustrate the adverse effects of recreation on ecosystems and wildlife. Here I briefly consider similar examples, in terms of the land ethic, to illustrate what outdoor recreation has become: hyperreality.

Consider fishing. Through the market and the associated development of recreational fisheries management, fish have become commodities, under tight control of civilization. The fish are no longer celebrated in song and ritual as legitimate members of the great economy of life. Neither are they part of a land community with which the angler identifies. Thoreau, from his nineteenth-century vantage point, wrote that the fisherman "does not make the scenery less wild, more than the jays and muskrats, but stands there as a part of it. . . . He belongs to the natural family of man, and is planted deeper in nature and has more root than the inhabitants of towns." The angler was a person who had, at least momentarily, slipped the leash of civilization and become a person-in-the-wilderness, someone more in tune with organic fundamentals than with social custom and convention. Do such authentic anglers yet exist? Or have they been replaced by the recreational consumer?

In Colorado, for example, the cutthroat trout is a native species adversely impacted by the introduction of game fish. When Anglo-Europeans first came to the Colorado River Basin, the cutthroat ranged throughout. Yet, even as early as 1888, nonnative trout were being stocked, first in the Gunnison River, and then elsewhere across the basin. Patrick Trotter (1987:155) states that, "By 1897, the Gunnison was famed for its rainbow trout fishing, and the native cutthroats had all but disappeared. In the other big rivers the story was the same: the native cutthroat could not coexist with introduced brown and rainbow trout. Nor could they compete with introduced brook trout in smaller streams at the higher elevations."

Those natives not driven from their habitat by competition usually hybridize, so that the pure cutthroat has disappeared in all but the most remote reaches of the basin. What's good for anglers and profitable for a state's tourist industry is not always good for indigenous species. Thus, an image displaces reality. Rainbow trout are fighting game fish that suit the tastes of Euroman. So, the Colorado Division of Wildlife, itself part of the modern world that Leopold called into question, caters to Euroman.

To learn the ways of the cutthroat is to revel in wildness, to celebrate its epochal migration across the face of the northwestern part of Turtle Island, now called the United States. Long before the Anglos came to name these places, there were the ancestral fish, slowly but steadily advancing with the successive waves of glaciation. To revere the cutthroat is to recall a time before human existence. Speciation, as is often pointed out, moves in a frame beyond ordinary human ken. The cutthroat is a living testimony to the imperceptible comings and goings of the ice ages, to the shaping of the landscape. Cutthroats bear witness, indeed, to the very processes of creation that produced *Homo sapiens*. Do you remember that cutthroats are fellows in the community of life?

Fish are easy marks since the quarry is so subject to manipulation. It's harder to fool with rock; perhaps climbing is a form of recreation impervious to commodification—an honest, authentic event. Or is it? Has mountaineering degenerated into technique, into routes with numbers that indicate "ego points," into "peak bagging," and "guided ascents" for urban alpinists who have about as much feel for a mountain community as corporate raiders for factory workers. With precious few exceptions, one cannot read a contemporary piece on mountaineering that reveals any sense of community, any sensibility remotely resembling a consciousness in the tradition of Muir. Again, the ersatz displaces the authentic. Climbing has become a business with multimillion dollar endorsements of heroes who free-climb the world's "8's" (peaks over 8,000 meters) without oxygen. Mountaineers of ordinary status flock out of Denver in droves to conquer Long's, Meeker, and Lady Washington in one day. International competitions with five-figure prizes are now held on artificial walls (to offer uniform conditions to the competitors) as they "rock climb." Spectators actually pay money to attend these events! Muir must be rolling in his grave.

Walking? Even if climbing is now part of the hyperreal, such is not the case with walking, which survives in a wild and undomesticated, indeed, Thoreauvian way. There is nothing illusionary about walking, as blisters on the hiker's feet reveal. Yet, the very act of walking in the wilderness begins to transform it, perhaps in ways less dramatic than agriculture, mining, etc., but nonetheless, changes accumulate. Paths trod by human feet become "manways" that mark high alpine terrain, almost permanently staking a human claim. There are the short-cuts, where hikers leave the trail, providing new channels for damage and erosion. Another subtle effect is seen in the habituation of wild creatures to the presence of human beings, converted by the lure of handouts into "beggars." Walking could take one on a path toward wildness, the wildness in which lies the preservation of the world; but walking has become guidebooks, signposts, and scenic vistas, as recreationists follow the yellow brick road laid down for them by the recreation industry. The true walker, as Thoreau suggests, has no need for a compass because intuition will guide those who are called to walking. Modern recreational hikers become distressed at even the hint that they have lost their place in the grid of marked trails and named places. Studies at the Matador Refuge in Texas show that visitors seldom travel more than a quarter mile from the road.

The Factory Managers, Reconsidered

To this point I have left factory managers, the professional elite, out of the frame of the land ethic. Taken seriously, the land ethic is an arrow aimed at the heart of the recreational profession in all its different guises. Michael Crichton's *Jurassic Park* (1990) is a delightful tale that, through hyperbole, makes clear the subtle (and therefore insidious) bureaucratic-capitalistic dynamic that I am trying to bring to the surface. How easy it is to be in charge, in control of an ecosystem, a word deliberately and consciously employed by A. G. Tansley (1935) to separate humans—as scientists, engineers, managers—from the land community; to create objective distance between observer and observed. The professions that inculcate recreational managers into their grasp, orient, frame, and provide a moral horizon that defines and simultaneously limits the ambit of possibility.

Tom Birch (1990) argues that wilderness managers, charged with incarcerating wildness, are more concerned with the advancement of their careers through achieving quantifiable goals (number of park visitors, total revenues) and developing park and forest amenities (roads, "scenic" turnouts, restrooms, paved trails, maps, campgrounds) than with perpetuating the land community of which they are a part. The moral horizon of many managers is circumscribed by the political economy of the market society (Hood 1993). The professionals who manage the wilderness areas, national forests, and national parks are either consciously or unconsciously creating a commodity for consumption by the masses.

A few professionals have discussed the subversive effect of the land ethic on their thinking and careers. Pister (1987:222) notes that his education as a recreational manager and the management programs that he moved into "were designed to meet the desires and demands of a public hungry for outdoor recreation following World War II. They were, in short, model 'utilitarian' management procedures." He found himself caught up in the recreation industry, supplying the ever-increasing demand of consumers. The activities of Pister and his fellow managers were socially legitimated, and during the recreational boom, the recreational bureaucracy grew even stronger and more powerful. The management program "was popular, it kept us fully employed, it fueled the tourist economy, and it was heresy to think otherwise" (Pister 1987:222).

In 1964, Pister came to rethink the implications of Leopold's land ethic. Suddenly, he realized he had become caught up in the recreation industry, part of California's put-and-take fishing industry. This, he heretically concluded, could not continue. "Any management program worthy of the name," he argued, "must begin with the integrity of the land and water. Using this as a foundation, the resource manager can then build his [or her] pyramid upward, adding to this base the flora and fauna, and then build further to accommodate, finally, the species of economic and political interest" (Pister 1987:223). Needless to say, as Pister makes clear, the response of California's recreational managers was (to borrow a phrase from the Zen masters) the sound of one hand clapping. I have not found, in my own interactions with the recreational bureaucracy, many exceptions.

Though the land ethic approach to management is "a path not taken" by the majority of our resource managers, Pister took the other road down which Leopold himself walked. Leopold's own experiences, his attempts to manage deer on utilitarian principles in the Gila National Forest, his observations of the failures of forest management in Germany, and others, gave him increasing reason to doubt the resource management approach institutionalized in both higher education and recreational bureaucracies. He came to reject the managerial model (see Hargrove 1989), in favor of an approach we now call therapeutic nihilism—nature knows best, and it is a vanity to think humans are in control.

Of course, such vanity or managerial arrogance comes to us through a long historical process, a process that has created the socially constructed reality in which we live. Most natural resource managers, though, remain blind to the institutionalized dynamic that shapes their behavior. A few do not. They follow in Leopold's wake, deconstructing the leading managerial philosophy (see Livingston 1981; Evernden 1992). The dominant outlook on the land is European; American attitudes are the offshoot of Euroculture. Within this context, place becomes the stage on which history is enacted. The land becomes a collection of legal entities, like towns and states. Outside these constructs, nature is viewed as a theoretical object for scientific inquiry, as environment, as playground, and as natural resource. This characteristic attitude is the consequence of events stretching into prehistory.

Recreation Recontextualized through the Land Ethic

Allow me to summarize an almost 500-page book that speaks to the evolution of the idea of history itself. In *The Idea of Wilderness* (Oelschlaeger 1991), I argue that, following the Neolithic revolution, (1) ancient peoples became increasingly adept and aggressive in their endeavors to humanize the land; (2) almost concomitantly, they became aware of themselves as partially dependent on but distinct from the larger community of life; (3) they devised increasingly abstract and complicated explanatory schemes (mythologies, theologies) to account for their relation to, domination of, and separation from the natural world; (4) these schemes recognized a limited mastery over the land through technology, while preserving the idea that some forces were beyond human control; (5) they also conceived of the landscape as divinely designed for human habitation, cultivation, and modification; (6) ultimately, the earth came to be conceived as valueless until humanized, no more than standing reserve for the purposes of human appropriation; and (7) the first scientific revolution gave Western culture the idea that Man was the master and possessor of nature. Everything that followed in the wake of modernism, such as the recreation industry, represented an unparalleled amplification of the exploitation of the land for utilitarian ends.

Recreation is a subtext in the cultural narrative that frames nature as a commodity. Whatever our illusions about recreation, there is no reason to think that it escapes the influence of the dominant cultural theme. I think of my own dawning awareness that I was a consumer of manufactured recreational experiences. The turning point came at the Grand Canyon, a place of great beauty that I now assiduously avoid, for I have come to see it as symbolizing the modern attitude toward place. It exposes me as a typical, rootless modern person coming to the Canyon as a consumer of a product packaged by the National Park Service and marketed by the Flagstaff Chamber of Commerce. I come; I see; I leave, with scenic photographs and a sense of disappointment at having been herded through a "canyon experience" with hundreds of others. The park is managed by a cadre of professionals and seasonals, all overworked and constantly harried by visitors, trained to make each visit a pleasant and safe encounter with scenery. The staff stay on the site longer than visitors, but are themselves caught up in the cultural narrative; successful managers move on in a few years to their next post, as they advance in their careers. Neither employees nor visitors seem to have a conscious awareness of the beauty around them.

Perhaps there is a lesson here for anyone who hopes to re-create themselves: the vast majority of us are rootless, so caught up in the routines and regimen of life that we are oblivious to place. It is only as we develop an affinity for place, cultivating what Leopold would call a sense of the land community, of which we are members rather than rulers, that a land ethic might be truly a practice. A land ethic is perhaps a marker on the road toward an alternative kind of culture, a culture of place that includes more than just humanity in a council of all beings. Try thinking of the land ethic as a guide to exploring the conceptual boundaries between nature and civilization. It invites us to cross over. Considered as re-creation, outdoor experience might open the doors of perception, disclose human connections to the earth, and help us begin to see strategies by which nature and recreation can coexist without undercutting the harmony, stability, and beauty of the land community.

Joseph Sax (1980) argued that the primary justification for our national parks was to provide citizens the opportunity for recreation, the time and place to erase the trace of culture and stand outside the domestic frame. Perhaps modern people cannot go home again, since at least ten thousand years of cultural history separate us from band society and any intuition that we are bound up with all life on earth. But recreation offers ample opportunity for contemplative encounters, occasions for human beings to reflect on life and cosmos, on meaning and significance that transcend the culturally relative categories of modern existence. Recreationists begin to sense the longer and slower rhythms of time with events measured on a grander scale. They bear witness to the interactions of geology and biology over millenia. Recreationists can realize the contingency of their socially constructed, Western selves.

Edward O. Wilson's writing on biophilia (1984) and the idea of wilderness (1992) bears powerfully on one's coming to grips with recreation. He believes that human beings are predisposed to feel the love of life, an affinity for the natural world. He also recognizes that humans are socially conditioned to behave in some ways and not others; primarily we are conditioned to be consumers. But recreation can lead us toward what Wilson calls "a deep conservation ethic" (1984:138–139). Such an ethic is predicated on the explicit knowledge and account of biophilia—the love of life and earth with which we are irretrievably bound.

Our connections to earth are concealed by history as the defining story of Man's conquest of nature; but, this is a potentially fatal illusion. The linguistic structures that overlay our genetic inheritances (the nucleic-acid code that defines the human genotype), and make us specifically humans, selfishly perpetuate themselves. They predispose us to think of outdoor experience as entertainment, diversion, scenery, escapism. It may be all these things, but it is also much more. Humans remain tied with the ongoing reality of life on earth. As Wilson (1992:349) notes, "Only in the last moment of human history has the delusion arisen that people can flourish apart from the rest of the living world." Through biophilia, and more generally through the idea of wilderness, we might develop a story of who we are, a deep ecological ethic that naturalizes history, that reconnects us with earth. "Wilderness," Wilson argues, "settles peace on the soul because it needs no help; it is beyond human contrivance" (1992:349). It reminds us of our place in the larger scheme of things.

Leopold's land ethic, then, is complemented by the work of scholars like Sax and Wilson. They collectively imply that any ethical concept of recreation limited to principles of prudence, utilitarian calculus of pleasure, and no harm to the environment, is far wide of the mark. Increasingly, given the ecological malaise that is engulfing the earth, there is evidence that only those communities that rekindle a sense of affiliation with the earth will endure. Recreation might help us find our way on a path toward an age of ecology. The inescapable truth, Wilson argues, "is that we have never conquered the world, never understood it; we only think we have control. We don't even know why we respond a certain way to other organisms, and need them in diverse ways, so deeply. The prevailing myths concerning our predatory actions toward each other and the environment are obsolete, unreliable, and destructive" (1984:139–140).

By using the land ethic to frame recreation as re-creation it is possible to see, however faintly, the first, light of the dawn, heralding a tomorrow where humans might once again become aware of their citizenship in a council of all beings. This awareness, that they are members of a land community, might enable them to reinhabit place and again take root in the green world.

Literature Cited

Abbey, E. 1988. Introduction. In *The Land of Little Rain, Mary Austin*. New York: Penguin Books.

Anderson, E. 1967. *Plants, Man and Life*. Berkeley, California: University of California Press.

Baudrillard, J. 1988. *America*, trans., Chris Turner. New York: Verso.

Berman, M. 1989. *Coming to Our Senses: Body and Spirit in the Hidden History of the West*. New York: Simon and Schuster.

Birch, T. 1990. The incarceration of wildness: wilderness areas as prisons. *Environmental Ethics* 12:3–26.

Botkin, D.B. 1990. *Discordant Harmonies: A New Ecology for the Twenty-First Century*. New York: Oxford University Press.

Bugbee, H.G. 1974. Wilderness in America. Pacific Northwest Regional Meeting of the American Academy of Religion.

Cohen, M.P. 1992. A brittle thesis: a ghost dance: a flower opening. In *The Wilderness Condition: Essays on Environment and Civilization*, ed., M. Oelschlaeger. San Francisco, California: Sierra Club Books.

Cobb, E. 1977. *The Ecology of Imagination in Childhood.* New York: Columbia University Press.

Crichton, M. 1990. *Jurassic Park.* New York: Ballantine Books.

Crosby, A.W. 1986. *Ecological Imperialism: The Biological Expansion of Europe.* New York: Cambridge University Press.

Evernden, N. 1992. *The Social Creation of Nature.* Baltimore, Maryland: Johns Hopkins University Press.

Hargrove, E.C. 1989. *Foundations of Environmental Ethics.* Englewood Cliffs, New Jersey: Prentice Hall.

Hood, R.L. 1993. Discursive horizons of human identity and wilderness in postmodern environmental ethics: a case study of the Guadalupe mountains of Texas. Unpublished masters thesis. Denton, Texas: University of North Texas.

Flader, S.L. 1974. *Thinking Like a Mountain: Aldo Leopold and the Evolution of an Ecological Attitude Toward Deer, Wolves, and Forests.* Columbia, Missouri: Univerity of Missouri Press.

Leopold, A. 1933. *Game Management.* Madison, Wisconsin: University of Wisconsin Press.

Leopold, A. 1949. *Sand County Almanac and Essays on Conservation from Round River.* New York: Oxford University Press.

Leopold, A. 1991. *The River of the Mother of God and Other Essays by Aldo Leopold,* eds., S.L. Flader and J.B. Callicott. Madison, Wisconsin: University of Wisconsin Press.

Livingston, J.A. 1981. *The Fallacy of Wildlife Conservation.* Toronto, Canada: McClelland and Stewart.

Meine, C. 1989. *Aldo Leopold: The Man and His Works.* Madison, Wisconsin: University of Wisconsin Press.

Oelschlaeger, M. 1991. *The Idea of Wilderness: From Prehistory to the Age of Ecology.* New Haven, Connecticut: Yale University Press.

Pister, E.P. 1987. A pilgrim's from group A to group B. In *Companion to A Sand County Almanac,* ed., J.B. Callicott. Madison, Wisconsin: University of Wisconsin Press.

Rosenau, P.M. 1992. *Post-Modernism and the Social Sciences: Insights, Inroads, and Intrusions.* Princeton, New Jersey: Princeton University Press.

Sax, J.L. 1980. *Mountains Without Handrails: Reflections on the National Parks.* Ann Arbor, Michigan: University of Michigan Press.

Shepard, P. 1978. *Thinking Animals: Animals and the Development of Human Intelligence.* New York: Viking Press.

Tansley, A.G. 1935. The use and abuse of vegetational concepts and terms. *Ecology* 16:284–307.

Trotter, P.C. 1987. *Cutthroat: Native Trout of the West.* Boulder, Colorado: Colorado Associated University Press.

Wilson, E.O. 1984. *Biophilia.* Cambridge, Massachusetts: Harvard University Press.

Wilson, E.O. 1992. *The Diversity of Life.* Cambridge, Massachusetts: Harvard University Press.

List of Scientific Names

Mammals

American opossum	*Didelphis marsupialis*
Asian black bear	*Selenarctos thibetanus*
Barking deer	*Muntiacus reevesi*
Beaver	*Castor canadensis*
Bighorn sheep	*Ovis canadensis*
Black bear	*Ursus americanus*
Caribou	*Rangifer tarandus*
Cheetah	*Acinonyx jubatus*
Cottontail	*Sylvilagus* spp.
Coyote	*Canis latrans*
Dall sheep	*Ovis dalli*
Deer mouse	*Peromyscus maniculatus*
Domestic cat	*Felis catus*
Domestic dog	*Canis familiaris*
Eastern chipmunk	*Tamias striatus*
Eastern gray squirrel	*Sciurus carolinensis*
Elk	*Cervus elaphus*
Ermine	*Mustela erminea*
Florida manatee	*Trichechus manatus*
Fox squirrel	*Sciurus niger*
Giant otter	*Pteronura brasiliensis*
Gray bat	*Myotis grisescens*
Grizzly bear	*Ursus arctos*
Harbor seal	*Phoca vitulina*
Harp seal	*Phoca groenlandica*
Howler monkey	*Alouatta palliata*
Hyena	*Crocuta crocuta*
Jaguar	*Panthera onca*
Kangaroo rat	*Dipodomys merriami*
Lion	*Panthera leo*
Mountain goat	*Oreamnos americanus*
Mountain sheep	*Ovis canadensis*
Moose	*Alces alces*
Mule deer	*Odocoileus hemionus*
Musk deer	*Moschus chrysogaster*
Olive baboon	*Papio cynocephalus*
Pig	*Sus scrofa*
Polar bear	*Ursus maritimus*

Raccoon   *Procyon lotor*
Red deer   *Cervus elaphus*
Red fox   *Vulpes vulpes*
Red panda   *Ailurus fulgens*
Reindeer   *Rangifer tarandus*
Roe deer   *Capreolus capreolus*
Serow   *Capricornis sumatrensis*
Svalbard reindeer   *Rangifer tarandus platyrhynchus*
Swamp rabbit   *Sylvilagus aquaticus*
Tahr   *Hemitgrasus jemlahicus*
Tapir   *Tapirus* spp.
Thomson's gazelle   *Gazella thomsoni*
Timber wolf   *Canis lupus*
Vole   *Microtus* spp.
White-tailed deer   *Odocoileus virginianus*
White-throated capuchin   *Cebus capucinus*
Wildebeest   *Connochaetes taurinus*
Woodchuck   *Marmota monax*
Zebra   *Equus burchelli*

Birds
American black duck   *Anas rubripes*
American crow   *Corvus brachyrhynchos*
American goldfinch   *Carduelis tristis*
American robin   *Turdus migratorius*
Anhinga   *Anhinga anhinga*
Arctic loon   *Gavia arctica*
Bald eagle   *Haliaeetus leucocephalus*
Barred ground dove   *Geopelia striata*
Bean goose   *Anser fabalis*
Black skimmer   *Rynchops niger*
Black-billed magpie   *Pica pica*
Blue-winged teal   *Anas discors*
Brant goose   *Branta bernicla*
Brewer's blackbird   *Euphagus cyanocephalus*
Brown noddy   *Anous stolidus*
Brown pelican   *Pelecanus occidentalis*
Brown-headed cowbird   *Molothrus ater*
Brunnich's guillemot   *Uria lomvia*
Canada goose   *Branta canadensis*
Canvasback   *Aythya valisineria*
Chestnut-backed chickadee   *Parus rufescens*
Clark's nutcracker   *Nucifraga columbiana*
Collared plover   *Charadrius collaris*
Common eider   *Somateria mollissima*
Common loon   *Gavia immer*
Common raven   *Corvus corax*
Common sandpiper   *Actitis hypoleucos*

Common tern	*Sterna hirundo*
Cooper's hawk	*Accipiter cooperii*
Crested tern	*Sterna bergii*
Double-crested cormorant	*Phalacrocorax auritus*
Eastern kingbird	*Tyrannus tyrannus*
Elegant tern	*Thalasseus elegans*
Emperor goose	*Chen canagica*
European kestrel	*Falco tinnunculus*
Fox sparrow	*Passerella iliaca*
Glaucus-winged gull	*Larus glaucescens*
Golden eagle	*Aquila chrysaetos*
Golden plover	*Pluvialis apricaria*
Gray jay	*Perisoreus canadensis*
Great black-backed gull	*Larus marinus*
Great blue heron	*Ardea herodias*
Greater prairie chicken	*Tympanuchus cupido*
Greater snow goose	*Chen caerulescens atlantica*
Greater white-fronted goose	*Anser albifrons*
Green-backed heron	*Butorides striatus*
Guillemot	*Cepphus* spp.
Heermann's gull	*Larus heermanni*
Herring gull	*Larus argentatus*
King shag	*Phalacrocorax albiventer*
Kittiwake	*Rissa* spp.
Lapland longspur	*Calcarius lapponicus*
Large-billed tern	*Phaetusa simplex*
Laysan albatross	*Diomedea immutabilis*
Least sandpiper	*Calidris minutilla*
Least tern	*Sterna antillarum*
Lesser prairie chicken	*Tympanuchus pallidicinctus*
Lesser white-fronted goose	*Anser erythropus*
Macaw	*Ara* spp.
Magellanic penguin	*Spheniscus magellanicus*
Mallard	*Anas platyrhynchos*
Mottled duck	*Anas fulvigula*
Mountain chickadee	*Parus gambeli*
Northern bobwhite	*Colinus virginianus*
Oregon junco	*Junco oreganus*
Osprey	*Pandion haliaetus*
Oystercatcher	*Haematopus palliatus*
Pacific black brant	*Branta bernicla nigricans*
Pacific eider	*Somateria mollissima v-nigra*
Peregrine falcon	*Falco peregrinus*
Pied lapwing	*Hoploxypterus cayanus*
Pink-footed goose	*Anser brachyrhynchus*
Pinyon jay	*Gymnorhinus cyanocephalus*
Piping plover	*Charadrius melodus*
Redhead	*Aythya americana*
Red-tailed hawk	*Buteo jamaicensis*

Red-winged blackbird	*Agelaius phoeniceus*
Resplendent quetzal	*Pharomachrus mocinno*
Ring-necked duck	*Aythya collaris*
Ring-necked pheasant	*Phasianus colchicus*
Rock ptarmigan	*Lagopus mutus*
Royal tern	*Sterna maxima*
Ruddy shelduck	*Tadorna ferruginea*
Ruffed grouse	*Bonasa umbellus*
Sand-colored nighthawk	*Chordeiles rupestris*
Sanderling	*Calidris alba*
Sandhill crane	*Grus canadensis*
Sooty tern	*Sterna fuscata*
Trumpeter swan	*Cygnus buccinator*
Tufted duck	*Aythya fuligula*
Vaux's swift	*Chaetura vauxi*
Western gull	*Larus occidentalis*
Willow ptarmigan	*Lagopus lagopus*
Yellow-billed tern	*Sterna superciliaris*

Reptiles

American alligator	*Alligator mississippiensis*
Black caiman	*Melanosuchus niger*
Bull snake	*Pituophis melanoleucus*
Canebrake rattlesnake	*Crotalus horridus atricaudatus*
Coachwhip	*Masticophis flagellum*
Crocodile	*Crocodilus niloticus*
Eastern diamondback rattlesnake	*Crotalus adamanteus*
Gopher snake	*Pituophis melanoleucus*
Glossy snake	*Arizona elegans*
Land iguana	*Conolphus subcristatus*
Leopard lizard	*Gambelia silus*
Massasauga	*Sistrurus catenatus*
Mojave rattlesnake	*Crotalus scutulatus*
Prairie rattlesnake	*Crotalus viridis viridis*
Puerto Rican coqui	*Eleutherodactylus coqui*
Racer	*Coluber constrictor*
Side-necked turtle	*Podocnemis unifilis*
Spadefoot toad	*Scaphiopus couchi*
Spectacled caiman	*Caiman crocodilus*
Timber rattlesnake	*Crotalus horridus*
Western diamondback rattlesnake	*Crotalus atrox*

Fish/Aquatic Invertebrates

Bluegill	*Lepomis macrochirus*
Dungeness crab	*Cancer magister*
Kokanee salmon	*Oncorhynchus netica*
Oppossum shrimp	*Mysis relicta*
Salmon	*Oncorhynchus* spp.

Vegetation

Beech	*Fagus grandifolia*
Cottonwood	*Populus sargentii*
Limber pine	*Pinus flexilis*
Sugar maple	*Acer saccharum*
Three-square bulrush	*Scirpus americanus*
Yellow birch	*Betula allegheniensis*

Index

Contributors

Stanley H. Anderson is the leader of the Wyoming Cooperative Fish and Wildlife Research Unit at the University of Wyoming.

Robert G. Anthony is the assistant leader of the Oregon Cooperative Wildlife Research Unit at Oregon State University.

Phillip C. Arena is a research assistant in the School of Veterinary Studies at Murdoch University in Perth, Western Australia.

Jean Bédard is a professor in the Department of Biology at Laval University in Québec City, Canada.

Luc Bélanger is a biologist with the Canadian Wildlife Service in Québec City, Canada.

Ann E. Bowles is a senior staff biologist at the Hubbs–Sea World Research Institute in San Diego, California.

James J. Brett is curator of the Hawk Mountain Sanctuary Association.

Joanna Burger is a professor of biology and was director (1978–1993) of the Graduate Program in Ecology and Evolution at Rutgers University–Piscataway.

David N. Cole is a research biologist at the Aldo Leopold Wilderness Research Institute of the U.S. Forest Service in Missoula, Montana.

H. Ken Cordell is project leader of the Outdoor Recreation and Wilderness Assessment Group of the U.S. Forest Service's Southeastern Forest Experiment Station in Athens, Georgia.

Daniel J. Decker is the department chairman, as well as co-leader of the Human Dimensions Research Unit, in the Department of Natural Resources at Cornell University.

David Duvall is an associate professor of zoology in the Department of Life Sciences at Arizona State University West in Phoenix.

Curtis H. Flather is a research wildlife biologist and wildlife and fish resource specialist for national assessments at the U.S. Forest Service's Rocky Mountain Forest and Range Experiment Station in Fort Collins, Colorado.

Geir W. Gabrielsen is a research scientist at the Norwegian Institute for Nature Research in Tromsö, Norway.

Kevin J. Gutzwiller is an associate professor in the Department of Biology and the Department of Environmental Studies at Baylor University.

Leslie HaySmith is a doctoral candidate in the Department of Resource Recreation and Tourism at the University of Idaho.

John D. Hunt is the head of the Department of Resource Recreation and Tourism at the University of Idaho.

Paul Kerlinger is director of the Cape May Bird Observatory and director of research for the New Jersey Audubon Society.

Richard L. Knight teaches wildlife conservation at Colorado State University.

Peter B. Landres is a research ecologist at the Aldo Leopold Wilderness Research Institute of the U.S. Forest Service in Missoula, Montana.

Richard A. Larson is the liaison between the Colorado Division of Wildlife and the State Board of the Great Outdoors Colorado Trust Fund.

Michael J. Manfredo is a professor in the Department of Natural Resources Recreation and Tourism at Colorado State University. He is also the leader of the Human Dimensions in Natural Resources Unit in the College of Natural Resources at Colorado State University.

Kevin McGarigal has completed a doctoral degree in the Department of Forest Science at Oregon State University.

Max Oelschlaeger is a professor in the Department of Philosophy and Religion Studies at the University of North Texas.

Thomas J. O'Shea is assistant director of the National Biological Survey's Midcontinent Ecological Science Center at Fort Collins, Colorado. He was formerly Sirenia Project Leader for the U.S. Fish and Wildlife Service in Florida.

E. Norbert Smith is the director of the Laboratory Animal Science Program at Redland Community College in El Reno, Oklahoma.

Robert J. Steidl has completed a doctoral degree in the Department of Fisheries and Wildlife at Oregon State University.

Stanley A. Temple is the Beers-Bascom Professor of Conservation at the University of Wisconsin–Madison.

Jerry J. Vaske is an associate professor in the Department of Natural Resources Recreation and Tourism at Colorado State University.

Clifford Warwick is director of the Institute of Herpetology at Worcester, England.

618020